Sustainable Development Goals Series

The **Sustainable Development Goals Series** is Springer Nature's inaugural cross-imprint book series that addresses and supports the United Nations' seventeen Sustainable Development Goals. The series fosters comprehensive research focused on these global targets and endeavours to address some of society's greatest grand challenges. The SDGs are inherently multidisciplinary, and they bring people working across different fields together and working towards a common goal. In this spirit, the Sustainable Development Goals series is the first at Springer Nature to publish books under both the Springer and Palgrave Macmillan imprints, bringing the strengths of our imprints together.

The Sustainable Development Goals Series is organized into eighteen subseries: one subseries based around each of the seventeen respective Sustainable Development Goals, and an eighteenth subseries, "Connecting the Goals", which serves as a home for volumes addressing multiple goals or studying the SDGs as a whole. Each subseries is guided by an expert Subseries Advisor with years or decades of experience studying and addressing core components of their respective Goal.

The SDG Series has a remit as broad as the SDGs themselves, and contributions are welcome from scientists, academics, policymakers, and researchers working in fields related to any of the seventeen goals. If you are interested in contributing a monograph or curated volume to the series, please contact the Publishers: Zachary Romano [Springer; zachary.romano@springer.com] and Rachael Ballard [Palgrave Macmillan; rachael.ballard@palgrave.com].

Glenn W. Muschert
Vijay Pereira • Vikash Ramiah
Asli Cansin Doker
Editors

Financial Inclusion

In Pursuit of UN Sustainable Development Goal 8.10

Springer

Editors
Glenn W. Muschert
Khalifa University of Science and Technology
Abu Dhabi, United Arab Emirates

Vijay Pereira
NEOMA Business School
Reims, France

Vikash Ramiah
University of Wollongong in Dubai
Dubai, United Arab Emirates

Asli Cansin Doker
Erzincan Binali Yildirim University
Merkez, Erzincan, Türkiye

ISSN 2523-3084 ISSN 2523-3092 (electronic)
Sustainable Development Goals Series
ISBN 978-3-031-68802-7 ISBN 978-3-031-68803-4 (eBook)
https://doi.org/10.1007/978-3-031-68803-4

Color wheel and icons: From https://www.un.org/sustainabledevelopment

This Springer imprint is published by the registered company Springer Nature Switzerland AG
The registered company address is: Gewerbestrasse 11, 6330 Cham, Switzerland

Contents

About the Contributors

Shehnaz Al-Helo is a Tutor for International Business Management and Employee Relations, among other subjects, at the University of Wollongong in Dubai. Driven by the desire to bridge inequality, she is striving to establish the link between expatriate entrepreneurship and Sustainable Development Goals, including Gender Equality, Quality Education, and Financial Inclusion. Other research interests include firm dynamics and resource allocation, monetary policy, and cryptocurrency. She employs context-appropriate methodologies, including systematic review and econometrics. She is keen on forging connections with others to disseminate information that will shape the future. She holds a 'Master's degree in International Business and was recognized as a top graduate from UOWD, where she also obtained her Bachelor's degree in Business Administration with distinction.

Abdulkader Aljandali holds a PhD in Finance from London Metropolitan University and is a Senior Fellow of the British Higher Education Academy. Prior to joining Africa Business School as the Head of Finance cluster, he worked across various institutions in the United Kingdom from 2012 to 2021. Dr. Aljandali's publications include textbooks on quantitative analysis with a research focus on the examination of financial time series in emerging markets. More recently, he has published in the area of Fintech with a particular focus on the impact of the latter on financial inclusion and MNCs resilience. Dr. Aljandali is a previous member of the CFA UK society and JCI London and a current member of the British Accounting and Finance Association. He is fluent in Arabic, French, and English with working knowledge in both German and Spanish. https://orcid.org/0000-0003-0880-6634

Maheshvari Appavoo is a PhD candidate in Development Finance at Stellenbosch Business School, Stellenbosch University, South Africa. Her thesis focuses on factors contributing to financial recovery from climate-related disasters in rural South Africa. Prior to enrolling in the PhD program, she graduated from the University of London with an MSc in Finance, majoring in Economic Policy. Her research interests include behavioral finance and climate change.

Ahmad Arslan is currently working as a Professor at the Department of Marketing, Management and International Business, University of Oulu, Finland. He also holds the position of Honorary at the Business School, University of Aberdeen, Scotland, UK. His earlier research has been published in prestigious academic journals like *British Journal of Management, Human Resource Management* (US), *IEEE Transactions on Engineering Management, International Business Review, International Marketing Review, Journal of Business Research, Journal of International Management, Journal of Knowledge Management, Production Planning and Control, Supply Chain Management, Technological Forecasting and Social Change*, and *Scandinavian Journal of Management*, among others. Finally, he holds several editorial board memberships and is currently a Senior Editor of the *International Journal of Emerging Markets* (Emerald).

Saranjam Baig teaches Political Economy at Sultan Qaboos University, Muscat Oman, and Economics and Public Policy at Karakoram International University, Gilgit, Pakistan. Dr. Baig has also been working with Mahbubul Haq Human Development Centre (MHDC) and University of La Verne, California. He received his education at the Claremont Graduate University, California, with a Master's in Economics and a Doctorate Degree in Economics and Political Science. He also holds an MS degree in Development Studies from the University of Glasgow, Scotland UK. He is an avid researcher, and his work has been published in many local and international journals. Saranjam does research in local economy of Gilgit-Baltistan, political economy, and Belt and Road Initiative.

Apoorva Vaishnavi Bhattaram is a second-year student pursuing Master's in Business Administration at Woxsen University. She received a Bachelor's degree in Engineering from Jawaharlal Nehru Technological University. She has worked on various projects related to strategy, market research, and financial analysis. She also worked with young entrepreneurs as a digital marketer to develop their business. She has a keen interest in marketing for financial institutions.

Reena Bhattu-Babajee is a Senior lecturer in finance at the University of Mauritius since 2006. She holds a Ph.D. in Finance from the University of Mauritius. Her fields of study are Intellectual Capital, Corporate Social Responsibility, and International Finance. r.bhattu@uom.ac.mu

Roshan Bhut is a working professional in Factset Systems India Pvt. Ltd. He is also a second-year MBA student at Woxsen University. His majors are Finance, Operations, and Data Science. He received his 'Bachelor's degree in Commerce from Ravenshaw University. He has 2 years of experience in the manufacturing industry where he was the manager of several projects. He is interested in Financial Analytics and is working on another book chapter, financial analytical tools and its importance in today's tech-oriented economy.

Kofi A. Boateng is a Senior Lecturer in the Department of Supply Chain and Information Systems at KNUST. He has a BA (Hons) in Publishing Studies from KNUST. His MSc in Analysis, Design and Management of Information Systems and PhD in Information Systems degrees were obtained from the London School of Economics and Political Science (LSE) in 2005 and 2010, respectively. His current research engagements include platform technologies, digital transformation, ICT-mediated interactions, and digital innovation. His publications have appeared in journals such as *Information and Management*, *Electronic Journal of Knowledge Management*, *Communications of the IIMA*, and *Information Knowledge Systems Management*.

Mercy A. Boateng obtained her Bachelor's degree in BA (Hons) Publishing Studies (2002) and her MSc in Development Management (2018), both at the Kwame Nkrumah University of Science and Technology, Kumasi, Ghana.

Debdutta Choudhury is an IIM Calcutta–NIT Jaipur Alumnus with expertise in Global Accreditations like AACSB, EFMD, and AMBA. Debdutta is an academician with 18 years of corporate experience spanning the domains of Manufacturing, BFSI, Technology, and Education. Debdutta has spent most of his career in sales and marketing with exposure to sales strategy, sales management, corporate strategy, integrated marketing communication, corporate finance, and general management. Debdutta has held several CXO positions in various organizations and has delivered tangible results under varied business environments and geographies in India.

Rosemary Boateng Coffie holds a DipEd (Management) (UCC), BEd (Management) (UCC)) MEd (Int. Mgt. and Policy) (UoB, UK), PGD (HRM) (WCC, UK), and PhD (Management) (KNUST) and is a Senior Lecturer in the Department of Human Resource and Organizational Development at the Kwame Nkrumah University of Science and Technology. She is also a member of the Institute of Human Resource Management Practitioners (Ghana). Since 2006 she has been engaged in teaching and research in the area of Management and Human Resource Management. Her research interests are in the areas of management, human resource management, strategic management, strategic human resource management, information systems and HRM, cross-cultural management, organizational climate, and institutional and societal coping.

D. Priya Dhatri is pursuing MBA from Woxsen University with a specialization in finance and marketing. She received her bachelor's degree in "Public Policy and Administration." She is part of the Centres of Excellence for Financial Inclusion and Global Development at Woxsen University. Her area of interest includes grassroots policy-making, structural inequalities, and socio-cultural anthropology. Her prior experiences of interning at local government bodies and startups have provided her with insights into the reality of institutionalized socio-economic inequalities and the reality of CSR and public policy implications.

Florian Gerth is an Associate Professor of Economics at the Asian Institute of Management in Manila, Philippines, and a Non-Resident Fellow for the Center of the Study of Global Economic Future, Dubai, UAE. During his career, he has been working for monetary policy institutions and universities all around Europe, the Middle East, and South-East Asia. Apart from having published articles in the field of financial economics, he also contributed to academic journals in the areas of monetary economics, financial stability, development economics, business cycle fluctuations, and productivity analysis. He holds an MSc in Econometrics and a PhD in Economics from the University of Kent, UK, and an MA in International Relations and Affairs from the University of Wollongong, Australia.

Maria Alejandra Gonzalez-Perez is a Full Professor of Management and Head of the Masters in Sustainability at Universidad EAFIT in Colombia, a Member of the Global Council for Sustainable Development Goal Number 1 (SDG 1: End Poverty), and a former Vice-President at the Academy of International Business (AIB) and regional president for Latin America and the Caribbean. She is an internationalization, sustainability, and CSR expert with 75 peer-reviewed scholarly papers and 18 books. She is a frequent business media contributor and a Visiting Professor at numerous universities worldwide. She has lived, studied, or worked in several countries and traveled to 97 countries. https://orcid.org/0000-0001-6338-4281; mgonza40@eafit.edu.co

Narvada Gopy-Ramdhany is a senior lecturer in finance at the University of Mauritius since 2006. She has a master's degree in finance from the University of Strasbourg, France, and an MPhil in finance from the same institution. She has a Ph.D. degree in Applied Finance from the University of Mauritius. One of her main research interests is in the field of real estate economics and finance.

Calvin Habasonda is currently pursuing a Doctor of Philosophy in Development Finance at the University of Stellenbosch, Cape Town, South Africa. He has over 16 years in financial sector policy and is currently Assistant Director—Policy and Research in Bank Supervision Department at the Bank of Zambia. His research interests include financial system regulation, financial inclusion, risk management, corporate governance, and anti-money laundering.

Morshadul Hasan is a PhD candidate at Murdoch Business School, Murdoch University, Australia (Prestigious Australian Government RTP scholarships fellow). His research focuses on Fintech, financial literacy, inclusive finance, corporate governance, and sustainable development. Besides, he also works as a Lecturer (Sessional Academic) at Murdoch Business School. His teaching focuses on Corporate Finance, International Finance, Financial Accounting, and Business Analytics in both undergraduate and postgraduate levels. He completed his 'Master's in Finance from Nanjing Audit University, China, with a number of prestigious awards, such as the Jiangsu Government Outstanding Student Awards, University Excellent Postgraduate Awards, and School Best Thesis Award (TSP 2018 Master of Finance Program). He presented his research at several international conferences across Asia and Europe. He has published his research in several prestigious ABS 3, ABDC A, and high-impact factor Q1 journals.

Carolina Herrera-Cano holds a Master's Degree in Global Development Studies from the Norwegian University of Life Sciences, a Master's in International Business from Universidad EAFIT (Colombia), and a Gender Diploma from Universidad Nacional (Colombia). She serves as an Adjunct Lecturer at Universidad EAFIT and as Senior Consultant at Aequales—a multilatina consultancy in gender equity within organizations. She has served as Associate Professor, Lecturer, and Researcher at different universities. She has done consulting for UNDP and the Colombian Ministry of Environment. Her research pieces on gender, sustainability, and international trade have been published and presented at academic journals and international conferences. https://orcid.org/0000-0003-4266-0077; cherre16@eafit.edu.co

Ariful Hoque is a Senior Lecturer of Finance at Murdoch Business School. He earned his PhD in Finance from Curtin University, Australia. Besides academic professions (almost 20 years), he has been working in different positions in the industry for 30 years. He has teaching experience in all areas of Finance, including research methodology and data analytics to undergraduate and postgraduate students at Curtin University, University of Dubai, University of Southern Queensland, University of South Australia, and Murdoch University. Also, he has authored one of the top five international finance textbooks entitled '*International Financial Management*' published by Cengage. His research interest areas (alphabetically) include but are not limited to Data Analytics, Derivatives, Financial Engineering, Fintech, and Market Microstructure. He has published his research works in several ABDC A* and ABDC A-ranked journals. He has completed six PhD supervisions, received several grants and awards for excellence in research, and has several academic leadership positions.

Haruna Issahaku is an Associate Professor of Applied Economics and Finance and the Vice-Dean of the School of Business, University for Development Studies, Ghana. He earned his PhD in Finance from the University of Ghana Business in 2017. Before undertaking the PhD program, he completed a one-year coursework in MPhil Finance in 2012. He also has an MPhil in Agricultural Economics from the University of Ghana. In addition, he has several publications in international peer-reviewed journals and serves as a visiting scholar at Covenant University, Nigeria. iharuna@uds.edu.gh

Leonie Jooste is an Associate Professor in the Faculty of Management at the Canadian University in Dubai. She started her career as a schoolteacher in South Africa and left the teaching profession to continue her career as a university academic. She is a Research Associate, and lectures both post- and undergraduate students. She served on the academic board and research colloquium of universities and was the academic head at a college in Malta. Leonie is research active and currently serves on an Accreditation and Quality Assurance committee. Leonie obtained her Doctorate in Accounting in 2004 at the University of Pretoria. She publishes articles in accredited journals, presents papers at internationally acclaimed conferences, reviews and serves on the editorial board of academic journals, and is an external examiner for PhD students. She acts as an expert witness on cash flows in High Courts, offers workshops, and consults with professionals in her field. Leonie1248@gmail.com

Paul de R. Jooste is an Accountant at Walkers Corporate Limited, part of Walkers Global. This leading international law firm provides legal, corporate, compliance, and fiduciary services to global corporations, financial institutions, capital markets participants, and investment fund managers, with a global presence spanning the Americas, Europe, the Middle East, and Asia. Walkers provide accessible advice on the jurisdictions of Bermuda, the British Virgin Islands, the Cayman Islands, Guernsey, Ireland, and Jersey. Paul started his career as a tax consultant at Grant Thornton before joining the KPMG group in 2015 as a Financial Services Auditor, specializing in the audits of internationally known banks, insurance companies and investment funds, and other industry-leading financial services companies such as VISA and Unionpay. In 2017 Paul became an Association of Charted Certified Accountants "ACCA" affiliate after completing his professional studies. Paul was admitted as a member of ACCA in 2018 and currently practices under the professional body. Paul has a passion for teaching and giving back to the community. He has given a few guest lectures on auditing at the University College of the Cayman Islands "UCCI" and, in his spare time, also coaches the under-10 rugby team at his local Rugby club in the Cayman Islands. Paul joined Walkers in 2021 and is no stranger to big names. Paul manages a portfolio of clients and investment vehicles, including internationally recognized clients such as Apollo, BlackRock, Royal Families, and High Net Worth Family offices. Paul ensures his clients comply with the laws and regulations imposed by the Cayman Islands Monetary Authority "CIMA" and the Department of International Tax Cooperation "DITC." jooste.paul654@gmail.com

Nuray Karaman is a Lecturer in the Department of Sociology at Uşak University, Turkey. She completed her Master's Degree at Florida Atlantic University and obtained her PhD from the University of Tennessee, Knoxville (UTK) under the mentorship of Dr. Michelle Christian (Dissertation title: "Experiences of Muslim Female Students in Knoxville: At the Intersection of Race, Gender, and Religion"). She is also a global fellow of the Center for the Study of Social Justice at UTK. Her research mainly focuses on racialization, whiteness, race, and gender. Her recent research article titled "My Hijab Is Like My Skin Color": Muslim Women Students, Racialization, and Intersectionality was published in *Sociology of Race and Ethnicity*.

Aftab Ahmed Khan received his BS degree in Information Technology from Pir Mehr Ali Shah Arid Agriculture University, Rawalpindi, MSc from the Technical University of Berlin, Germany, and PhD degree in 3D Geoinformatics from the Technical University Munich, Germany, in 2015. From 2013 to 2015, he worked as a Research Associate at the Technical University of Munich. Currently, Dr. Aftab works as an Associate Professor at the Computer Science Department, Karakoram International University, Gilgit, Pakistan. He is the author of two book chapters and more than 10 peer-reviewed articles. His research interests include 3D geoinformatics and machine learning using earth science informatics.

Amal Laaribi is a Doctor in Management Science and graduate of the ISCAE Group. Amal is a Professor at ENCG Casablanca, the business school of Hassan II University. Her doctoral thesis was nominated for the third edition of the research prize of the Order of Chartered Accountants. It dealt with aspects of public management, management control, and performance management in Moroccan public universities. Dr. Laaribi is the author of several publications/communications at national and international level. She is also a project manager and consultant who intervenes on issues related to higher education and continuing education.

Katia Lopez is a full-time PhD student at the University of Wollongong in Dubai, and she specializes herself in Fintech and Behavioral Finance. Katia holds a MEconSc in International Finance from the National University of Galway, Ireland, and a Bachelor's degree in Economics from the Universidad Autonoma de Nuevo Leon, Mexico. She is a member of Researchers sans Frontiers, a research association that aims to contribute to society through quality research and building a bridge between industries and academia. https://orcid.org/0000-0001-8829-1241

Imad A. Moosa is currently a Professor of Economics at Kuwait University. Before taking on the present position, he had held professorial positions at RMIT, Monash University, and La Trobe University. Prior to becoming an academic in 1991, he was a professional economist in investment banking and a financial journalist for over ten years. He also worked as an economist at the Bureau of Statistics of the IMF and served as an advisor for several organizations, including the US Treasury. Professor Moosa has published 31 books and 270 papers in scholarly journals. His most recent books are *Fintech: A Revolution or a Transitory Hype* and *Financialisation: Measurement, Driving Forces and Consequences*.

Carlos Felipe Munera-Alzate is an economist from Universidad EAFIT and holds a Master's in Applied Economics from the same university. He has worked for more than three years on research in economics, social, and sustainable development topics. At present, he is part of the Bien+ research group and has worked developing SROI-related methodologies for the private sector in Colombia, for which he and the group his is part of have been awarded the "Nicanor Restrepo Santamaría" award on research with social impact. Additionally, he has worked as a Research Analyst at Universidad EAFIT for the last two years. https://orcid.org/0000-0002-2134-0165; cmunera7@eafit.edu.co

Glenn W. Muschert is a professor of sociology based in the Department of Humanities and Social Sciences at Khalifa University of Science and Technology (جامعة خليفة) in Abu Dhabi, UAE. He previously served as a Research Advisor to the Abu Dhabi Department of Community Development and is appointed as the Len Jessup Professor of Leadership and Organizational Change at Woxsen University, India. Previously he served appointments on the Law and Society faculty at Purdue University (USA) and the Sociology, Social Justice Studies, and Comparative Media Studies faculties at Miami University (USA). His research focuses on the metaverse, digital inequalities, sustainable development, and the ethical resolution of social problems. Dr. Muschert earned a PhD (2002) in Sociology from the University of Colorado (USA) and a BS (1992) in International Studies from Drexel University (USA). He has been a visiting scholar at various universities worldwide, including Seoul City University (South Korea), Erzincan University (Turkey), and Atatürk University (Turkey). He is proficient in various qualitative and quantitative methodologies of social research inquiry and frequently serves as reviewer and editorial board member for dozens of academic journals. With over 25 years of experience as a university instructor, he is certified by the UAE Ministry of Education as an External Reviewer for the Council on Academic Accreditation. Dr. Muschert is a member of international professional associations such as the Academy of International Business (AIB) and the Society for the Study of Social Problems (SSSP), where he is a lifetime member and has served since 2010 as SSSP Secretary. He has written or edited over a dozen academic volumes/special issues of journals, dozens of academic research articles, and dozens of chapters in academic volumes. https://orcid.org/0000-0003-3748-4961

Biju A. V. Nair is an Assistant Professor at the Department of Commerce, School of Business Management and Legal Studies, University of Kerala. He is a research supervisor in the Faculty of Commerce and Faculty of Management. He is a member of Research Sans frontier (RSF) in Dubai; this research association aims to contribute to society through quality research and building a bridge between industry and academia. He is taking lectures at the University of Wollongong, Dubai, and Jain University, Cochin. He serves as the managing editor of the *Journal of Commerce and Business Researcher*, published by the Department of Commerce, University of Kerala. He also serves as a review member of springer, Wiley, Inderscience, and IGI Global journals such as the *SN Business and Economics*, *International Journal of Applied Management Science*, *International Journal of E-business Research* (IJEBR), and *International Journal of Business Analytics* (IJBAN). He received the best paper award from the IIM Nagpur conference in 2022. https://orcid.org/0000-0001-5583-6495; bijuav@keralauniversity.ac.in

Sabu K. Nair is a postdoctoral fellow in the Department of Commerce at the University of Kerala, India. His fields of expertise include Marketing, Retail Management, and Operations Research.

Alina R. Oxendine (PhD, University of Minnesota) is a Professor and Chair of the Department of Political Science at Hamline University in St. Paul, MN. Her research and teaching focus on American politics, public policy, inequality, and political psychology. Professor Oxendine's work has been published in numerous collections and journals, has been supported by the Centennial Center for Political Science and Public Affairs, and recognized with an International Society of Political Psychology Best Dissertation Award. She has also been a Visiting Researcher at the University of York and taught at the University of Minnesota's Humphrey Institute of Public Affairs.

Arley Pino-Villegas holds a Bachelor of Business Administration (BBA) degree from Universidad EAFIT, Colombia. He served as Coordinator of the Research Incubator of Administration and Organisations (2020), Excellence Intern at the Secretariat of Economic Development of the Mayor's Office of Medellin (2021), and Analyst and Co-coordinator of the Trade, Investment and Development Observatory (2021-2022). At present, he is a Researcher at the Imagining Futures Centre and a member of the Young Researchers Program sponsored by the Colombian Ministry of Science. His academic and professional interests span the social sciences, sustainable development, business management, learning sciences, and futures studies. https://orcid.org/0000-0003-3549-9537; apinov@eafit.edu.co

Dimitrios Reppas is an Assistant Professor of Economics at Khalifa University (UAE). He has also served as a Lecturer in the Department of Economics at the University of Manchester (UK). He received his PhD and MSc in Agricultural and Resource Economics (as a Fulbright scholar) from the University of Maryland (USA). His research focuses on mathematical applications of microeconomic theory to environmental problems, and the potential of alternative forms of payment (mobile money) to achieve economic and social sustainability in marginalized economies. His research has been awarded twice (2020 and 2022) by the Academy of International Business.

Mandira Sarma is a faculty member at Jawaharlal Nehru University (JNU), Delhi, India, where she teaches, researches, and mentors research scholars. Her research intersects banking and financial sector issues with those of economic opportunities, growth, inclusiveness, and financial stability. She has published in reputed journals and written chapters for many books. She has been a recipient of several fellowships including the Fulbright fellowship. She is well-traveled and enjoys working in a multicultural and diverse environment.

Boopen Seetanah is a Professor at the University of Mauritius (UoM) with research interests in tourism and transport, international trade and finance, and development economics. He is currently the Co-Chair of the WTO Chair (UoM) and the Director of Research at the International Centre for Sustainable Tourism and Hospitality (ICSTH) at the UoM. He is currently the Director of the Doctoral School at the University of Mauritius. Boopen is an editorial board member and a reviewer for numerous high-rated journals. He has been consulting with the government and also numerous international organizations, including the UNEP, UNDP, UNECA, UNCTAD, IISD, World Bank, ADB, ILO and RMCE, amongst others.

Asima Shirazi formerly Assistant Professor at the University of Wollongong in Dubai, has extensive experience in teaching economics at both the undergraduate and postgraduate levels. She is an active researcher with a keen interest in Sustainable Development Goals including Financial Inclusion. Her other research interests include child poverty, development economics, and macroeconomic issues including economic growth and foreign direct investment. She is interested in the Capability Approach as a framework for discussing social justice. Her research has focused on disadvantaged women and children in Pakistan with particular reference to women's corporeal mobility and children's enrolment in schools. She has presented papers at various international and local conferences and colloquia.

Md. Nur Alam Siddik is working in the Department of Finance and Banking, Begum Rokeya University, Rangpur, Bangladesh. He obtained his BBA and MBA degrees (major in Finance and Banking) from the University of Rajshahi, Bangladesh, and obtained a PhD degree in Finance from Dongbei University of Finance and Economics, Dalian, China. His current research interests include financial inclusion, FinTech, school banking, mobile banking, inclusive development, and sustainable development goals. He can be reached at mnas@brur.ac.bd; nasiddikbru@gmail.com

Kerry Slade is an educator and facilitator in the areas of innovation, emerging markets, and social entrepreneurship. She formerly was an Assistant Professor at Temple University's Fox School of Business and Assistant Academic Director of the Innovation and Entrepreneurship Institute. As an attorney, she assists clients with contract, property, employment, and other disputes. She owns Slade Unlimited, an ethical venture that imports products from African countries. Ms. Slade received a BA in Sociology from Brown University with honors, a JD from Northwestern University cum laude, and an MBA from Temple University.

About the Editors

Glenn W. Muschert, a sociologist of social problems, is a Professor of Sociology at Khalifa University of Science and Technology in Abu Dhabi (UAE). His research focuses on social problems, most frequently related to the use of digital technologies. He has authored dozens of academic articles and chapters and has been author or editor of over a dozen academic volumes and special issues of academic journals.

Vijay Pereira, a scholar of international business, is a Professor of International and Strategic Human Capital Management in the People and Organizations Department at NEOMA Business School, France. Professor Pereira is currently the elected Vice President of the Academy of International Business (Middle East and North Africa) and the Secretary of the Indian Academy of Management (an affiliate of the Academy of Management, USA).

Vikash Ramiah is an Associate Professor of Finance at the University of Wollongong in Dubai. He has published in reputable academic journals and supervises numerous PhD students. He specializes in applied finance and his research areas include financial markets, behavioral finance, and environmental/sustainable finance.

Aslı Cansın Doker, an economist, is an Assistant Professor at Erzincan Binali University (Turkiye). Her research interests include experimental economics, game theory, economic growth and development, and applied econometrics.

Part I

Introductory Matters

On Financial Inclusion and Global Development: An Introductory Remark

1

Glenn W. Muschert, Sabu K. Nair, and Biju A. V. Nair

1.1 Introduction

As stated in the Sustainable Development Goal (SDG) 8.10 of the UN's Agenda 2030, financial inclusion is a crucial piece of the puzzle for achieving economic empowerment. Financial inclusion allows people to manage risks, invest in their future, and build businesses (Ade'Soyemi et al., 2020). The UN's SDG 8.10 targets financial inclusion, a critical economic empowerment aspect (Baffoe et al., 2021). This goal recognizes that financial services like banking, insurance, and credit are essential for full economic participation (Rai et al., 2019). SDG 8.10 calls on countries to strengthen their financial institutions to serve all citizens better. Achievement of this goal involves expanding access to bank accounts and digital financial services, creating financial products for low-income people and improving financial literacy and consumer protection. By promoting financial inclusion, countries can work toward reducing poverty and building a sustainable global economy.

Financial inclusion is a cornerstone of global development, deeply intertwined with the aspirations of sustainable economic growth, social equity, and inclusive prosperity. This handbook dives deep into the complexities of financial inclusion and its linkages with the SDG. It provides an overview of how banking, insurance, and credit access empower individuals and communities. The authors contributed chapters to this book to examine the challenges that prevent many from accessing services and analyze strategies that governments, financial institutions, and nongovernmental organizations (NGOs) can use to promote broader financial inclusion (Chibba, 2009). Ultimately, by closing the gap in financial access, SDG 8.10 helps pave the way for a more just and prosperous world for everyone.

At its core, this volume embodies the spirit of collaboration and inquiry, bringing together scholars and practitioners from diverse disciplines to engage in a collective dialogue on one of the most pressing challenges of our time. Through its interdisciplinary approach and rich empirical insights, Financial Inclusion: In Pursuit of UN SDG 8.10 seeks to inform, inspire, and catalyze action, driving us closer to realizing a more inclusive and sustainable global economy.

1.2 The Salience of Financial Inclusion

Financial inclusion, a pivotal aspect of global development initiatives, ensures that all individuals and businesses, particularly those marginalized

G. W. Muschert (✉)
Khalifa University of Science and Technology, Abu Dhabi, United Arab Emirates
e-mail: glenn.muschet@ku.ac.ae

S. K. Nair · B. A. V. Nair
University of Kerala, Thiruvananthapuram, Kerala, India

G. W. Muschert et al. (eds.), *Financial Inclusion*, Sustainable Development Goals Series,
https://doi.org/10.1007/978-3-031-68803-4_1

or underserved, can access affordable and suitable financial services (see Pesqué-Cela et al., 2021). Financial inclusion aims to create inclusive financial systems that empower individuals, enhance livelihoods, and foster economic growth. Financial inclusion encompasses a broad spectrum of financial services, including but not limited to banking, credit, insurance, and savings. It goes beyond mere access to encompass the availability, affordability, and usability of financial products and services tailored to meet the diverse needs of users. Financial inclusion strives to bridge the gap between the financially excluded and the formal financial system, promoting financial resilience, stability, and prosperity.

Measuring financial inclusion involves assessing various dimensions, including access, usage, quality, and outcomes (Demirguc-Kunt et al., 2018). Access indicators gauge the reach of financial services, such as the availability of bank branches in remote areas or the barriers individuals encounter in accessing formal financial institutions. Usage indicators delve into how individuals utilize financial services over time, providing insights into savings behavior, transaction patterns, and credit usage. Quality measures evaluate the suitability and effectiveness of financial products and services, ensuring they align with users' needs and preferences. Finally, outcome indicators assess the impact of financial inclusion on households and businesses, shedding light on economic empowerment, entrepreneurship, and human capital development.

To effectively measure and monitor financial inclusion, policymakers and stakeholders rely on a combination of demand-side and supply-side data surveys (see Beck, 2016; Park & Mercado, 2021). Demand-side surveys capture insights directly from users of financial services, offering a nuanced understanding of their needs, behaviors, and socioeconomic characteristics. On the other hand, supply-side surveys provide data from regulated financial institutions, offering insights into geographical access, pricing, and product penetration. By leveraging both data types, countries can develop comprehensive strategies to address financial exclusion, identify critical barriers, and tailor policies to specific population groups.

Financial inclusion is a cornerstone of inclusive economic development, empowering individuals, fostering entrepreneurship, and driving sustainable growth. Through collaborative efforts and data-driven insights, stakeholders can work toward building more inclusive and resilient financial systems, thereby unlocking opportunities for all segments of society.

1.3 Introductory Matters

The volume's Part I, Introductory Matters, contains three chapters. The academic journey begins with a comprehensive overview of SDG 8.10, setting the stage for subsequent discussions. Following this introduction is Chap. 2, "Unveiling SDG 8.10: Financial Inclusion for Sustainable Development," in which authors Leonie Jooste and Paul de R. Jooste emphasize strengthening domestic financial institutions to expand access to banking, insurance, and financial services for everyone. The chapter explores the rationale behind this target and its significance for achieving sustainable development. The chapter highlights the ongoing need for research in this area and emphasizes the critical role of financial inclusion in achieving sustainable development, particularly in the face of challenges like COVID-19 (Tay et al., 2022).

Key areas covered in the chapter include:

- Background and purpose of SDG 8.10 and its indicators (8.10.1 and 8.10.2)
- The role of effective financial structures in fostering economic development and inclusion
- The link between financial inclusion and mobile money services. (Ahmad et al., 2020)
- Analysis of progress made toward achieving SDG 8.10, including successes and limitations
- The impact of the COVID-19 pandemic on financial inclusion efforts

Chapter 3, "Forward and Backward Linkages of Financial Inclusion, Fintech, and Sustainable Development Goals: A Bibliometric Analysis and Concept Mapping," traverses the forward and backward linkages of financial inclusion, financial technologies (fintech), and the SDGs. Authors Biju A. V. Nair and Katia Lopez explore the crucial role of financial inclusion in achieving the SDGs, particularly SDG 8: decent work and economic growth. It investigates how fintech bridge financial inclusion and reach these global development targets (Gabor & Brooks, 2020). The chapter establishes the connection between financial inclusion and social and economic development. It highlights how widespread access to financial services improves living standards, health outcomes, and poverty reduction. Fintech can be a crucial driver of financial inclusion.

Through bibliometric mapping, the chapter delves into how fintech solutions like mobile money can facilitate financial access for a broader population. The researchers investigate how fintech-enabled financial inclusion contributes to SDG 8, particularly the sub-indicator 8.10.2, which focuses on enhancing access to financial services. A bibliometric analysis using tools like VOSviewer and R Studio is employed to explore the existing research landscape. This analysis examines keyword co-occurrence and thematic mapping to identify research trends and knowledge gaps. The chapter aims to identify fruitful areas for further research in financial inclusion, fintech, and sustainable development. Such topics include exploring the impact of fintech on financial inclusion across different countries and at the individual level, ultimately contributing to achieving the SDGs. It aims to provide valuable insights into financial inclusion mechanisms and highlight areas requiring further investigation to achieve the UN's SDGs by 2030. Overall, it emphasizes the critical role of fintech in promoting financial inclusion and its positive impact on achieving sustainable development goals.

1.4 Conceptual Concerns

Part II of the volume presents conceptual inquiries into the myths and realities surrounding financial inclusion and pragmatic examinations of frugal innovation and fintech's transformative potential. Each chapter traverses diverse aspects of the conceptual terrain, shedding light on critical issues and emerging trends.

Chapter 4 is "Financial Inclusion Reduces Inequality and Poverty: Myth or Reality?" Authors Asima Shirazi and Shehnaz Al-Helo illuminate the role of financial inclusion in reducing inequality and poverty. This chapter explores the connection between financial inclusion and achieving SDGs, particularly those focused on poverty reduction and inequality (Roy et al., 2018). Expanding access to financial services like banking and insurance is crucial to financial inclusion, as stated in SDG 8.10. The chapter argues that financial inclusion can be a tool for achieving broader development goals. The chapter investigates the impact of financial inclusion on poverty, income inequality, and other non-income inequalities like health and education. Using panel data regression, researchers analyzed data from 138 developed and developing countries over 9 years, from 2010 to 2018, allowing them to examine changes within and between countries over time. The study finds a strong negative association between financial inclusion and poverty (Erlando et al., 2020), suggesting that financial inclusion can effectively reduce poverty. However, the evidence for its impact on income inequality and other non-income inequalities is less conclusive. The chapter highlights the need for policymakers to design financial services that promote equal opportunities and integrate financial inclusion strategies with broader development initiatives to achieve the SDGs. The authors suggest that financial inclusion is a crucial factor in poverty alleviation but may require additional measures to address broader aspects of inequality (Omar & Inaba, 2020).

In Chap. 5, "Driving Financial Inclusion through Frugal Innovation to Achieve UN SDG Target 8.10," Kerry Slade examines how mobile money, a financial service offered through mobile phone companies, can increase financial inclusion, which is the accessibility of financial services (Okello et al., 2018). It signifies that mobile money services address this challenge by making financial services more accessible through a technology many people already have: a mobile phone. This chapter uses the case study of M-Pesa, the world's first mobile money service (Mugambi et al., 2014), to illustrate how a frugal innovation approach can help design financial services that meet the needs of the unbanked (Burns, 2018). Frugal innovation focuses on reducing complexity and cost, making it an ideal approach for developing solutions in resource-constrained environments. The chapter argues that frugal innovation can be a transformative tool for developing and developed economies as they seek to create more inclusive financial systems (Altamirano & van Beers, 2018).

Imad A. Moosa explores the concept of fintech, or financial technology, and analyzes its impact on society. In Chap. 6, "The Benefits and Costs of Fintech," the author argues for a balanced perspective on financial inclusion, pointing out that many uncritically celebrate fintech's potential to promote financial inclusion and economic growth. The chapter begins by highlighting the factors that have accelerated its development, such as big data and advancements in artificial intelligence (Hoang et al., 2022). It then examines the supposed benefits of fintech, including increased financial inclusion and economic growth. However, the authors caution that these claims may overlook potential drawbacks. The chapter identifies two critical areas of concern: operational risks and the "war on cash" (Ferreira & Dickason-Koekemoer, 2019). Operational risks encompass the possibility of fraud and technological failures. The "war on cash" refers to the potential societal consequences of phasing out physical currency in favor of digital transactions. The author points out that fintech enthusiasm often overshadows these potential downsides. They call for a more critical assessment of fintech's costs and benefits, urging policymakers to consider the impact on financial stability, privacy, and consumer protection. The chapter concludes by emphasizing the need for a cautious approach to adopting fintech solutions. Although technological advancements offer exciting possibilities, it is crucial to acknowledge and mitigate the potential risks associated with them (Buckley et al., 2019).

Haruna Issahaku contributes Chap. 7, "Digital Finance for Inclusive Development in Agriculture," discussing the application of digital finance for the inclusive development of agriculture. The author identified the potential of digital finance to transform agriculture and promote inclusive development. The chapter defines digital finance as using digital channels to access and deliver financial products and services (Durai & Stella, 2019). It highlights the knowledge gap regarding the concrete impact of digital finance on agricultural development and poverty reduction and emphasizes the need for rigorous research to move beyond anecdotal evidence.

The urgency of exploring this topic stems from several factors. First, digital finance offers a promising avenue to address financial exclusion, a significant barrier for smallholder farmers, most of whom live in poverty. Second, digital finance can enhance agricultural productivity, potentially lifting millions out of poverty (Xu & Wang, 2023). Finally, the COVID-19 pandemic underscores the importance of digital financial inclusion (Machasio, 2020), which minimizes physical contact, reducing the spread of the virus. The chapter then discusses the rise of digital financial tools driven by technological advancements, mobile money services, and fintech companies. These innovations have led to new financial products specifically designed for the agricultural sector, including digital payments, savings platforms, credit and insurance solutions, crowdfunding opportunities, and e-commerce platforms.

While the chapter acknowledges mixed results, it points toward evidence suggesting a positive influence of digital finance on agricultural output, productivity, consumption patterns, financial inclusion, and poverty reduction. It also highlights the potential of digital finance to empower women by narrowing the gender finance gap (Kofman & Payne, 2021). However, the chapter recognizes challenges that hinder the widespread adoption of digital financial tools. These challenges include security concerns, regulatory uncertainties, limited access to affordable and reliable internet connectivity, potential disruption of existing business models, and a need for more financial products tailored to the specific seasonal needs of smallholder farmers. It concludes by emphasizing the need to address these challenges through enhanced security features, stricter regulations to combat fraud, capacity building for agribusinesses to navigate digital disruption, and the development of financial products catering to smallholder farmers' unique requirements. By overcoming these hurdles, digital finance has the potential to unlock a new era of agricultural development and inclusive growth (Anshari et al., 2019).

In Chap. 8, "Financial Literacy, Economic Inequality, and Trust in Government," Alina R. Oxendine outlined the critical issue of economic inequality and its impact on political attitudes. The chapter highlights the importance of financial literacy in promoting trust in government and a healthy democratic society, aligning with the UN Sustainable Development Goals for economic empowerment and reduced inequality. The chapter examined existing research on how exposure to economic inequality can erode public trust in government. It then investigates the role of financial literacy in shaping these political beliefs. The author uses survey data to analyze the connection between financial knowledge, trust in government, and people's views on inequality.

The key finding is that financial literacy strengthens positive democratic attitudes, such as confidence in government and belief in one's ability to influence politics, which is political efficacy. Furthermore, financial literacy acts as a buffer, mitigating the adverse effects of economic inequality on trust in government (Hung et al., 2009). Those with lower financial literacy are more likely to lose faith in government when faced with high inequality. This finding underscores the importance of promoting financial literacy to combat economic inequality's negative political consequences. The chapter argues for increased access to financial education and further research on the connections between financial knowledge and political attitudes. By fostering financial literacy, we can create a more resilient and trusting citizenry equipped to address the challenge of economic inequality (Lusardi & Mitchell, 2014).

In the section's final Chap. 9, "Financial Literacy, Access to Microfinance, and Sustainable Development," authors Morshadul Hasan and Ariful Hoque evaluated the crucial role of financial literacy in enabling access to microfinance in Chap. 9. Microfinance, a powerful tool for poverty reduction, offers financial services to underserved communities (Bakhtiari, 2006). Financial literacy empowers individuals to understand and utilize these services effectively. The author pinpointed the role of microfinance as a driver of financial inclusion and economic development. The chapter emphasizes microfinance's role in poverty alleviation, promoting gender equality, and fostering economic growth, all of which contribute to achieving SDGs. The author discusses the importance of financial literacy in enabling informed decisions about financial products and services like microfinance. Financial literacy empowers individuals to manage money effectively, utilize microfinance for productive purposes, and avoid financial risks (Suprapti et al., 2024). The chapter acknowledges existing financial literacy and microfinance research but identifies a gap in understanding the impact of technology-based microfinance access (Singh & Padhi, 2015). The chapter mainly bridges this gap by examining the relationship between financial knowledge and access to microfinance.

The author employed an experimental methodology using probabilistic models to analyze the impact of specific financial knowledge areas (savings, loans, and interest rates) on microfinance access. The findings confirm a positive and significant correlation between financial literacy and microfinance access (Okello et al., 2020). The chapter concludes by underscoring the practical implications of the research for promoting financial inclusion, particularly in rural areas. It highlights the need for further research on measuring and developing financial literacy programs for underserved communities (Jacob et al., 2000).

Overall, this chapter emphasizes the importance of financial literacy as a critical factor in ensuring adequate access to microfinance and promoting financial inclusion for sustainable development.

1.5 Global and Regional Topics

Part II meticulously examines the global and regional financial inclusion landscape, with chapters delving into the challenges and opportunities across various geographical contexts. From the financial inclusion of rural women in the Global South to initiatives and hurdles encountered in the BRICS countries, the volume offers a panoramic view of the multifaceted nature of financial inclusion efforts.

Chap. 10, "Financial Inclusion of Rural Women in the Global South," discusses rural women's financial inclusion in the Global South. Carolina Herrera-Cano, Arley Pino, Carlos Felipe Munera-Alzate, and Maria Alejandra Gonzalez-Perez evaluate the critical role of financial inclusion for rural women in achieving sustainable development, particularly within the framework of SDG 8.10. It emphasizes the importance of a gendered perspective within the 2030 Agenda (Rai et al., 2019) and analyzes the current limitations on rural women's access to essential financial services. The chapter identifies entrenched gender norms, socioeconomic segregation, and a lack of financial literacy as key barriers to financial inclusion for rural women. It explores the negative consequences of this exclusion, including limited economic opportunities, vulnerability during crises, and hindered progress toward other SDGs like poverty reduction and gender equality (Razavi, 2016). The chapter then highlights the potential benefits of financial inclusion for rural women, including increased economic autonomy, bargaining power, and agency. It showcases successful historical and current programs that promote financial inclusion in the Global South, offering recommendations for businesses, governments, and civil society to implement similar initiatives. The chapter argues that by promoting financial inclusion for rural women, we can achieve a more inclusive and sustainable future, fostering economic growth and empowering women to contribute to personal wealth-building and community development.

Chapter 11 of the collection, "Assessing the Impact of Financial Inclusion on Economic Growth in Island Economies," investigates the link between financial inclusion and economic growth, specifically focusing on the effect of *financial inclusion on economic growth in island economies*. The authors Narvada Gopy Ramdhany, Reena Bhattu Babajee, and Boopen Seetanah, while investigating the positive relationship between these two factors, generally acknowledged that more evidence is needed, particularly for small island developing states (SIDS) economies (Rudge, 2021). Key points noted in the chapter are financial inclusion and access to formal financial services, which are believed to drive economic growth, especially for underserved populations (Lyons & Kass-Hanna, 2021). The authors have noted that existing research primarily focuses on developed economies, with limited studies on developing countries and none specifically on SIDS. This chapter aims to fill this gap by examining the impact of financial inclusion on economic growth in a sample of SIDS using a multidimensional financial inclusion index and advanced econometric techniques. The study also explores the possibility of a reverse causal effect where economic growth influences. Findings from this research contribute to understanding financial inclusion's role in achieving SDG 8.10 on promoting decent work and economic growth and provide valuable policy recommendations for SIDS governments. This chapter offers valuable insights for policymakers, academics, and anyone interested in promoting inclusive and sustainable economic growth in SIDS (Ghina, 2003).

In Chap. 12, "Financial Inclusion: Initiatives and Challenges in the BRICS Countries," authors Debdutta Choudhury, Apoorva Vaishnavi Bhattaram, D. Priya Dhatri, and Roshan Bhut studied the financial inclusion initiatives undertaken by the BRICS economies such as Brazil, Russia, India, China, and South Africa. These developing economies aim to empower citizens by promoting financial inclusion, reducing poverty and inequality, and fostering economic growth (Narayan, 2002). The chapter delves into the specific strategies employed by each BRICS nation. For instance, it is crucial to examine Brazil's focus on expanding physical access to banking channels and its innovative digital solutions. Then, Russia's emphasis on financial literacy programs highlights its efforts to create a barrier-free financial ecosystem for vulnerable groups. India's success story in leveraging Aadhaar and mobile technology to drive financial inclusion through schemes like PMJDY and UPI is analyzed. China's targeted approach, which uses a combination of monetary, tax, and regulatory policies to promote financial inclusion, particularly for underserved segments like small businesses and rural residents, is explored. Finally, South Africa's efforts to increase access to financial services through stakeholder collaboration and address gender and racial disparities in financial inclusion are discussed. By comparing and contrasting the approaches of these BRICS economies, the chapter aims to provide valuable insights for policymakers and financial institutions across the globe who are striving to achieve financial inclusion in their own countries. Key themes explored in the chapter include:

- The importance of financial inclusion for economic development and poverty reduction
- Strategies for increasing physical and digital access to financial services
- The role of government policies and regulations in promoting financial inclusion
- The importance of financial literacy and consumer protection (Rutledge, 2010)
- Addressing gender (Roy & Patro, 2022) and racial disparities in financial inclusion
- Lessons learned from the BRICS economies for other developing countries

Author Calvin Habasonda uses a newly developed financial inclusion index (FII) to explore financial inclusion in African countries in Chap. 13, "Assessing Financial Inclusion in 41 African Countries." The FII considers three critical dimensions of financial inclusion: access, usage, and availability. The chapter highlights the following key points: Financial inclusion is critical for economic growth, poverty reduction (Menyelim et al., 2021), and achieving the SDGs. Despite the perceived positive impact of mobile banking, progress in financial inclusion across African countries has been uneven. The chapter proposes a novel FII that considers access, usage, and availability to measure financial inclusion progress in 41 African countries from 2010 to 2019. The analysis reveals that most countries fall under the low financial inclusion category, highlighting the need for continued efforts to expand financial services. The study finds a positive correlation between financial inclusion and the availability of financial services, suggesting that policies promoting accessibility may also indirectly increase usage (Yadav et al., 2021). The chapter emphasizes the importance of measuring financial inclusion and highlights areas for policy intervention to achieve financial inclusion goals in Africa.

1.6 Country Studies

Part IV contains country-specific studies that enrich the discourse, offering insights from on-the-ground experiences and localized challenges. These chapters offer valuable insights into the intricate dynamics shaping financial inclusion strategies and outcomes from Bangladesh to South Africa and from Turkey to Morocco.

In Chap. 14, "The Role of School Banking in Promoting Financial Inclusion: Evidence from Bangladesh," author Nur Alam Siddik investigates the potential of school banking programs to promote financial inclusion in Bangladesh. Notably, this chapter investigated the potential power of access to financial services for all citizens, which is crucial for achieving SDGs, particularly SDG 8.10 on sustainable economic growth. The chapter begins by highlighting the importance of financial inclusion for Bangladesh's development goals. It then introduces school banking as a strategy to bring young people into the formal financial system. School banking programs aim to educate students on essential financial management skills and banking practices, potentially fostering responsible financial behavior in the future (Johnson et al., 2018).

Despite the limited recognition of school banking in Bangladesh, the chapter argues that it is gaining traction among students and banks. It examines the potential benefits of school banking programs, including educating children on financial management skills, acclimatizing students to banking products and services, and promoting savings behavior among youth. The chapter presents original research using data from 2011 to 2019 to analyze the impact of school banking on financial inclusion in Bangladesh. The findings suggest that school banking activities positively influence all three dimensions of financial inclusion: access, usage, and depth. The chapter concludes by emphasizing the need for policymakers to create a more conducive environment for school banking, develop consistent reporting systems for banks, and encourage nationwide program adoption. Bangladesh can make significant strides toward achieving its sustainable development goals by fostering financial inclusion through school banking initiatives (Mondal et al., 2015).

In Chap. 15, "Microenterprises in the Agricultural Sector and (Lack of) Insurance: Exploratory Case Studies from Fresh Fruit Farmers and Retailers from Northern Pakistan," Ahmad Arslan, Saranjam Baig, and Aftab Ahmed Khan examined the financial inclusion landscape for microentrepreneurs in the agricultural sector, mainly focusing on the lack of access to microinsurance schemes. The study occurred in Gilgit-Baltistan, a mountainous and remote region in Pakistan. Here are the key takeaways for readers while reading through this chapter. The authors noted that, despite some knowledge of micro-health insurance, none of the interviewed micro-entrepreneurs (producers and retailers of fresh fruits) had micro-business or agricultural insurance. There is a need for increased awareness about the existence and benefits of such schemes. Without formal insurance, microentrepreneurs rely heavily on social networks like family and village communities for financial support in case of losses, a dynamic also observed in many other developing economies.

The study suggests a correlation between education level and awareness about microinsurance options (Asmare & Worku, 2018; Mohammed & Mukhtar, 2012). Similar to findings in other financial inclusion contexts, those with higher education exhibited a better understanding of the importance of insurance for business sustainability. The mountainous terrain and harsh weather conditions in Gilgit-Baltistan lead to spoilage of fresh fruits due to limited storage facilities and frequent blockages of the Karakorum Highway, the primary access route, highlighting the vulnerability of these businesses and the need for risk mitigation strategies like insurance. Despite the lack of microinsurance, all respondents expressed interest in such schemes if they were made available and accompanied by proper guidance. This positive sentiment aligns with their existing use of private health insurance, indicating a basic understanding of insurance products. The income and product availability of these microentrepreneurs highly depend on the season. The lack of proper storage facilities creates additional challenges, highlighting the need for infrastructure development alongside financial inclusion initiatives.

While some microentrepreneurs operate solely in cash, others, especially those experiencing business growth, recognize the increasing necessity of formal banking channels and financial services. Overall, the chapter underlines the critical role of micro-insurance in enhancing the financial security (Habib et al., 2016) and sustainability of microentrepreneurs in the agricultural sector, particularly in geographically challenging regions like Gilgit-Baltistan. It highlights the need for targeted interventions to raise awareness, improve access to microinsurance products, and develop infrastructure to address the limitations.

In Chap. 16, "Financial Inclusion for Climate Resilience in Rural South Africa: A Qualitative Analysis of Opportunities and Barriers," Maheshvari Appavoo evaluated the link between financial inclusion and climate resilience (Haworth et al., 2016) in rural South Africa, particularly for vulnerable populations. It examines how access to financial services can bolster communities' ability to cope with climate-related disasters like floods, droughts, and storms. The study's key findings are that climate change disproportionately impacts impoverished populations in developing countries, especially those reliant on natural resources. Financial inclusion can strengthen resilience to climate shocks through access to formal financial products and services. While South Africa has made strides in financial inclusion, significant challenges remain, particularly for rural communities. The chapter investigates the potential of financial services in managing crises, minimizing risk exposure, and supporting economic recovery after climate disasters (Ghesquiere & Mahul, 2010). It emphasizes the importance of designing financial products that consider the needs of vulnerable populations, including low-income households, migrants, and women.

The chapter contributes to understanding financial inclusion as a tool for climate change adaptation, particularly in developing countries. It highlights the limitations of traditional disaster relief programs and underscores the need for long-term solutions. The research encourages policymakers to prioritize financial inclusion strategies to improve the financial resilience of vulnerable communities (Hussain et al., 2021). The chapter suggests integrating indigenous knowledge with scientific approaches to develop more effective climate adaptation policies. It emphasizes the need for further research on financial products and services tailored to address the specific risks of climate change for low-income populations.

In Chap. 17, "Gendered Social Norms and Financial Inclusion in Turkey, Nuray Karaman examines the gender gap in financial inclusion. While financial inclusion has become a key development strategy globally, current initiatives in Turkey often overlook the specific challenges women face (Antonijević et al., 2022). The authors examine the reasons behind the gender gap. The author elucidates the factors that determine gender imbalances in financial inclusion. First, social and cultural factors (Rink et al., 2021) are examined, where the author analyzes how cultural norms and patriarchal structures in Turkey can limit women's access to financial resources and participation in the financial system. Subsequently, the author discussed policy shortcomings, arguing that current financial inclusion policies in Turkey do not adequately address the unique needs of women.

The research presented in this chapter highlights the importance of understanding gendered social norms to achieve equitable financial inclusion (Waldron et al., 2021). It proposes that incorporating a gendered lens into policymaking can lead to more effective solutions that bridge the gap in women's access and usage of financial services in Turkey. This chapter contributes to the field of financial inclusion by providing policymakers with a fresh perspective for crafting future regulations that specifically address the needs and contributions of women.

In Chap. 18, "Fintech and Financial Inclusion in the Kingdom of Morocco: Demographics and Socioeconomic Characteristics," Abdulkader Aljandali and Amal Laaribi examined the role of fintech in financial inclusion in Morocco. It highlights the challenges of financial exclusion in Morocco, including low bank account penetration and a cash-dominant economy. The chapter then discusses the rise of fintech solutions in Morocco (Naz et al., 2022), particularly mobile banking and payments and their potential to address these challenges. The chapter covers the need for financial inclusion: Financial inclusion is essential for economic development and individual well-being. However, many members of the Moroccan population lack access to formal financial services. Several factors hinder financial inclusion in Morocco, including limited access to bank branches, high costs of financial services, and a lack of financial literacy, particularly among women and rural populations.

The authors find that fintech presents an opportunity to expand financial inclusion in Morocco. Mobile banking and payments offer a convenient and affordable way for people to access financial services. The chapter touches on the governance initiatives of the Moroccan government, which is actively supporting the development of the fintech sector (Abdelkhalek et al., 2021). It has implemented regulations that allow non-bank institutions to offer financial services and is investing in digital infrastructure. The strategy of the Moroccan government outlines a roadmap for achieving financial inclusion in Morocco. It emphasizes the role of fintech in reaching underserved populations and developing alternative financial models.

This chapter contributes to understanding fintech's role in promoting financial inclusion in developing economies. It provides a specific case study of Morocco and highlights the opportunities and challenges associated with leveraging fintech for financial inclusion. The authors concluded that fintech has the potential to revolutionize the financial landscape in Morocco and promote inclusive economic growth. However, much work remains concerning challenges such as consumer protection and ensuring the sustainability of financial inclusion gains (Jagtiani & John, 2018).

In Chap. 18, "Mobile Money Transactions for Financial Inclusion in Socio-economic Development: An Interpretative Account of Two Ghanaian Rural Communities," Kofi A. Boateng, Rosemary Boateng Coffie, and Mercy A. Boateng examined the growing significance of mobile money (MoMo) transactions in fostering financial inclusion and driving socioeconomic development within rural communities. The authors highlighted that the increasing ubiquity of mobile phones and information communication technology (ICT) has paved the way for innovative financial tools like MoMo, particularly in developing nations. The chapter utilizes a qualitative approach, drawing on the experiences of MoMo users and merchants in two Ghanaian communities to understand how these transactions contribute to financial inclusion. Through this lens, the research sheds light on a previously under-examined facet of development literature: the role of technology-driven financial inclusion in rural settings.

The chapter highlights the positive impact of MoMo on socioeconomic development (Agrawal, 2016; Nan, 2019) and acknowledges potential challenges. Security concerns, particularly around fraud and limitations in rural infrastructure, such as network connectivity (Senso & Venkatakrishnan, 2013), are identified as areas requiring further attention. Finally, the chapter proposes directions for future research, including investigating potential barriers to financial inclusion in urban areas and the often-held assumption that urban centers hold all the advantages for development. By challenging this notion, further research can provide valuable insights for policymakers aiming to bridge the development gap between rural and urban communities.

1.7 Driving Research Forward

As the volume concludes, it begins to look ahead, examining how financial inclusion affects the economy, the small and large economic factors that affect the use of mobile money, and how the UN's Sustainable Development Goals can be used to measure financial inclusion.

In Chap. 20, "Nexus between Financial Inclusion and Economic Activity: A Study about Traditional and Non-Traditional Financial Service Indicators Determining Financial Outreach," Florian Gerth investigates the relationship between financial inclusion and economic activity. The author argues that financial inclusion is vital for economic development and that nontraditional financial services such as mobile money and non-branch retail agent outlets are becoming increasingly important in promoting financial inclusion. The chapter also highlights the importance of a healthy financial sector for financial inclusion. The chapter makes several contributions to the literature. First, it uses a new measure of financial inclusion that considers nontraditional financial services (Gerth, 2023). Second, it uses a more extended sample period and more countries than previous studies. Third, it finds a positive relationship between financial inclusion and economic activity (Sethi & Acharya, 2018), even when controlling for nontraditional financial services. Fourth, it finds that the financial sector's health is an essential determinant of financial inclusion. The chapter concludes with several policy implications. It recommends that policymakers design policies to promote financial inclusion, such as increasing financial literacy (Grohmann et al., 2018) and ensuring the financial sector is safe and sound.

"Micro- and Macro-Economic Factors of Mobile Money Adoption in the Middle East and North Africa," in Chap. 21 by Dimitrios Reppas, examines how mobile money has emerged as a powerful tool for financial inclusion, especially in sub-Saharan Africa. While mobile money adoption is rising globally, the Middle East and North Africa (MENA) region lags. This chapter delves into the reasons behind this disparity. It is a digital currency stored on mobile phones, enabling cashless transactions like transfers, deposits, and bill payments. It bypasses traditional banking systems (Broby, 2021). Mobile money has seen significant growth worldwide, particularly in Africa. However, MENA has a lower penetration rate with significant variation between countries (Siddika & Sarwar, 2023). The chapter aims to identify the factors influencing mobile money adoption in MENA. It analyzes existing research, which primarily focuses on Africa, and highlights the need for studies specific to the MENA region.

The chapter explores potential micro- and macroeconomic factors affecting Mobile Money adoption in MENA. These factors could include:

- Existing banking infrastructure
- Population demographics (e.g., young, tech-savvy population)
- Regulatory environment
- Income inequality
- Recent digitalisation efforts

By understanding these factors, researchers can design better studies to promote financial inclusion through mobile money in MENA. The chapter emphasizes the need for further research to validate the proposed drivers.

In Chap. 22, "UN SDG Indicators 8.10 for Measuring Financial Inclusion: An Assessment," Mandira Sarma explores the concept of financial inclusion and how it is measured, focusing on the limitations of current methods used by the United Nations SDGs. Financial inclusion ensures that people can easily access and use formal financial services, like bank accounts, credit, and insurance (Krishnan et al., 2012). It is crucial for economic development and poverty reduction. The chapter also highlighted the problem with current SDG indicators: The SDG sub-target 8.10 aims to expand access to financial services. However, the current indicators used to measure progress (number of bank branches, ATMs per capita, and account ownership) are limited. The author discussed the limitations of current indicators because of the following:

The author suggests the need for an index-based, more comprehensive approach. Such measurement would combine various indicators into a single score, capturing a more holistic view of financial inclusion. The chapter proposes additional indicators to consider: The number of microfinance outlets, mobile money, and insurance companies—usage measures like outstanding deposits, loan volume, and transaction value. The chapter concludes and argues for a more comprehensive approach to measuring financial inclusion. By using broader indicators (Tram et al., 2023) and index-based methods, we can better understand a country's progress toward financial inclusion.

1.8 Conclusion

This book delves into the intersection and mapping of financial inclusion, fintech, and SDG. The readers of this book experience the pertinent areas of financial inclusion and its thematic interconnection, a noted research gap. The chapters underscore the role of financial inclusion in sustainable development and its impacts on reducing poverty. Some chapters have empirically proven the interlinkages of financial inclusion, digital finance, and fintech. Some chapters provide insight into financial literacy, financial inclusion, and sustainable development linkages. Chapters cover the significant contributions of regional and national evidence of financial and digital financial inclusion and its impacts on economic prosperity. This book provides tools for measuring financial inclusion that are more useful to readers looking to understand the quantum of financial inclusion. After traversing through chapters, this handbook underscores the significant contribution of financial inclusion and its technological additions to poverty eradication. This handbook confirms that overall financial inclusion significantly contributed to the UN agenda of SDG.

References

Abdelkhalek, T., Ajbilou, A., Benayad, M., Boccanfuso, D., & Savard, L. (2021, November). *How can the digital economy benefit Morocco and all Moroccans? Economic Research Forum (ERF)*. https://bit.ly/3I2PXUX

Ade'Soyemi, K., Olowofela, O. E., & Yunusa, L. A. (2020). Financial inclusion and sustainable development in Nigeria. *Journal of Economics and Management, 39*(1), 105–131. https://sciendo.com/article/10.22367/jem.2020.39.06

Agrawal, R. (2016). Mobile money empowering people living at the bottom of pyramid and boosting socio-economic development in a big way. *Economic Analysis, 49*(1–2), 1–23. https://library.ien.bg.ac.rs/index.php/ea/article/view/325/321

Ahmad, A. H., Green, C., & Jiang, F. (2020). Mobile money, financial inclusion, and development: A review with reference to African experience. *Journal of Economic Surveys, 34*(4), 753–792. https://doi.org/10.1111/joes.12372

Altamirano, M. A., & Beers, C. P. V. (2018). Frugal innovations in technological and institutional infrastructure: Impact of mobile phone technology on productivity, public service provision, and inclusiveness. *The European Journal of Development Research, 30*, 84–107. https://doi.org/10.1057/s41287-017-0115-2

Anshari, M., Almunawar, M. N., Masri, M., & Hamdan, M. (2019). Digital marketplace and FinTech to support agriculture sustainability. *Energy Procedia, 156*, 234–238. https://doi.org/10.1016/j.egypro.2018.11.134

Antonijević, M., Ljumović, I., & Ivanović, Đ. (2022). Is there a gender gap in financial inclusion worldwide? *Journal of Women's Entrepreneurship and Education, 1–2*, 79–96. https://doi.org/10.28934/jwee22.12.pp79-96

Asmare, A., & Worku, A. (2018). Determinants of micro-insurance demand in Jimma zone. *International Research Journal of Business Studies, 11*(3), 145–157. http://download.garuda.kemdikbud.go.id/article.php?article=848109&val=11160&title=Determinants%20of%20Micro-Insurance%20Demand%20in%20Jimma%20Zone

Baffoe, G., Zhou, X., Moinuddin, M., Somanje, A. N., Kuriyama, A., Mohan, G., Saito, O., & Takeuchi, K. (2021). Urban-rural linkages: Effective solutions for achieving sustainable development in Ghana from an SDG interlinkage perspective. *Sustainability Science, 16*(4), 1341–1362. https://doi.org/10.1007/s11625-021-00929-8

Bakhtiari, S. (2006). Microfinance and poverty reduction: Some international evidence. *International Business & Economics Research Journal (IBER), 5*(12). https://doi.org/10.19030/iber.v5i12.3550

Beck, T. (2016, November). Financial Inclusion–Measuring progress and progress in measuring. In *This paper was written for the Fourth IMF Statistical Forum "Lifting the Small Boats: Statistics for Inclusive Growth"*. Cass Business School, City, University of London, CEPR, and CESifo.

Broby, D. (2021). Financial technology and the future of banking. *Financial Innovation, 7*(1), 47. https://doi.org/10.1186/s40854-021-00264-y

Buckley, R. P., Arner, D. W., Zetzsche, D. A., & Selga, E. (2019). The dark side of digital financial transformation: The new risks of fintech and the rise of tech risk. *UNSW Law Research Paper, 19–89*. https://doi.org/10.2139/ssrn.3478640

Burns, S. (2018). M-Pesa and the 'market-led' approach to financial inclusion. *Economic Affairs, 38*(3), 406–421. https://doi.org/10.1111/ecaf.12321

Chibba, M. (2009). Financial inclusion, poverty reduction, and the millennium development goals. *The European Journal of Development Research, 21*, 213–230. https://doi.org/10.1057/ejdr.2008.17

Demirguc-Kunt, A., Klapper, L., Singer, D., & Ansar, S. (2018). *The Global Findex Database 2017: Measuring financial inclusion and the Fintech revolution*. World Bank Publications.

Durai, T., & Stella, G. (2019). Digital finance and its impact on financial inclusion. *Journal of Emerging Technologies and Innovative Research, 6*(1), 122–127. https://www.researchgate.net/publication/330933079

Erlando, A., Riyanto, F. D., & Masakazu, S. (2020). Financial inclusion, economic growth, and poverty alleviation: Evidence from eastern Indonesia. *Heliyon, 6*(10) https://www.cell.com/heliyon/pdf/S2405-8440(20)32078-8.pdf

Ferreira, S., & Dickason-Koekemoer, Z. (2019). A conceptual model of operational risk events in the banking sector. *Cogent economics & finance, 7*(1), 1706394. https://doi.org/10.1080/23322039.2019.1706394

Gabor, D., & Brooks, S. (2020). The digital revolution in financial inclusion: International development in the fintech era. In *Material cultures of financialisation* (pp. 69–82). https://doi.org/10.1080/13563467.2017.1259298

Gerth, F. (2023). Nexus between financial inclusion and economic activity: A study about traditional and non-traditional financial service indicators determining financial outreach. https://mpra.ub.uni-muenchen.de/119265/.

Ghesquiere, F., & Mahul, O. (2010). Financial protection of the state against natural disasters: A primer. *World Bank Policy Research Working Paper, 5429*. https://ssrn.com/abstract=1680722

Ghina, F. (2003). Sustainable development in small Island developing states. *Environment, Development and Sustainability, 5*, 139–165. https://doi.org/10.1023/A:1025300804112

Grohmann, A., Klühs, T., & Menkhoff, L. (2018). Does financial literacy improve financial inclusion? Cross-country evidence. *World Development, 111*, 84–96. https://doi.org/10.1016/j.worlddev.2018.06.020

Habib, S. S., Perveen, S., & Khuwaja, H. M. A. (2016). The role of micro health insurance in providing financial risk protection in developing countries—A systematic review. *BMC Public Health, 16*, 1–24. https://doi.org/10.1186/s12889-016-2937-9

Haworth, A., Frandon-Martinez, C., Fayolle, V., & Simonet, C. (2016). Climate resilience and financial services. https://www.cakex.org/sites/default/files/documents/10316.pdf

Hoang, T. G., Nguyen, G. N. T., & Le, D. A. (2022). Developments in financial technologies for achieving the sustainable development goals (SDGs): FinTech and SDGs. In *Disruptive technologies and eco-innovation for sustainable development* (pp. 1–19). IGI Global. https://doi.org/10.4018/978-1-7998-8900-7.ch001

Hung, A., Parker, A. M., & Yoong, J. (2009). *Defining and measuring financial literacy*. https://doi.org/10.2139/ssrn.1498674

Hussain, A. B., Islam, M., Ahmed, K. J., Haq, S. M. A., & Islam, M. N. (2021). Financial inclusion, financial resilience, and climate change resilience. In *Handbook of climate change management: Research, leadership, transformation* (pp. 2085–2107). Springer International Publishing. https://doi.org/10.1007/978-3-030-57281-5_19

Jacob, K., Hudson, S., & Bush, M. (2000). *Tools for survival: An analysis of financial literacy programs*. Woodstock Institute. https://search.issuelab.org/resources/30348/30348.pdf

Jagtiani, J., & John, K. (2018). Fintech: The impact on consumers and regulatory responses. *Journal of Economics and Business, 100*, 1–6. https://doi.org/10.1016/j.jeconbus.2018.11.002

Johnson, L., Lee, Y., Njenga, G., Kieyah, J., Osei-Akoto, I., Rodriguez Orgales, C., et al. (2018). School banking as a strategy for strengthening youth economic participation in developing countries: Lessons from YouthSave. *Global Social Welfare, 5*, 265–275. https://doi.org/10.1007/s40609-017-0109-1

Kofman, P., & Payne, C. (2021). Digital financial inclusion of women: An ethical appraisal. *Handbook on ethics in finance*, 133–157. https://doi.org/10.1007/978-3-030-29371-0_34

Krishnan, J., Mondal, S., & SEM, I. (2012). *Financial inclusion* (p. 716). Emerging Strategies for Sustainable Growth & Competence. https://www.researchgate.net/profile/Mahadev-Patil/publication/

Lusardi, A., & Mitchell, O. S. (2014). The economic importance of financial literacy: Theory and evidence. *American Economic Journal: Journal of Economic Literature, 52*(1), 5–44. https://doi.org/10.1257/jel.52.1.5

Lyons, A. C., & Kass-Hanna, J. (2021). Financial inclusion, financial literacy, and economically vulnerable populations in the Middle East and North Africa. *Emerging Markets Finance and Trade, 57*(9), 2699–2738. https://doi.org/10.1080/1540496X.2019.1598370

Machasio, I. N. (2020). *COVID-19 and digital financial inclusion in Africa: How to leverage digital technologies during the pandemic*. https://hdl.handle.net/10986/34637.

Menyelim, C. M., Babajide, A. A., Omankhanlen, A. E., & Ehikioya, B. I. (2021). Financial inclusion, income inequality, and sustainable economic growth in sub-Saharan African countries. *Sustainability, 13*(4), 1780. https://doi.org/10.3390/su13041780

Mohammed, A. H., & Mukhtar, S. (2012). The prospects of micro-insurance in the rural areas of Nigeria. *European Scientific Journal, 8*(3) http://download.garuda.kemdikbud.go.id/article.php?article=848109&val=11160&title=Determinants%20of%20Micro-Insurance%20Demand%20in%20Jimma%20Zone

Mondal, U., Hossain, M. S., & Khan, M. A. M. (2015). School banking: A new idea of banking operation in Bangladesh. *International Journal of Science and Research, 4*(11), 1926–1931. https://www.researchgate.net/profile/Ujjal-Mondal-3/publication/334084358

Mugambi, A., Njunge, C., & Yang, S. C. (2014). Mobile-money benefits and usage: The case of M-PESA. *IT Professional, 16*(3), 16–21. https://doi.org/10.1109/MITP.2014.38

Nan, W. V. (2019). Mobile money and socioeconomic development: A cross-country investigation in sub-Saharan Africa. *Journal of International Technology and Information Management, 27*(4), 36–65. https://doi.org/10.58729/1941-6679.1392

Narayan-Parker, D. (Ed.). (2002). *Empowerment and poverty reduction: A sourcebook*. World Bank Publications. https://doi.org/10.1596/0-8213-5166-4

Naz, F., Karim, S., Houcine, A., & Naeem, M. A. (2022). Fintech growth during COVID-19 in MENA region: Current challenges and prospects. *Electronic Commerce Research, 1–22*. https://doi.org/10.1007/s10660-022-09583-3

Okello Candiya Bongomin, G., Ntayi, J. M., Munene, J. C., & Malinga, C. A. (2018). Mobile money and financial inclusion in sub-Saharan Africa: The moderating role of social networks. *Journal of African Business, 19*(3), 361–384. https://doi.org/10.1080/15228916.2017.1416214

Okello Candiya Bongomin, G., Mpeera Ntayi, J., & Akol Malinga, C. (2020). Analyzing the relationship between financial literacy and financial inclusion by microfinance banks in developing countries: Social network theoretical approach. *International Journal of Sociology and Social Policy, 40*(11/12), 1257–1277. https://doi.org/10.1108/IJSSP-12-2019-0262

Omar, M. A., & Inaba, K. (2020). Does financial inclusion reduce poverty and income inequality in developing countries? A panel data analysis. *Journal of Economic Structures, 9*(1), 37. https://doi.org/10.1186/s40008-020-00214-4

Park, C. Y., & Mercado, R. V. (2021). Financial inclusion: New measurement and cross-country impact assessment 1. In *Financial inclusion in Asia and beyond* (pp. 98–128). Routledge.

Pesqué-Cela, V., Tian, L., Luo, D., Tobin, D., & Kling, G. (2021). Defining and measuring financial inclusion: A systematic review and confirmatory factor analysis. *Journal of International Development, 33*(2), 316–341. https://doi.org/10.1002/jid.3524

Rai, S. M., Brown, B. D., & Ruwanpura, K. N. (2019). SDG 8: Decent work and economic growth–A gendered analysis. *World Development, 113*, 368–380. https://doi.org/10.1016/j.worlddev.2018.09.006

Razavi, S. (2016). The 2030 agenda: Implementation challenges to attain gender equality and women's rights. *Gender and Development, 24*(1), 25–41. https://doi.org/10.1080/13552074.2016.1142229

Rink, U., Walle, Y. M., & Klasen, S. (2021). The financial literacy gender gap and the role of culture. *The Quarterly Review of Economics and Finance, 80*, 117–134. https://doi.org/10.1016/j.qref.2021.02.006

Roy, P., & Patro, B. (2022). Financial inclusion of women and gender gap in access to finance: A systematic literature review. *Vision, 26*(3), 282–299. https://doi.org/10.1177/09722629221104205

Roy, J., Tscharket, P., Waisman, H., Abdul Halim, S., Antwi-Agyei, P., Dasgupta, P., … Suarez Rodriguez, A. G. (2018). *Sustainable development, poverty eradication, and reducing inequalities*. https://centaur.reading.ac.uk/81048/

Rudge, P. (2021). Beyond the blue economy: Creative industries and sustainable development in small Island developing states. *Routledge*. https://doi.org/10.4324/9781003011514

Rutledge, S. L. (2010). Consumer protection and financial literacy: Lessons from nine country studies. *World Bank Policy Research Working Paper, 5326*. https://doi.org/10.1596/1813-9450-5326

Senso, N. C., & Venkatakrishnan, V. (2013). Challenges of mobile-phone money transfer services' market penetration and expansion in Singida District, Tanzania'. *International Journal of Research in Management & Technology, 3*(6), 205–215. https://citeseerx.ist.psu.edu/document?repid=rep1&type=pdf&doi=224cc98492d44f0625c48ce49e6cd07479565066

Sethi, D., & Acharya, D. (2018). Financial inclusion and economic growth linkage: Some cross country evidence. *Journal of Financial Economic Policy, 10*(3), 369–385. https://doi.org/10.1108/JFEP-11-2016-0073

Siddika, A., & Sarwar, A. (2023). Socio and macroeconomic determinants of mobile money adoption: A cross-cultural study of selected MENA and sub-Saharan countries. *Digital Policy, Regulation and Governance.* https://doi.org/10.1108/DPRG-09-2023-0138

Singh, V., & Padhi, P. (2015). Information and communication technology in microfinance sector: Case study of three Indian MFIs. *IIM Kozhikode Society & Management Review, 4*(2), 106–123. https://doi.org/10.1177/227797521560

Suprapti, I. A. P., Harsono, I., Sutanto, H., Chaidir, T., & Arini, G. A. (2024). Financial inclusion strategies (exploring the landscape through systematic literature review). *Accounting Studies and Tax Journal (COUNT), 1*(1), 101–110. https://doi.org/10.62207/qfpspk76

Tay, L. Y., Tai, H. T., & Tan, G. S. (2022). Digital financial inclusion: A gateway to sustainable development. *Heliyon, 8*(6) https://www.cell.com/heliyon/pdf/S2405-8440(22)01054-4.pdf

Tram, T. X. H., Lai, T. D., & Nguyen, T. T. H. (2023). Constructing a composite financial inclusion index for developing economies. *The Quarterly Review of Economics and Finance, 87*, 257–265. https://doi.org/10.1016/j.qref.2021.01.003

Waldron, D., Siek, H., Mattern, M., & Tukahiirwa, W. (2021). *Addressing gender norms to increase financial inclusion*. https://prd-control-multisite.maneraconsult.com/media/jzvjs2im/2021_cgap_technical_guide_gender_norms.pdf.

Xu, S., & Wang, J. (2023). The impact of digital financial inclusion on the level of agricultural output. *Sustainability, 15*(5), 4138. https://doi.org/10.3390/su15054138

Yadav, V., Singh, B. P., & Velan, N. (2021). Multidimensional financial inclusion index for Indian states. *Journal of Public Affairs, 21*(3), e2238. https://doi.org/10.1002/pa.2238

An Overview of SDGs 8.10.1 and 8.10.2

2

Leonie Jooste and Paul de R. Jooste

Abstract

SDG 8.10 is a relatively new and unresearched topic. Therefore, this chapter briefly overviews target SDG 8.10 and the indicators to strengthen financial institutions and expand access to banking, insurance, and financial services. This chapter reviews the reports, agendas, and surveys on SDG indicators 8.10.1 and 8.10.2. A brief introduction appears on the development of SGD targets and the SDG indicators of SDG 8.10. The chapter then summarizes the rationale and measurements for the indicators, including their impact on financial inclusion and micro- and macroeconomics. Finally, the chapter discusses discrepancies, limitations, and the impact of the COVID-19 pandemic on the achievement of the indicators. The conclusion will include possible relevant topics on SGD 8.10 for future research.

L. Jooste (✉)
Canadian University Dubai,
Dubai, United Arab Emirates
e-mail: Leonie.jooste@cud.ac.ae

P. de R. Jooste
Walkers Professional Services,
George Town, Cayman Islands

Keywords

SGD targets · SGD indicators · United Nations Goals · Sustainable development · SDG 8.10 · ATM · Mobile money · Mobile wallet · Financial inclusion · Financial institutions · Financial services

2.1 Introduction

The Sustainable Development Goals (SDGs) emerged in 2012 at a United Nations (UN) Conference on Sustainable Development in Rio de Janeiro. The conference's objective was to produce goals to meet environmental, political, and economic challenges in the world. The primary goal, Goal 1, of the United Nations (UN) is to end poverty in all its forms everywhere (UN, 2018a). In total, 17 goals emerged to be achieved by 2030 (UN, 2018b). Goal 8, set by the Department of Economic and Social Affairs (IMF, 2018), promotes inclusive and sustainable economic growth, employment, and decent work for all. Sustained and inclusive economic growth was identified as the drive to progress, create decent jobs, and improve living standards worldwide (SDG 8).

One aspect of economic development and justice involves the extent to which all population

G. W. Muschert et al. (eds.), *Financial Inclusion*, Sustainable Development Goals Series,
https://doi.org/10.1007/978-3-031-68803-4_2

segments can access financial resources and services. Goal 8.10 aims to strengthen domestic financial institutions' capacity to encourage and expand access to banking, insurance, and financial services for all. The indicators for goals 8.10.1 and 8.10.2 are to increase bank branches, automated teller machines (ATMs), and accounts at financial institutions or with a mobile service provider.

SDG 8.10 (SDG, 2018) is a new topic requiring more research. Therefore, the purpose of this chapter is to review the available literature and summarize the background and purpose of SDG 8.10.1 and 2. The chapter appears in several sections, such as delving into the literature available on SDG 8.10.1 and 2, the role of effective banking structures, and the weaknesses of the financial system, financial inclusion, and mobile money. The last sections include conclusions, suggestions, and a discussion of this study's significance.

2.1.1 Literature Study

Sustainable development is recognized to have origins in the 1972 UN Conference on the Human Environment. The idea of sustainable development was relatively new. It has now taken form and has become a principal policy goal of many of the world's foremost institutions, including the UN, the World Bank, and the World Trade Organization. All goals are committed to better and more equitable outcomes in health, gender, housing, and sanitation that directly affect poorer groups. Poverty and inequality are recognized as significant causes and effects of global environmental problems and are central concerns of sustainable development. Climate change, economic recession, rising food and fuel prices, and commodity prices impact society's poorest people the hardest. Therefore, finding ways to address and prevent this is at the core of sustainable development (Elliott, 2013).

2.1.1.1 Background to Sustainable Development Goals, Global Findex, and Financial Inclusion

In 2011, the Bill and Melinda Gates Foundation funded the World Bank to launch the Global Findex database (World Bank, 2014, 2020), the most comprehensive database on adult savings, borrowings, spending, and risk management. This Global Findex Database includes over 140 economies, representing over 150,000 adults aged 15 and above, and is updated every 3 years. Furthermore, in July 2015, at the third International Conference on Financing for Development, held in Addis Ababa (an outcome document), the Addis Ababa Action Agenda was adopted on mobilizing financial and technical resources in support of sustainable development, including the Sustainable Development Goals (SDGs) covering the period up to 2030 (UN, 2015).

Sustainable development continued, and the Global Partnership for Financial Inclusion (GPFI, 2017) developed indicators to assess the state of financial inclusion and digital financial services globally. The indicators in the Global Financial Inclusion (Global Findex) database (World Bank, 2020) are derived from survey data covering and representing more than 97% of the world's population. Gallup, Inc. conducted a survey in 2014 as part of the Gallup World Poll. Samples were from nationally representative samples, and the target population included the entire civilian, noninstitutionalized population aged 15 and above. Since its inception in 2005, Gallup has continually conducted surveys of approximately 1000 people in each of more than 160 economies and over 140 languages.

The global indicator framework for Sustainable Development Goals emerged from the work of the Inter-Agency and Expert Group on SDG Indicators, set during the 48th session of the United Nations Statistical Commission held in March 2017 (UN, 2017).

The purpose of the SDGs of the United Nations was to end poverty, protect the planet, and ensure that by 2030, all people would enjoy peace and prosperity. In this regard, 17 goals emerged to transform the world and promote sustainable growth and prosperity for all (UN, 2022c).

The 2017 Global Findex database includes and has access to formal and informal financial services. It has different financial technology (or fintech) data, including using mobile phones and the Internet to conduct financial transactions. Furthermore, the data reveal opportunities to access people without bank accounts and information on the digital financial services of people with bank accounts (WB, 2017, WB, 2020a, WB, 2020b).

2.1.1.2 Overview of SDG 8 and Its Purpose

SDG 8 (UN, 2019) is one of the 17 Sustainable Development Goals established by the United Nations General Assembly in 2015. The main goal of SDG 8 is to promote inclusive and sustainable economic growth, employment, and decent work for all. It includes higher productivity levels, technological innovation, encouraging entrepreneurship and decent job creation, access to financial services, and the protection of labor rights.

SDG 8 promotes sustained, inclusive, sustainable economic growth, full and productive employment, and decent work for all (UN, 2022c). The goal has twelve targets that aim to be achieved by 2030. Some targets are for 2020 and others for 2030. The first 10 goals are outcome targets:

- Sustainable economic growth
- Diversify, innovate, and upgrade for economic productivity
- Promote policies to support job creation and growing enterprises
- Improve resource efficiency in consumption and production
- Full employment and decent work with equal pay
- Promote youth employment, education, and training
- End modern slavery, trafficking, and child labor
- Protect labor rights and promote safe working environments
- Promote beneficial and sustainable tourism
- Universal access to banking, insurance, and financial services

The last two targets are for means of achieving, which are to:

- Increase aid for trade support
- Develop a global youth employment strategy

Simonetti (2018) stated at the International Trade Union Confederation that promoting decent work is one of the main objectives of trade unions in their 2030 Agenda. Human and labor rights, freedom of association, collective bargaining, and social dialogue are essential for sustainable economic growth, development, and democracy. Similarly, social dialogue fosters socioeconomic progress to facilitate sustainable development. Bringing workers and employers together when making decisions that impact social, economic, and environmental conditions reinforces institutional stability. Sustained and inclusive economic growth can drive progress, create decent jobs, and improve living standards. However, with all this in mind, no policy could envision COVID-19, which has disrupted billions of lives and endangered the global economy (IMF, 2021).

2.1.1.3 A Brief Introduction to SDG 8.10 as Financial Inclusion Goals

In 2009, the Financial Access Survey (FAS) of the International Monetary Fund was launched (IMF, 2014). The FAS is a supply-side dataset on access to and use of financial services, supporting policymakers to measure and monitor financial inclusion and benchmark progress. Target SDG 8.10, therefore, aims to strengthen the capacity of domestic financial institutions to encourage and expand banking, insurance, and financial services for all. Subsequently, the subgoals of SDG 8.10.1 and 2 stress the importance of the availability of commercial bank branches, ATMs, and adults with an account at a financial institution or with a mobile money service provider (UN, 2019).

SDG 8.10 identified indicators 8.10.1 (UN, 2022c) as

(a) The number of commercial bank branches per 100,000 adults
(b) The number of ATMs per 100,000 adults

SDG 8.10 identifies indicators 8.10.2 (UN, 2022a) as the proportion of adults (15 years and older) with an account at a bank, other financial institution, or a mobile-money-service provider.

The SDG 8.10.1 and 2 indicators are included in the Financial Access Survey FAS database, collected and updated annually since 2009, covering 2004–2019. The information derives from central banks or other principal financial access survey regulators for 189 jurisdictions (IMF, 2014).

2.2 Overview of SDG 8.10

SDG 8.10 (UN, 2019) is a new and unresearched topic. Therefore, this section briefly overviews target SDG 8.10 and the indicators to strengthen financial institutions, encourage, and expand access to banking, insurance, and financial services.

This section aims to review the reports, agendas, and surveys available on SDG 8.10. A brief introduction is given on the development of SGD targets and the SDG indicators of SDG 8.10. The rationale and measurements for the indicators appear in the summary, including their impact on financial inclusion and micro- and macroeconomics. Finally, the discrepancies and limitations and the impact of COVID on the achievement of the indicators will be discussed. The conclusion will include possible relevant topics on SGD 8.10 to research in the future.

The commercial bank branches in SDG 8.10.1 refers to the total number of commercial bank branches in a country that are reported annually by the central bank or the primary financial regulator of a country and submitted to the Financial Access Survey (IMF, 2018, 2014). Hence, the number of commercial bank branches is scaled per 100,000 adults to make the indicator meaningful for cross-country comparison. Similarly, the number of ATMs refers to the number of ATMs in the country for all types of financial institutions, such as commercial banks, nondeposit-taking microfinance institutions, deposit-taking microfinance institutions, credit unions and credit cooperatives, and other deposit takers. The country's central bank, or top financial regulator, also submits this data annually to the Financial Access Survey. Furthermore, the number of ATMs is scaled per 100,000 adults to make the indicator meaningful for cross-country comparison.

Indicator SGD 8.10.2 (UN, 2019) indicates the proportion of adults (25 years and older) with an account at a bank, other financial institution, or a mobile money service provider.

SGD 8.10.1 (2019) states that access to and use of formal financial services is essential for sustainable development. The availability of financial services such as savings, insurance, payments, credit, and remittances allows people to manage their lives, budget, pay expenses, grow their businesses, and improve their overall welfare. Commercial banks remain one of the key institutions for accessing formal financial services. Therefore, having access to a bank branch is an important starting point for initiating and using all available financial services. Financial institutions offer services such as ATMs to assist customers with banking facilities, which can extend financial services to remote locations and after-hours use.

2.2.1 Financial Inclusion

The importance of financial inclusion intersects with several Sustainable Development Goal targets. SDG 8.10 calls explicitly to "strengthen the capacity of domestic financial institutions to improve access to banking, insurance, and financial services for all." The Addis Ababa Agenda (2015) stresses the role of micro-, macroeconomic, and structural policies of institutions, governance, and capacity building at all levels. It supports the need for financial inclusion and regulatory frameworks to encourage access to stable financial structures (WB, 2017).

Financial inclusion (GPFI, 2017) includes the availability of financial products and services that all sections of society need. This definition covers society in general, including vulnerable and weaker groups and low-income groups. Moreover, financial institutions need to make financial services and products available at an affordable cost and fairly and transparently. Therefore, individuals and businesses may have access to valuable and affordable financial products and services to meet their needs responsibly and sustainably. These products and services may include payments, savings, credit, and insurance transactions.

The ability to have access to a bank account is a first step toward broader financial inclusion since a transaction account allows people to store money and send and receive payments. A transaction account serves as a gateway to other financial services, so ensuring that people worldwide can access a transaction account is the focus of the World Bank Group's Universal Financial Access Initiative (WB, 2020a, b).

The Global Findex Database 2017, Measuring Financial Inclusion and the Fintech Revolution (WB, 2017), reports how adults in more than 140 economies access accounts, make payments, save, borrow, and manage risk. The data includes the successes, challenges, and opportunities regarding financial inclusion in the respective economies. Furthermore, the report includes and tracks the advances in each country, such as reducing poverty, hunger, and gender inequality, and progress toward the Sustainable Development Goals. This aspect reflects the continuous evolution of financial inclusion.

Global and national efforts and policies in the respective countries have appeared to expand financial inclusion. In 2017, more than 69% of adults had account ownership, a 7% increase since 2014 and 18% since 2011, when ownership was 52%. This figure means that 515 million adults have gained access to financial services, but 1.5 billion remain unbanked. Furthermore, the Financial Access Survey of the IMF indicates that mobile money is growing, and there are twice as many mobile money accounts as bank accounts per 1000 adults (GPFI, 2017).

The recent progress in financial inclusion was due to digital payments, government policies, and the power of financial technology to expand and access financial inclusion through mobile phones and the Internet. Sub-Saharan Africa reports that 21% of adults have mobile money accounts, which has doubled since 2014. It is also the highest of any country in the world. The 2017 Findex data reports that 52% of adults globally sent or received digital payments in 2016. Adults now have a more profound knowledge of financial services, and a broad range of financial services have emerged from the technology sector. Payments made through technology platforms increased account use in emerging economies, such as China, where 57% of account owners use mobile phones or the Internet to make purchases and pay bills (WB, 2017).

Some of the advances include helping women in India access financial services. In 2014, men were the primary users of accounts, which decreased dramatically because their government pushed ownership through biometric identification cards. However, women still lag globally, where 65% of women have an account compared to 72% of men. Furthermore, equality of ownership has yet to be reached between the rich and the poor. Ownership in the 60% of wealthiest households within economies is 13% higher than in the poorest 40%. Furthermore, urban populations still have access to more financial services than rural communities. Around 200 million rural adults in China remain outside the formal financial structure (IMF, 2014).

Businesses need to progress, expand financial inclusion, and pay wages to unbanked adults electronically to include them in the formal financial sector. In 2017, 230 million unbanked adults worldwide received wages in cash. Furthermore, around one billion financially excluded adults own mobile phones, and 480 million have access to the Internet (WB, 2017).

The quest for savings and borrowings has stagnated, and the impact of SDG 8.10 will be lost. Therefore, in support of SDG 8.10, the private sector, governments, and development organizations are called upon to focus on using financial services.

2.2.2 Measuring Financial Inclusion

Financial inclusion is a critical enabler in the fight against poverty. Therefore, reliable data are needed to focus on financial policies and initiatives. At the Cannes Summit in 2011, the Group of Twenty (G20) agreed to support global and national financial inclusion data efforts, called the G20 Basic Set of Financial Inclusion Indicators of the work of the Global Partnership of Financial Inclusion (GPFI) (GPFI G20). A wide range of payment instruments are covered. Payments include payments held with a bank or other financial institution or an e-money account, made on the Internet, through a mobile phone or device connected to a Wi-Fi network, or using a point of sale (POS) terminal. Government statistical agencies gather indicators to assess accomplishments and disparities using digital financial services and the technology and supportive environment required to take advantage of them (GPFI G20). These financial indicators measure access to and usage of financial services and the quality of the products and service delivery.

Since 2009, 189 economies have contributed to the FAS, covering financial inclusion from 2004 to 2019. All the contributing countries' central banks report annual information on the SDG 8.10 indicators. These indicators are calculated based on data collected directly from the central bank or the primary financial regulator in the country (UN, 2019).

The FAS contacts the respondent countries to initiate the annual survey process. The countries compile data sent and submitted by the IMF through the Integrated Collection System (ICS) or National Summary Data Page (NSDP) for processing and validation by the FAS. The reported data undergoes a round of validation checks and careful review. Inconsistencies are spotted and corrected, if any, and made available on the FAS website.

2.3 SDG 8.10.1—Metadata—08-10-01 (UN, 2022b)

Commercial banks are the key institutions for formal financial services such as savings, insurance, payments, credit, and remittances. It is essential and allows people to plan and manage their lives, grow businesses, and improve their welfare. Bank branches and access to automated teller machines at all financial institutions are complementary and can extend financial services to remote locations. Commercial banks refer to the banks reported by the country's central bank. ATMs will include all financial institutions, for example, commercial banks, deposit and nondeposit microfinance institutions, credit unions and cooperatives, and other deposit takers.

Indicator SDG 8.10.1 aims to strengthen financial inclusion in a country and increase the accessibility of branches and ATMs for banking and financial services. The increase in the number of commercial bank branches and automatic teller machines per 100,000 adults will serve as a gauge of SDG 8.10.1's success.

2.3.1 Measuring SDG 8.10.1

The indicators are calculated based on data collected directly from the country's central bank or primary financial regulator. The formula to obtain these indicators is

$$\begin{aligned}&\text{The number of commercial bank}\\&\text{branches per } 100{,}000 \text{ adults}_{it}\\&= \frac{\text{Number of commercial bank branches}_{it}}{\dfrac{\text{Adult population}_{it}}{100{,}000}}\end{aligned}$$

$$\begin{aligned}&\text{The number of automated teller}\\&\text{machines}\left(\text{ATMs}\right)\text{per } 100{,}000 \text{ adults}_{it}\\&= \frac{\text{Number of automated teller machines}\left(\text{ATMs}\right)_{it}}{\dfrac{\text{Adult population}_{it}}{100{,}000}}\end{aligned}$$

where "*i*" indicates the country and "*t*" indicates the year. The FAS is the source of information for the number of commercial bank branches and ATMs. In contrast, the source of information for the adult population is the World Development Indicators or the CIA Factbook.

2.3.2 Limitations of Indicators

The data collection is centralized at the central bank, which sources data from financial institutions and service providers. The aggregate for the total economy is reported to the FAS. The FAS is a supply-side database that reports data from an administrative source, which tends to be more accurate than demand-side surveys. It provides country-level metadata, which explains the financial inclusion of each reporting economy. Data from the FAS may differ from household-based surveys because of possible differences in coverage, scope, or concept definitions.

2.4 SDG 8.10.2: Metadata-08-10-02 (UN, 2022a)

Access to formal financial services is essential for people, regardless of income level, gender, age, education, or domicile, to manage their lives, build their futures, and grow businesses. A banking account is the starting point for various financial services such as savings, insurance, payments, credit, and remittances.

A respondent's account includes an account at a bank or other financial institution such as a credit union, microfinance institution, cooperative, or the post office (if applicable) or a debit card in the respondent's name. SDG 8.10.2 includes respondents who report receiving wages, government transfers, or agricultural product payments into a financial institution account in the past 12 months, including respondents who have paid utility bills or school fees from an account at a financial institution in the past 12 months or received wages or government transfers into a card in the past 12 months. Mobile money accounts include respondents who report personally using GSM Association (GSMA) Mobile Money for the Unbanked (MMU) services in the past 12 months to pay bills or to send or receive money. Furthermore, it includes respondents who reported receiving wages, government transfers, or payments for agricultural products through a mobile phone in the past 12 months.

2.4.1 Measuring SDG 8.10.2

Indicator 8.10.2 measures the proportion of adults, 15 years of age and older, with an individual bank account or someone else at a financial institution or with a mobile money service provider in the past 12 months.

The indicator is based on data collected through individual-level surveys in each country with representative samples. Appropriate sampling weights are used to calculate country-level aggregates, the percentage of adults owning accounts multiplied by the number of adults in the country. In order to calculate the regional values, divide the total amount across the region's countries by the total number of adults living there. Similarly, for global figures, the same calculation is repeated for all countries.

2.4.2 Limitations of Indicators and Trends

The Global Findex is an individual-level survey measuring the individual's ownership of accounts. Therefore, financial inclusion is an individual-level concept and is the appropriate measure. Other surveys done at the household level may measure access to finance through another household member, which may overestimate financial inclusion.

2.5 Financial Inclusion Supports Sustainable Development, Growth, and the Economy

Financial inclusion is an enabler for 7 of the 17 sustainable development goals and is vital to reducing extreme poverty and boosting prosperity to be reached by 2020. Since 2010, more than 55 countries have committed and more than 60 countries have launched or developed national strategies for financial inclusion (WB, 2022).

2.5.1 Macroeconomic Effects of Financial Inclusion

Demirguc-Kunt et al. (2017) confirm the nexus of finance and growth and, thus, economic development. This finding is supported by new data

Sahay et al. accessed (2015). An analysis by the IMF (Sahay et al., 2015) shows that financial inclusion and microfinancing can positively impact economic growth, stability, and inequality. Microfinancing contributed to financial inclusion by building robust, durable financial institutions to bring valuable services to millions of needy people. Supportive evidence exists regarding the benefits of access to basic payments and savings, especially among poor households. Furthermore, the findings (Sahay et al., 2015) indicated that financial sectors providing broader access to finance are more conducive to economic growth.

Sectors more dependent on external finance grow more rapidly in countries with greater financial depth and inclusion. Financial inclusion enhances stability directly if more people use bank deposits during periods of stress. Financial inclusion may provide clients of financial firms with better risk management tools, which will boost their resilience. However, this must happen at high levels of supervision. Gender gaps in account usage and increasing diversity in the depositor base will help to improve growth without impairing financial stability. Inclusive financial sector governance will also contribute to financial stability.

Bertuch-Samuels (2019) states that saving is indispensable for sustained investment and growth. Greater access to financial services in all economic sectors contributes to mobilizing savings and the availability of sustainable financing, especially in rural and economically weaker areas. Furthermore, credit increases the basis for economic growth and development at the local and macroeconomic levels. However, credit alone is not sufficient. The provision of increased credit opportunities must be managed prudently and requires an adequate legal framework, good governance, oversight, supervision, and robust financial infrastructure.

A competitive, diverse financial sector will improve financial inclusion and foster more inclusive growth. Attention must go to different types of financial institutions or banks and associated business models. Some are more effective than others at achieving safe and reliable financial inclusion for banking partners and Micro, Small and Medium Enterprises (MSME). Decentralized and locally anchored banking structures can be more resilient and have countercyclical effects.

A weak labor market is one reason for inequality. Therefore, a comprehensive national employment policy needs implementation for SDG 8 on sustainable growth, development, and decent work. Governments must develop and implement pro-employment macroeconomic strategies with the support of progressive trade, industrial, tax, and infrastructure policies, investments in education and skills development, youth employment, equality, and the care economy (Simonetti, 2018).

A minimum wage needs implementation to guarantee an income that allows people to live with dignity and is essential to reducing poverty. The cost of living should be regularly reviewed and adjusted. Declining wages in many countries affect aggregate demand, detrimental to growth, employment, and the global economy. Social protection should be available for those without access to social security. The potential of social protection for supporting employment, creating jobs, fostering skill development, and contributing to overall economic growth must be considered when assessing budgetary implications (Simonetti, 2018).

Governments are responsible for providing fiscal space to support social policies. Tax havens, avoidance, and evasion must be eliminated. Redistributive tax systems must be in place, and public spending must be relocated to support formal employment to increase tax and social security contributions.

In many countries, the economic models favor corporate greed, which influences the fundamental rights of workers and freedoms. Global markets and supply chains increased competition and caused firms to cut labor costs through restructuring, outsourcing, and offshoring. As a result, wages and working conditions decreased. The International Monetary Fund Annual Report (IMF, 2018) confirms that the decline in unions contributed to the rise of income at the top. This trend, in turn, has increased downward pressure on wages, working conditions, the deregulation and support of labor markets, and collective bargaining. Governments must ensure "due diligence" in supply chains with effective grievance procedures to ensure protection for human and labor rights. Governments need to endorse specific criteria for the private sector. These criteria must respect and implement due diligence, international labor standards, and assessments of environmental development at the country level of any operation (Simonetti, 2018).

Countries must switch to environmentally friendly production methods, transforming economies and industries' operations. Changes must start at the level of labor. Governments must strengthen "Just Transition" methods, an inclusive approach with social dialogue between workers, communities, employers, and governments to drive concrete plans, policies, and investments for a fast and fair transformation towards a low-carbon economy. It is an approach to building social protection systems, providing skills training, redeployment, labor market policies, and community development. Governments must strengthen their capacity to deliver these "Just Transition" measures (Simonetti, 2018).

The challenge of digitization and the impact of new technologies pose primary issues for developing countries. The issues range from new jobs in a country's organizational forms to the demand for new skills in the labor market.

2.5.2 Weaknesses in the Global Financial System

The United Nations interagency task force on financing for development and sustainable development stated in their 2019 report (UN, 2019) that the weaknesses in the global financial system may pose heightened risks to achieving the Sustainable Development Goals. These risks include the volatility of international capital flows. This results from the short-term nature of many elements of international capital markets: persistent global imbalances, debt sustainability challenges in the public and private sectors, and growing monopoly power and less effective competition policies. The high debt levels of public and private entities raise vulnerabilities and feed boom-bust cycles. The compression of the wage share of income has worsened inequality, while the rapid pace and technological changes may have provided new remedies but also increased global systemic risks (UN 2022c, 2019).

Well-run National Development Banks (NDBs) should be holistically aligned with the SDGs and integrated with national financing frameworks. Collaboration between well-run NDBs and multilateral banks through cofinancing or on-lending arrangements can enhance SDG-related finance through global resources and local market knowledge. The collaboration of members of the United Nations and the international community may strengthen NDB's risk management. Furthermore, more research is needed on tailoring and applying regulatory frameworks to NDBs to protect their financial sustainability and their investment's practical, sustainable development (UN, 2019).

Declining correspondence in the banking sector remains a concern. Maintaining stringent anti-money laundering and related standards is one of the costs and risks driving this. Well-managed technological solutions may address the costs and risks of operating correspondent banking relationships. Therefore, members of the United Nations may work together to adopt know-your-customer utilities and the Legal Entity Identifier (ILO, 2015; UN, 2019).

Institutions, policies, and regulatory frameworks have not kept pace with science, technology, innovation, and capacity-building changes at national and international levels. These new and emerging changes have the potential to support the goals of the SDGs but may also raise new challenges. Recent developments in automation, artificial intelligence, and other technologies may make the labor of millions in developed and developing countries redundant. In this regard, governments should take action to be better prepared by encouraging innovation that uses technologies to create new products, services, and jobs; be sensitive to the differential impact on women and men; ensure social protection and extend social security mechanisms to compensate for the loss of working hours and jobs; and invest in people's capabilities in order to enable them to benefit from new technologies, with attention to the different needs of different groups, such as the young, older, persons with disabilities, women, men, and others (UN, 2019).

Financial inclusion has increased through access to the mobile Internet, cryptography, distributed computing, and financial innovations (fintech). However, they also led to new risks and challenges for financial markets that need regulation without stifling financial innovation. Appropriate regulatory frameworks must be designed through dialogue between policymakers, regulators, and new service providers to find the right balance for all financial institutions. The international community needs to support developing countries in keeping up with technological changes and progressing toward achieving the SDG goals. The constraints must be addressed, such as absorptive capacities, the digital skills gap, a lack of economic incentives, social and cultural factors, and intellectual property rights (IPRs). Increasing economic digitalization and connectivity create innovation possibilities and new challenges, especially for the poorest countries (UN, 2019).

Finally, the COVID-19 pandemic has created the world's most significant global economic crisis since the 1929 Great Depression. The UN Task Force urged governments to address the COVID-19 pandemic's economic and financial havoc and prevent a potentially devastating debt crisis in the Financing for Sustainable Development Report (UN, 2020). The report outlined measures to address the impact of the unfolding global recession and financial turmoil, especially in the world's poorest countries. All countries have been called upon to "rise to the challenge to save lives and safeguard livelihoods in our response and recovery. We have one chance to build better together for people and for the planet" (UN, 2020, 2021, WB 2021).

2.6 Conclusion

The Sustainable Development Goals (SDGs) of the United Nations were a call to action to end poverty, protect the planet, and ensure that by 2030, all people will enjoy peace and prosperity. Seventeen goals emerged to drive the transformation of our world. SDG 8 promotes decent work for all and sustained economic growth. Subsequently, SDG 8.10 strengthens the capacity of domestic financial institutions to encourage and expand access to banking, insurance, and financial services for all. In light of this, Indicators 1 and 2 stress the importance of the availability of commercial bank branches, ATMs, and adults with an account at a financial institution or with a mobile money service provider.

Research (Simonetti, 2018) shows that there has been a steady increase in bank branches as well as in ATMs since 2015 (SDG 8.10.1). However, the Groupe Speciale Mobile Association (GSMA, 2014) reports that mobile money (SDG 8.10.2) exceeded expectations, and accounts grew to 1.2 billion in 2020, double the projected forecast. Their latest report on mobile internet connectivity (2022) indicated that 4.3 billion people were using mobile Internet, almost a 300 million increase since the end of 2020 (GSMA, 2022).

The International Labour Organization (ILO, 2015) noted that the number of workers living in poverty in 2015 has decreased despite the economic recession of 2008 and that the middle class makes up more than 34% of total employment. However, despite the decreased poverty, economic growth was slower, inequalities were widening, and not enough jobs emerged to keep up with the growing labor force.

Sub-Saharan Africa had the highest unemployment rates in the world in 2020. Therefore, economic growth (capital and technology) needs to be improved. Technology and innovations are primarily generated in developed countries rather than developing countries. The World Bank agrees that economic growth is the biggest driver of success, but growth stimulates income-generated opportunities for people with low incomes. This approach involves raising the value of agricultural products or generating better jobs (ILO, 2022).

In light of the current COVID pandemic, unemployment increased, poverty increased, and economic growth declined. Mobile money services increased with the fastest growth in markets where governments provided relief during the COVID-19 pandemic. Very little support went to low-income countries. Some suggest the UN return to the drawing board and find critical indicators to overcome and support their 17 goals while living with COVID (ILO, 2022).

Had the paradigm shift envisioned by the 2030 Agenda for Sustainable Development been fully embraced over the past 6 years, the world would have been better prepared to face this crisis, with more robust health systems, expanded social protection coverage, the resilience that comes from more equal societies, and a healthier natural environment. Regrettably, the SDGs were already off track even before COVID-19 emerged. Progress has occurred in poverty reduction, maternal and child health, access to electricity, and gender equality, but not enough to achieve the Goals by 2030. Progress had either stalled or reversed in other vital areas, including reducing inequality, lowering carbon emissions, and tackling hunger (UN, 2021, WHO, 2021).

The question remains whether unemployment, noneconomic growth, and poverty need financial inclusion, even at a reasonable cost. Furthermore, sustainable economic growth may be achieved in general and in the current COVID-19 pandemic, which may eventually result in an endemic. Furthermore, the World Bank (WB, 2020a, b) indicated that economic growth is responsible for improving income-generating opportunities for low-income people. On the other hand, what drives long-run economic growth? A country's growth depends on the percentage of economic growth that comes from capital, labor, and technology.

Sustainable development Goals are now more critical than ever. Now is the time to secure the well-being of people, economies, societies, and our planet. The global community is at a critical point in pursuing the Sustainable Development Goals (SDGs). The human and economic toll has been unprecedented in the global pandemic, where millions of lives were lost. Recovery efforts so far have been uneven, inequitable, and insufficiently geared toward achieving sustainable development. The crisis is threatening decades of development gains, further delaying the urgent transition to greener, more inclusive economies, and throwing progress on the SDGs even further off track.

References

Bertuch-Samuels, A. (2019). *Financial sector development—the role of effective local banking structures*. Available at https://doi.org.developmentfinance.un.org

Demirguc-Kunt, A., Klapper, L., Singer, D., Ansar, S., & Hess, J. (2014). *Global Findex Database 2014: Measuring financial inclusion around the world*. Available at http://www.worldbank.org/globalfindex

Demirguc-Kunt, A., Klapper, L., Singer, D., Ansar, S., & Hess, J. (2017). *The Global Findex Database 2017: Measuring financial inclusion and the Fintech revolution*. Available at https://openknowledge.worldbank.org

Elliott, J. (2013). *An introduction to sustainable development*. Routledge.

GSM Association – (GSMA). (2014). *The State of Mobile Money Usage – How many people use mobile money globally?* Available at https://www.gsma.com/mobilefordevelopment/programme/mobile-money/the-state-of-mobile-money-usage-how-many-people-use-mobile-money-globally/

GSM Association – (GSMA). (2022). *The State of Mobile Money Internet Connectivity Report 2022*. Available at https://www.gsma.com/mobilefordevelopment/gsma_events/the-state-of-mobile-internet-connectivity-report-2022/

Global Partnership for Financial Inclusion – (GPFI). (2017). *G20 Financial Inclusion Indicators*. Available at https://databankfiles.worldbank.org/public/ddpext_download/g20fidata/G20_Financial_

International Labour Organization – (ILO). (2015). *World employment and social outlook – trends 2015. Geneva*. Available at https://www.ilo.org/global/research/global-reports/weso/2015/lang%2D%2Den/index.htm

International Labour Organization – (ILO). (2022). *ILO Modelled Estimates and Projections (ILOEST Database)*. Available at https://ilostat.ilo.org/resources/concepts-and-definitions/ilo-modelled-estimates/

International Monetary Fund – (IMF). (2014). *People's Republic of China*. Available at http://www.imf.org/external/pubs/ft/scr/2014/cr14235.pdf

International Monetary Fund – (IMF). (2018). *IMF annual report 2018. Building a shared future*. Available at https://www.imf.org/external/pubs/ft/ar/2018/eng/assets/pdf/imf-annual-report-2018.pdf

International Monetary Fund – (IMF). (2021). *Access to Macroeconomic & Financial Data – Financial Access Survey*. Available at https://data.imf.org/?sk=E5DCAB7E-A5CA-4892-A6EA-598B5463A34C&sId=1496861011734

Sahay, M. R., Cihak, M., N'Diaye, M. P., Barajas, M. A., Mitra, M. S., Kyobe, M. A., & Yousefi, M. R. (2015). *Financial inclusion: Can it meet multiple macroeconomic goals?* Available at https://www.imf.org/external/pubs/ft/sdn/2015/sdn1517.pdf

Sustainable Development Growth Tracker – (SDG Tracker). (2018). *Measuring progress towards the Sustainable Development Goals*. Available at https://sdg-tracker.org/

Simonetti, P. (2018). *International Trade Union Confederation*. Available at https://doi.org.www.2030spotlight.org/sites/default/files/spot2018/chaps/Spotlight_Innenteil_2018_sdg8_simonetti.pdf

United Nations – (UN). (2015). *Addis Ababa Actions Agenda of the Third International Conference on Financing for Development*. Available at https://sustainabledevelopment.un.org/content/documents/2051AAAA_Outcome.pdf

United Nations – (UN). (2017). *Economic and Social Council Statistical Commission Forty-eight Session Agenda*. Available at https://unstats.un.org/unsd/statcom/48th-session/documents/2017-29-unsystem-e.pdf

United Nations – (UN). (2018a). *Department of Economic and Social Affairs. Sustainable Development. Goals 8. Promote sustained, Inclusive and sustainable economic growth, full and productive employment and decent work for all*. Available at https://sdgs.un.org/goals/goal8

United Nations – (UN). (2018b). *United Nations Global indicator framework for the Sustainable Development Goals and Targets of the 2030 Agenda for Sustainable Development*. https://unstats.un.org/sdgs/indicators/Global%20Indicator%20Framework%20after%202020%20review_Eng.pdf

United Nations – (UN). (2019). *United Nations, Inter-agency Task Force on Financing for Development, Financing for Sustainable Development Report 2019*. Available at https://developmentfinance.un.org/fsdr2019

United Nations – (UN). (2021). *The Sustainable Development Goals Report 2021*. Available at https://unstats.un.org/sdgs/report/2021/

United Nations – (UN). (2022a). *The Sustainable Development Goals Report*. Available at https://unstats.un.org/sdgs/report/2022/

United Nations – (UN). (2022b). *SDG Indicator Metadata*. Available at https://unstats.un.org/sdgs/metadata/files/Metadata-08-01-01.pdf

United Nations – (UN). (2022c). *Inter-agency Task Force on Financing for Development, Financing for Sustainable Development Report 2022*. Available at https://developmentfinance.un.org/fsdr2022

World Bank. (2014). *Global Financial Development Report 2014: Financial Inclusion*. Available at https://openknowledge.worldbank.org/entities/publication/184063e6-52bd-59dd-bbb4-7ec8f840e009

World Bank. (2017). *The Global Findex Database 2017: Measuring Financial Inclusion and Opportunities to Expand Access to and Use of Financial Services*. Available at https://documents.worldbank.org/en/publication/documents-reports/documentdetail/491271624538346557/the-global-findex-database-2017-measuring-financial-i

World Bank. (2020a). *Universal Financial Access 2020*. Available at https://ufa.worldbank.org/en/ufa#:~:text=Universal%20Financial%20Access%202020&text=The%20UFA%20goal%20envisions%20tha

World Bank. (2020b). *The World Bank Economic Review, 34*(Supplement), 2020. Available at https://documents.worldbank.org/en/publication/documents-reports/documentdetail/717571624546779164/the-world-b

World Bank. (2021). *World Bank Global Findex Database 2021: Financial Inclusion, Digital Payments, and Resilience in the Age of COVID-19*. Available at https://www.worldbank.org/en/publication/globalfindex

WorldBank.(2022).*FinancialInclusion*.Availableathttps://www.worldbank.org/en/topic/financialinclusion/overview

World Health Organization – (WHO). (2021) *WHO Global Coordination Mechanism on the Prevention and Control of Noncommunicable Diseases*. Available at https://www.who.int/groups/gcm/about

Forward and Backward Linkages of Financial Inclusion, Fintech, and Sustainable Development Goals: A Bibliometric Analysis and Concept Mapping

3

Biju A. V. Nair and Katia Lopez

Abstract

We aim to analyze financial inclusion's forward and backward linkages through financial technology (fintech) and to scope this linkage toward sustainable development goals using a bibliometric analysis and concept mapping. We use platforms such as VOSViwer and R studio to analyze 779 articles collected from the Web of Science database, including keyword co-occurrence analysis and thematic mapping. Our findings show a significant impact of fintech and mobile money on financial inclusion and sustainability, with a focus on the current literature on sustainable development goals, fintech, and mobile money. We also find a substantial gap in the empirical research on cross-country and individual analysis, connecting fintech to financial inclusion toward the Sustainable Development Goals (SDGs). This study will help policymakers, regulators, and academic researchers know the nuts and bolts of financial inclusion and identify the relevant areas that need further research toward reaching the SDGs set by the United Nations.

Keywords

Financial inclusion · Fintech · Mobile money · Sustainable Development Goals · SDG 10 · Sustainability · Keyword co-occurrence · Thematic mapping

3.1 Introduction

The United Nations has set the Sustainable Development Goals (SDGs) as its key targets by 2030. Out of these goals, SDG 8 promotes sustainable economic growth for everyone, efficient and thorough use of labor resources, and decent work for all (United Nations, 2015). Financial technology (fintech)-led financial inclusion (FI) plays a vital role in reaching SDG 8, particularly in the sub-indicator 8.10.2, which focuses on enhancing the efficiency of domestic financial institutions to grant access to financial products and services. Fintech prompts digital transformation as a possible way to reach Sustainable Development Goals (Arner et al., 2020).

Financial inclusion refers to individuals and businesses having equal access to valuable and affordable financial products and services that

B. A. V. Nair (✉)
Department of Commerce, School of Business Management & Legal Studies, University of Kerala, Thiruvananthapuram, Kerala, India
e-mail: bijuav@keralauniversity.ac.in

K. Lopez
School of Business, University of Wollongong in Dubai, Dubai, UAE

G. W. Muschert et al. (eds.), *Financial Inclusion*, Sustainable Development Goals Series,
https://doi.org/10.1007/978-3-031-68803-4_3

meet their needs: transactions, payments, savings, credit, and insurance—delivered responsibly and sustainably (World Bank, 2019). A financially included country could be a good measure of its overall social and economic development (Sarma & Pais, 2011), which could help reach the SDGs. Through a deep analysis of past research and considering all the issues and bottlenecks of the present financial inclusion systems, we expect to understand better the factors that prevent a society from reaching SDG 8.

The levels of human development and financial inclusion in a country move closely with each other. Countries with high degrees of financial inclusion and relatively lower income tend to experience a significant trade-off between financial inclusion and output stability (Cavoli et al., 2020). Low financial inclusion causes low living standards, poor health, high illiteracy, and deprived well-being, along with a spillover effect of low human development in frontier market countries (Ababio et al., 2021). Additionally, there is a positive relationship between financial inclusion and economic growth, which is even more vital in countries with low income and a lower degree of financial inclusion (Van et al., 2021).

Through bibliometric mapping, we analyze the role of financial inclusion in SDG 8.10.2 with fintech as the primary mediator to serve the financial needs of more significant proportions of the population and, thereby, toward global upliftment and economic sustainability. We examine why governments and regulators should focus on fintech, how it closely relates to financial inclusion, and how fintech relates to SDG achievement.

3.2 Literature Review

3.2.1 Literature Evolution and Status of Financial Inclusion and Exclusion

About 1.7 billion adults worldwide are unbanked without an account with a financial institution or through a mobile money provider. Globally, 69% of adults have a bank account, and 56% of all unbanked adults are women (Demirguc-Kunt et al., 2018). The applied or policy literature positively perceives financial inclusion as a tool to reduce the costs to excluded people.

However, the scholarly literature is critical of financial inclusion (Prabhakar, 2019). The concept and literature on financial inclusion have also evolved and expanded. Before the 2008 crisis, the main questions were how the financial system discriminated against poor people that could not afford loans and would not be aware of other financial products that could be within their reach (Gabor & Brooks, 2017). The study of financial inclusion has shifted toward accessibility and equal opportunities for the poor to be part of the financial system.

After the 2008 global economic crisis, the concern for financial inclusion was a matter of the World Bank and other financial organizations, developing countries, and others, such as the Alliance for Financial Inclusion (AFI). The Pittsburgh Summit of 2009 established the Financial Inclusion Experts Group (FIEG) and acknowledged digital finance solutions as the route toward financial inclusion. The system becomes enhanced by providing alternatives to achieve financial goals and overcoming financial problems by reducing financial exclusion. In the same way, providing formal financial services to the unbanked urban and rural poor could lead to diversified risk (Aron, 2018). These ideas are embraced and supported by numerous countries globally, such as the United States, France, India, the United Kingdom, South Africa, and Germany, which have adopted measures to reduce financial exclusion. Through an overview of the literature on financial inclusion, we found the concept of financial inclusion to focus on the technological aspects considering artificial intelligence and blockchain-driven financial services tools to reach the UN SDGs. We can also see this shift in the literature toward digital finance through the evolution of new keywords in the theme of financial inclusion. We observe from the past literature and bibliometric coupling that technology, mobile banking, and fintech are the key factors that contribute to financial inclusion.

3.2.2 Role of Technology in Financial Inclusion

Financial technology (fintech) is a tool with significant potential to provide access to financial services and to help escape poverty's clutches (Lagna & Ravishankar, 2021). Technology enables the reconsideration of existing systems to balance market integrity, financial inclusion, and economic growth while meeting international financial standards (Arner et al., 2020). For example, in the COVID-19 pandemic crisis, the interaction between technology, finance, and sustainability has been of great importance and has urged all countries to rethink traditional models and rely more on technology and sustainability (Macchiavello & Siri, 2020). Therefore, we consider fintech an engine for sustainable development in developing economies.

By embracing fintech, commercial banks can improve their traditional business model by reducing bank operating costs, improving service efficiency, strengthening risk control capabilities, creating an enhanced customer-oriented business model, and consequently improving comprehensive competitiveness (Yang et al., 2021). The design of digital financial markets and systems supports broader access to finance and investment. Such access grants a tremendous potential to transform finance, economies, and societies and promotes sustainable development through the linkage of fintech and financial inclusion (Gai et al., 2018). An example of these possible transformations is the case of Indian states with short-run and unidirectional long-run causal relationships between financial inclusion, mobile telephony, and economic growth (Pradhan & Sahoo, 2021).

3.2.3 Fintech-Led Financial Inclusion

The fintech revolution has seemed a threat to the traditional commercial banking business because as technology takes over, services migrate to an online service, and physical contact declines (Agarwal & Zhang, 2020; Lee & Shin, 2018). Hence, governments are in a difficult phase where they must decide whether to grant the whole region or only selected ones (Jiao et al., 2021).

Some studies support fintech-led financial inclusion and sustainability (Moro-Visconti, 2021; Sarma & Pais, 2011; Kong & Loubere, 2021; Arner et al., 2020; Gomber et al., 2018). The fintech industry's success depends on the attitude toward adopting the technology (Shin & Choi, 2019). Therefore, the role of governments is vital as they could prepare the road for easier adoption and greater financial inclusion. Sarma and Pais (2011) describe the essentialities of fintech, such as telecom infrastructure, mobile connectivity, and usage; these factors could promote the development of such services and, therefore, reach FI. According to Arner et al. (2020), Macchiavello and Siri (2020), and Zetzsche et al. (2019), fintech is the best way to reach FI and SDGs through services such as online financing, credit, and crowdfunding, especially the SDG 8.10 goal (Gargano & Rossi, 2024). On the other hand, Moro-Visconti et al. (2020) found fintech companies to lead with a considerable premium over banks and to be less risky.

3.2.4 Role of M-Money in Financial Inclusion

Fintech services offer sustainable finance to microfinance ventures through inclusive ownership and provide solutions to international investments and digital platforms (Moro-Visconti et al., 2020). Fintech is promoting financial development through mobile money as evidenced by the success stories of Kenia and Khana (Senyo et al., 2020.; Asongu et al., 2020; Della Peruta, 2018; Gomber et al., 2018; Arner et al., 2020). M-money can provide access to formal financial services such as saving accounts, loan accounts, and insurance to the previously unbanked using a mobile phone (Lashitew et al., 2019).

Although M-money is accessible to everyone, mainly low-income groups use it and obtain more benefits, as experienced in Nairobi (Kim et al., 2020). Furthermore, M-money could enable small businesses, informal workers, and informal remittances to use formal channels leading to the formalization of the economy (Aron, 2018). It could also provide easy access to financial services to people in rural areas, solving geographical constraints and lack of infrastructure (Kim et al., 2018).

There have been some suggestions (Pal et al., 2021) to use policy interventions to nudge consumers to opt for mobile payments to increase their welfare. Some examples are M-Pesa in Kenya, India Stack, similar developments in Russia, China, and other countries in East Africa (Macchiavello & Siri, 2020; Arner et al., 2020; Asongu et al., 2020; Kim et al., 2020). Additionally, Phuc Nguyen et al. (2020) showed that increased mobile usage significantly affects financial inclusion.

From the theories discussed above, this is an excellent time to review the literature on financial inclusion, fintech, and sustainable development by analyzing the research done in those areas and developing linkages among them. We have been unable to identify an academic work that concludes the role of fintech and mobile money toward financial inclusion and Sustainable Development Goals through a coupling of past literature. Hence, we aim to address the following aspects:

(a) Analyze the growth pattern of research in financial inclusion, fintech, and SDG.
(b) Identify authors, research institutions, and journals contributing to the area of emerging fintech-based financial inclusion and SDG.
(c) Synthesize the findings evolved from research and understand the research trends.
(d) Identify the research gap prevailing in the area and provide insight into future research.

3.3 Methodology

3.3.1 The Rationale of Bibliometric Analysis

Social science disciplines widely adopt bibliometric reviews (Aria & Cuccurullo, 2017; Ellegaard & Wallin, 2015). Bibliometrics provides a structural analysis of the past literature over a period, identifies the shifts in the boundaries of disciplines, and detects the most prolific scholars and institutions (Crane, 1972). Scholars use different qualitative and quantitative literature reviewing approaches to understand and organize earlier findings. The bibliometrics analysis has the potential to introduce several approaches for a systematic and transparent review of the previous literature and give the synthesis based on a topic to the readers (Broadus, 1987; Diodatao, 1994).

3.3.2 Data

We collected 779 documents from the Web of Science using the "topic" search option. The search words are "financial inclusion," or "fintech," or "SDG 8," or "financial inclusion and sustainability," or "fintech and sustainability," or "financial inclusion and sustainable development goals," or "fintech and sustainable development goals." The period of research work collected was from 2004 to 2021. As shown in Fig. 3.1, we have used 779 documents, which consist of 313 sources. There are 2039 authors' keywords, and we reported the work of 1838 authors with such keywords.

3.3.3 Tools for Analysis

The bibliometric information from data extracted using the keywords helps us understand several aspects, including the research evolution, citation and co-citation, keyword co-occurrence, author, and coauthor networks (Bastian et al., 2009). Recently, Citation and Network analysis has evolved as a tool for analyzing bibliometric data, providing insights to readers and researchers across the globe. In this study, we used the keyword co-occurrence in VOSviewer. We have used the modularity-based clustering algorithm (Blondel et al., 2008; Newman & Girvan, 2004) to identify various linkages among the financial inclusion, fintech, and Sustainable Development gGals. This algorithm programmed in VOSviwer computes the difference between the fraction of

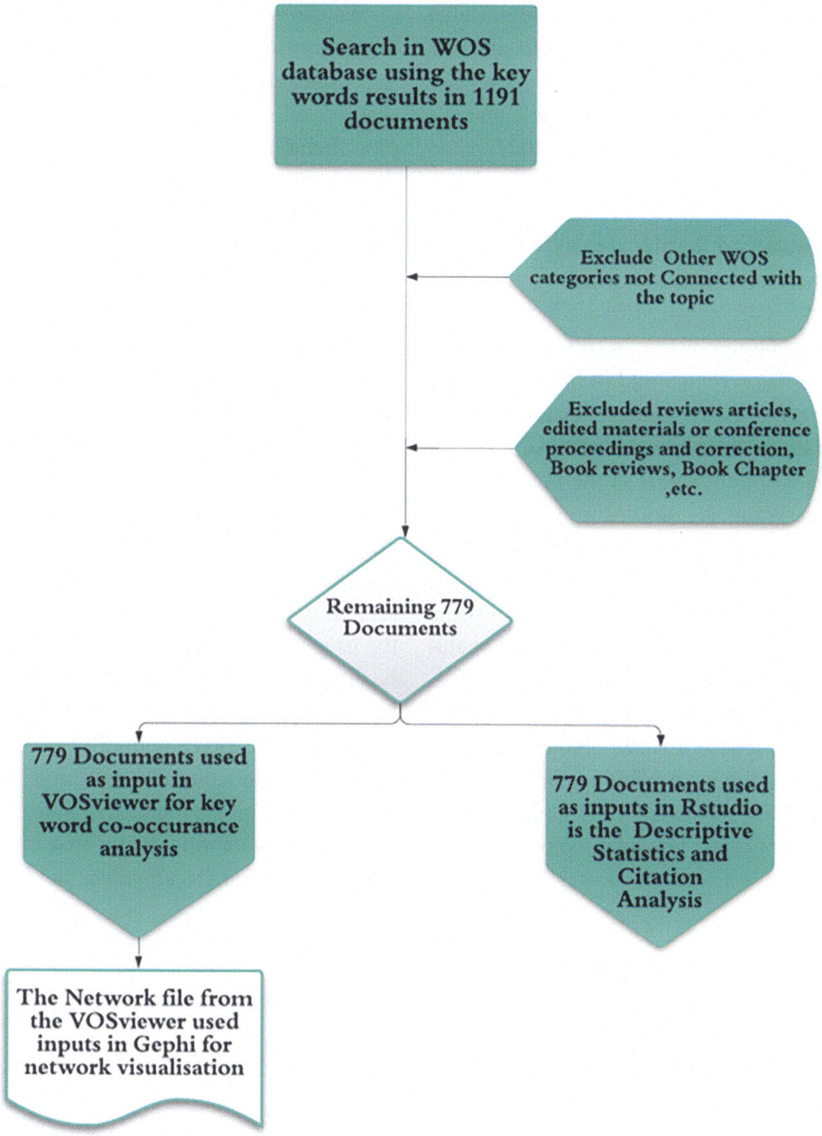

Fig. 3.1 Representation of the design followed in this study

the links within the specific cluster and will lead to the clustering of keywords.

3.4 Analysis and Discussion

We have used two programs (Vosviwer and Bibliometrix) to obtain an overview of past literature using keywords that connect financial inclusion, fintech, and Sustainable Development Goals. Vosviwer generated the keyword clustering, keyword co-occurrence, and author coauthor analyses. R studio generated citations, publications, journals, and inferential analyses.

3.4.1 Annual Publications and Citation

Table 3.1 presents a summary of data relating to citations and publications in the selected keywords. We can see a constant growth in papers and citations from 2004 to 2021. Table 3.1 also reports the average citation per article and the average citation per year. Figure 3.2 shows the mean growth of total citations per article and the number of publications from 2004 to 2021.

3.4.2 Growth of Publications, Journals, and Authors

We can see at the top of Fig. 3.3 the authors that started the research in the field of financial inclusion and the fintech revolution toward sustainability as well as the authors that have some of the most recent publications in the field, such as Asongu SA, Wang X, Koomson I, Zang YH, Chang V, and Liu C. Asongu S, who have done some of the latest published research, including mobile money and its role in financial inclusion. Recent work also covers digital financial inclusion. For example, Wang X's paper covered digital financial inclusion, cryptocurrency, and Industry 4.0. Likewise, Brooks studied the revolutionary role of fintech in the financial inclusion process, and Liu C (Hu et al., 2021) constructed a fintech index to predict financial inclusion.

We can see how the past research is more on the side of the formal commercial banking system rather than financial technology, as the current ones do. The authors who contributed to the past financial inclusion literature now work on the role of technology and digital financial inclusion, financial deepening with technology, and financial inclusion for sustainability matters. Johnson's work showed the level of financial inclusion through the opening of more formal bank accounts. Klapper L talked about the connectivity and linkages of financial literacy and financial inclusion. Guerini's work investigated the arguments against governments' formal financial inclusion policy framework, where he criticizes the discrimination attitude of the government toward the rural population. Cull R explored the role of mobile money in African countries for the successful inclusion of people in the usage of financial transactions. Bernards analyzed the role of technology in global financial inclusion.

Table 3.2 reports a journal-wise analysis of the publication and citation data and the authors' impact based on the h index, m index, and g index from the publications on the area of the selected keywords. *World Development* and *Sustainability*

Table 3.1 Annual publication and citation data

Year	N	MeanTCperart	MeanTCperyear	Citableyears	Number of citations
2004	1	54	3.18	17	2
2005	0	0	0	0	0
2006	0	0	0	0	0
2007	2	30	2.14	14	20
2008	2	22.5	1.73	13	26
2009	1	86	7.17	12	28
2010	2	31	2.82	11	30
2011	8	29	2.9	10	30
2012	11	30.9	3.43	9	42
2013	18	31.1	3.89	8	57
2014	17	19.6	2.8	7	126
2015	18	16.9	2.82	6	188
2016	42	18.1	3.61	5	258
2017	53	18.3	4.57	4	649
2018	123	15.6	5.21	3	1532
2019	149	8	4	2	2926
2020	225	3.4	3.41	1	2048

Notes: This tables presents the mean of the total citation per article (meanTCperArt), mean of the total citations per year (meanTCperYear) and the citable years annually between 2004 and 2020 and N represents number of papers

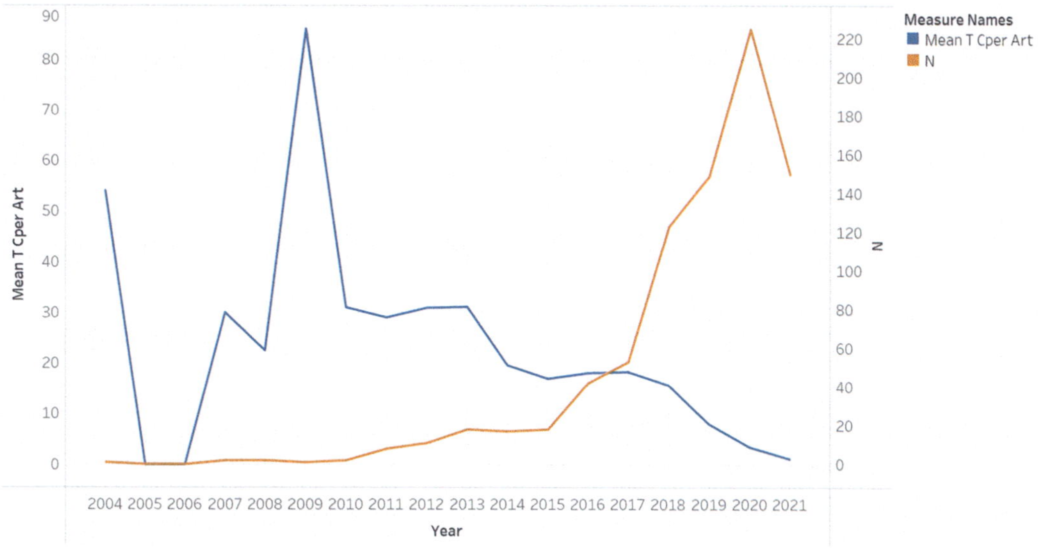

Note: The figure represents the growth of the Mean of total citations per article (Mean TC per Art) and the number of publications (N) from 2004 to 2021.

Fig. 3.2 Growth curve of publications. (Note: The figure represents the growth of the Mean of total citations per article (Mean TC per Art) and the number of publications (N) from 2004 to 2021)

journals have the highest number of publications, with 23 and 60, and total citations of 171 and 457, respectively. The top authors are Kauffman R J and Klapper, with 6 articles each and 144 and 317 citations, respectively (see Fig. 3.3). Figure 3.4 displays a mind mapping of the most relevant papers regarding citations.

3.4.3 Keyword Co-occurrence Analysis

Figures 3.5 and 3.6 represent the keyword co-occurrence analysis for which we have used the authors' keywords. The counting method is total counting, and of the 2048 keywords, out of the 789 articles selected, only 86 meet the minimum number of 5 occurrences. We calculated the strength of the co-occurrence link with other keywords for each of the 86 keywords, and then the keywords with the greatest total link strength were selected. Out of the selected keywords, "financial inclusion" has 272 occurrences, and the total link strength is 352. Similarly, fintech was repeated 165 times with a link strength of 217. For this calculation, we used the method of normalization as the LinLog modularity.

Cluster #1: Financial Access (Represented in Red)

The first theme in the keyword network is Financial Access. Most of the studies cover topics related to financial access and case studies in countries like India, Bangladesh, Africa, Kenya, and other developing countries. The most common keywords in the middle of the cluster are financial inclusion, financial development, financialization, mobile banking, and India.

Cluster #2: Technology (Represented in Green)

The second cluster shown in the keyword network is Technology, with the keyword *fintech* in its center. The most common keywords in this cluster are *blockchain*, *cryptocurrency*, *artificial intelligence*, and *mobile payment*. China has a higher reoccurrence in the keywords related to financial innovation and digital transformation topics.

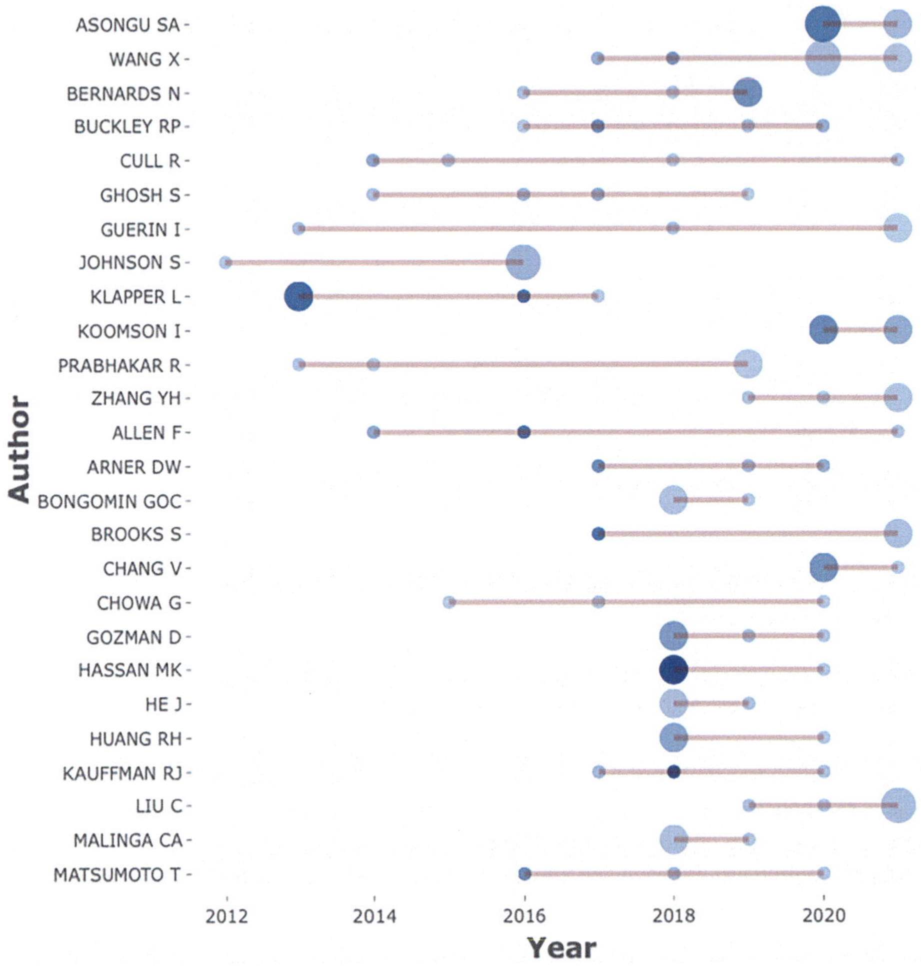

Note: This figure shows a thematic map of the authors' production from 2012 to 2020. We can observe that the authors have the latest work on the selected keywords and the pioneers who contributed from the inception using their keywords.

Fig. 3.3 Authors production over time (2012–2020). (Note: This figure shows a thematic map of the authors' production from 2012 to 2020. We can observe that the authors have the latest work on the selected keywords and the pioneers who contributed from the inception using their keywords)

Cluster #3: Economic Structure (Represented in Blue)

We can observe that clusters #1 and #2 are well-defined and have emerged more recently, whereas clusters #3 and #4 remain dispersed. The third cluster identified in the keyword network is Economic Structure which lacks a clear center. In this cluster, we can see the following words: *economic development*, *infrastructure*, *access to finance in developing countries*, and *remittances*.

Cluster #4: Sustainability (Represented in Yellow)

The fourth cluster is sustainability, with the fewest keywords. Looking at Figs. 3.5 and 3.6, we can see that even though clusters #3 and #4 are the smallest, they are also part of the recently explored topics, implying upcoming research in these themes. The keywords included in this cluster are *SDG*, *sustainable development*, *sustainability*, and *sustainable development goals*.

Table 3.2 Publication and citation status of journals based on h index, m index, g index, and total citations

Sl No	Source	h_index	g_index	m_index	TC	N
1	Sustainability	7	10	1.4	171	60
2	World development	11	21	1.2	457	23
3	Technological forecasting and social change	6	10	1	109	19
4	Electronic commerce research and applications	8	15	0.8	234	17
5	International journal of finance & economics	2	2		8	17
6	Emerging markets finance and trade	7	10		112	16
7	Financial innovation	6	10	1.2	113	16
8	International journal of bank marketing	5	8		73	15
9	European journal of finance	3	4		32	13
10	Journal of international development	6	13		224	13
11	Applied economics	4	7	0.7	63	11
12	European business organization law review	5	7	1.3	61	11
13	New political economy	7	11		161	11
14	Review of international political economy	6	10		114	11
15	Telecommunications policy	6	11	0.7	176	11
16	Finance research letters	5	9	1.3	91	10
17	International journal of Islamic and middle eastern finance and management	4	5	0.7	31	10
18	Journal of management information systems	7	9	1.4	276	9
19	Research in international business and finance	5	9	1	121	9
20	Singapore economic review	3	7		50	9

Note: The h-index measures the production and impact of journals, g-index is the most significant number such that the top g articles received (together) at least g2 citation. The m-Index is the h-index divided by the number of years, TC represents total citations, and N represents the number of papers

Note: Author's mind mapping using the mind space shows the most significant papers having more citations in financial inclusion through Fintech.

Fig. 3.4 Mind mapping of most relevant and cited papers in the fintech literature. (Note: Authors mind mapping using the mind space shows the most significant papers having more citations in financial inclusion through fintech)

Figure 3.6 shows the evolution of the keywords from 2010 to 2020. We can see the color change of the circles and links, showing the emerging research area. These areas are machine learning, artificial intelligence, crowdfunding, entrepreneurial finance, bitcoin, blockchain, and mobile payment.

3.4.4 Thematic Mapping of Keywords

Figure 3.7 is an intuitive plot where the meaning of topics reflects in the quadrant where they appear. Each quadrant represents a theme of topic mapping and classification based on their relevance and magnitude depending on the density and centrality. The upper-right quadrant contains the most significant and developed research areas or topics in financial inclusion. The lower-right quadrant contains basic themes with topics well-studied, developed, and seen as fundamentals of financial inclusion. The lower-left quadrant encloses emerging or disappearing themes. The upper-left quadrant includes very specialized/niche themes. The center represents the most relevant themes in which the research is currently taking place, where we can see the relevance that constitutes the Sustainable Development Goals and sustainability and and as a better and clearer connection between financial inclusion, fintech, and sustainability appears.

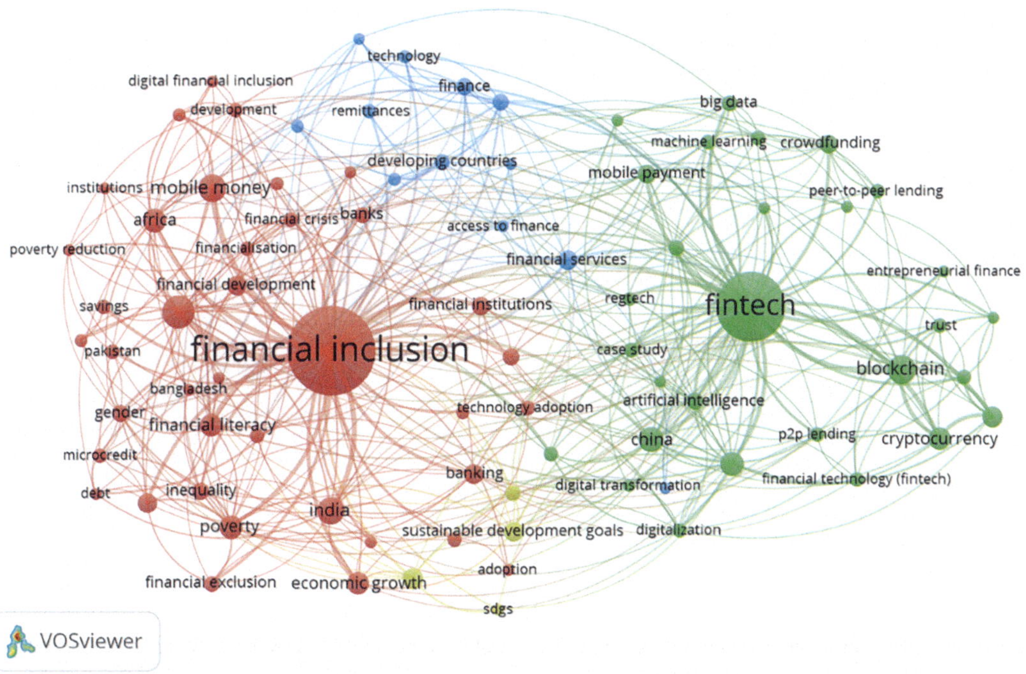

Fig. 3.5 Keyword co-occurrence network, the clusters: Cluster #1—red (41.56%) Cluster #2—green (31.03%) cluster #3—Blue (14.18%) Cluster #4—Pink (3.33%). (Note: Fig. 3.5 illustrates the network of keywords among the 779 documents analysed in this study. The diagram shows four clusters differentiated by colors where the size of the bubble represents the degree of association between the keywords. Clusters are arranged based on the number of keywords interconnected with a minimum of five reoccurrences)

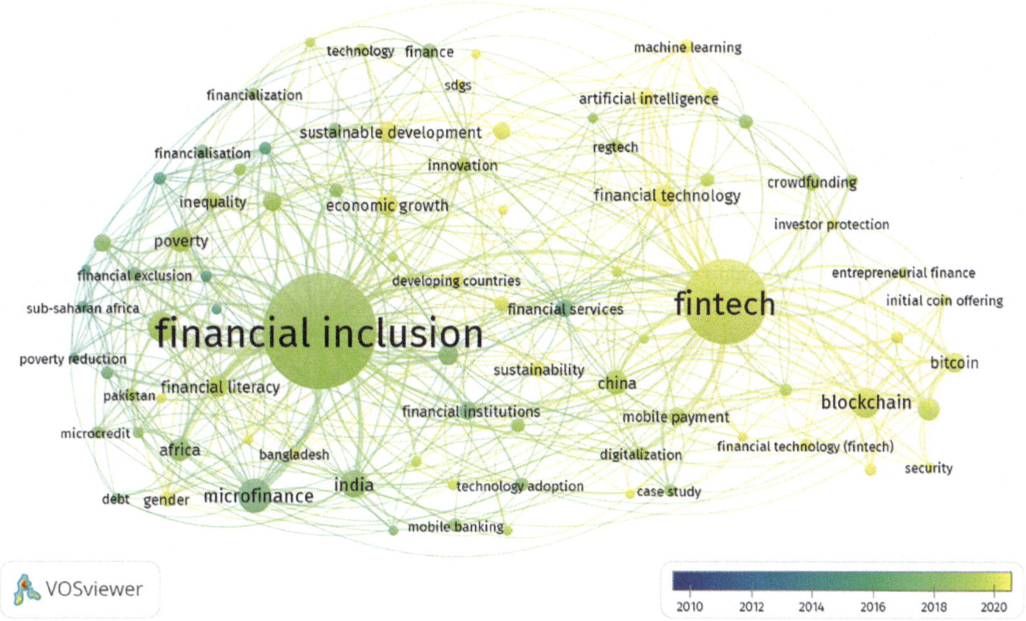

Fig. 3.6 Overlay visualisation of evolution of keyword co-occurrence from 2010 to 2020

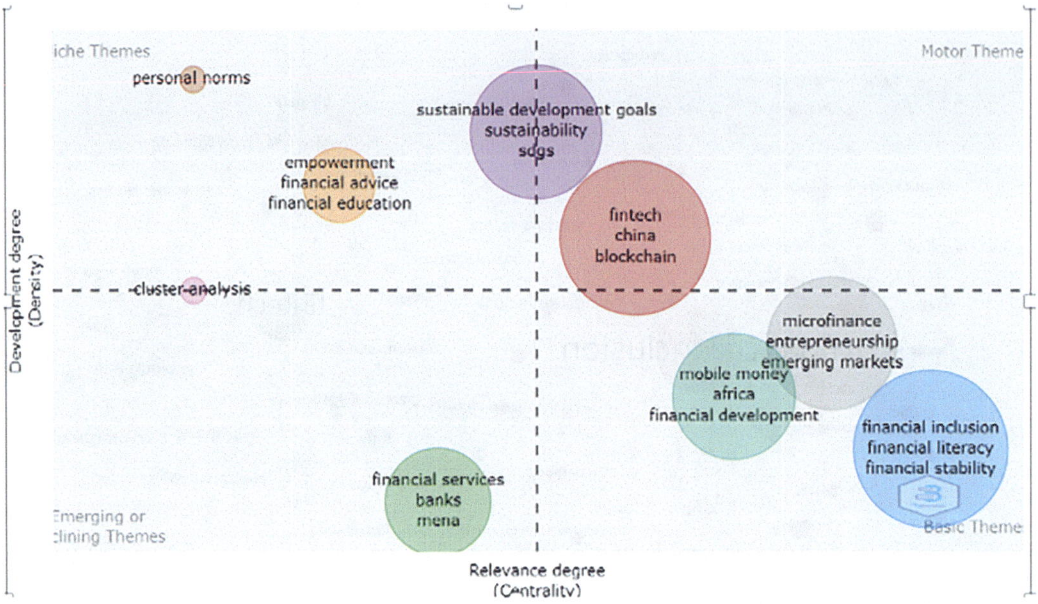

Note: Thematic map of the literature on Fintech for financial inclusion and SDG from the R studio "bibliometrix" package analytics.

Fig. 3.7 Thematic Mapping of keywords. (Note: Thematic map of the literature on fintech for financial inclusion and SDG from the R studio "bibliometrix" package analytics)

3.5 Financial Inclusion and Sustainability

Sustainability and financial inclusion are two sides of the same goal (Arner et al., 2020). The bibliometric analysis shows the powerful potential of Fintech-led financial inclusion toward sustainability. When financially excluded people receive opportunities to access financial services, they have a greater chance to reach longer-term solutions to their financial problems. The thematic map and keyword co-occurrence analysis confirm how the current literature focuses on the role of financial technology in financial inclusion.

3.6 Future Research Opportunities and Research Gap

There has been much research on fintech, but more focus is needed on financial inclusion. One of the exceptions is Information and Communication Technologies for Development (ICT4D), which focuses its research on financial inclusion in four crucial topics: mobile-enabled financial services, ICT-enabled banking, ICT and microfinance, and economic impact of ICT development and financial inclusion (Lagna & Ravishankar, 2021). On the other hand, there is much research about sustainability, financial inclusion, and fintech, but only separately (Arner et al., 2020). The European Commission's Sustainable Finance Action Plan needs more work on fintech. There has been much research on the legal aspects of financial technologies but not on fintech achieving financial inclusion and sustainability. We also found a substantial gap in the empirical research on the cross-country and individual analysis, which connects the fintech lead financial inclusion towards Sustainable Development Goals.

3.7 Conclusion

We understand that fintech is becoming a foundational framework for achieving SDG 8, promoting sustainable economic growth and productive employment. We recognize how it can ultimately improve the macroeconomic stability of the state as it lowers the transaction cost, expands access to bank deposits, and encourages savings (Coffie et al., 2021). The understanding of fintech business models remains in its infancy, with a need for cross-country comparisons (Laidroo et al., 2021). Digital technology has helped several low-income countries improve their financial inclusion, but this remains the same for other regions' low-income countries (Shen et al., 2021). We can see how users still need to be more hesitant to adopt and use fintech, ultimately affecting its growth. In order to achieve sustainable development and growth of fintech, an in-depth investigation of fintech continuance intentions is required (Ryu & Ko, 2020). The administrators of the nations would need to answer the following questions:

1. Is the global population financially literate to complement technological and financial developments?
2. Are the financial markets stable enough to utilize the productivity of these innovations?
3. Are fintech innovations secured enough with regulatory and legal provisions to enhance and change the financial landscape?

The above questions are part of the challenges of fintech to be addressed by nations and central banks worldwide. However, it is essential to consider what digital transformation is doing today by introducing new technologies for payments, funds transfers, and other financial transactions. Even though fintech provides excellent opportunities for financial inclusion, there are some possible disadvantages. Criminals might use and abuse financial technologies for fraud, extortion, money laundering, and financing activity in the criminal underground (Nikkel, 2020). Even in the legitimate sphere, the hypothesis remains that the growth of financial technology negatively influences bank performance (Phan et al., 2020). Hence, while promoting fintech services, the nations should have to provide sound regulation for better administration while protecting, at the same time, traditional commercial banking and cooperative banking businesses.

References

Ababio, J. O. M., Attah-Botchwey, E., Osei-Assibey, E., & Barnor, C. (2021). Financial inclusion and human development in frontier countries. *International Journal of Finance & Economics, 26*(1), 42–59. https://doi.org/10.1002/ijfe.1775

Agarwal, S., & Zhang, J. (2020). Fintech, lending and payment innovation: A review. *Asia-Pacific Journal of Financial Studies, 49*(3), 353–367. https://doi.org/10.1111/ajfs.12294

Aria, M., & Cuccurullo, C. (2017). Bibliometrix: An R-tool for comprehensive science mapping analysis. *Journal of Informetrics, 11*(4), 959–975. https://doi.org/10.1016/j.joi.2017.08.007

Arner, D. W., Buckley, R. P., Zetzsche, D. A., & Veidt, R. (2020). Sustainability, Fintech and financial inclusion. *European Business Organization Law Review, 21*, 1–29.

Aron, J. (2018). Mobile money and the economy: A review of the evidence. *World Bank Research Observer, 33*(2), 135–188. https://doi.org/10.1093/wbro/lky001

Asongu, S. A., Biekpe, N., & Cassimon, D. (2020). Understanding the greater diffusion of mobile money innovations in Africa. *Telecommunications Policy, 44*(8), 102000. https://doi.org/10.1016/j.telpol.2020.102000

Bastian, M., Heymann, S., & Jacomy, M. (2009, July). Gephi: An open source software for exploring and manipulating networks. In *BT – International AAAI conference on weblogs and social. International AAAI conference on weblogs and social media* (pp. 361–362). https://doi.org/10.13140/2.1.1341.1520

Blondel, V. D., Guillaume, J.-L., Lambiotte, R., & Lefebvre, E. (2008). Fast unfolding of communities in large networks. *Journal of Statistical Mechanics: Theory and Experiment, 2008*(10), P10008. https://doi.org/10.1088/1742-5468/2008/10/P10008

Broadus, R. N. (1987). Early approaches to bibliometrics. *Journal of the American Society for Information Science, 38*(2), 127–129. https://doi.org/10.1002/(SICI)1097-4571(198703)38:2<127::AID-ASI6>3.0.CO;2-K

Cavoli, T., Gopalan, S., & Rajan, R. S. (2020). Does financial inclusion amplify output volatility in emerging and developing economies? *Open Economies Review, 31*(4), 901–930. https://doi.org/10.1007/s11079-019-09568-0

Coffie, C. P. K., Hongjiang, Z., Mensah, I. A., Kiconco, R., & Simon, A. E. O. (2021). Determinants of Fintech payment services diffusion by SMEs in sub-Saharan Africa: Evidence from Ghana. *Information Technology for Development, 27*(3), 539–560. https://doi.org/10.1080/02681102.2020.1840324

Crane, D. (1972). *Invisible colleges: Diffusion of knowledge in scientific communities*. University of Chicago Press.

Della Peruta, M. (2018). Adoption of mobile money and financial inclusion: A macroeconomic approach through cluster analysis. *Economics of Innovation and New Technology, 27*(2), 154–173. https://doi.org/10.1080/10438599.2017.1322234

Demirguc-Kunt, A., Klapper, L., Singer, D., Ansar, S., & Hess, J. (2018). The global Findex database 2017: Measuring financial inclusion and the Fintech revolution. In *The global Findex database 2017: Measuring financial inclusion and the Fintech revolution*. https://doi.org/10.1596/978-1-4648-1259-0

Diodato v. (1994). *Dictionary of bibliometrics*. Haworth Press.

Ellegaard, O., & Wallin, J. A. (2015). The bibliometric analysis of scholarly production: How great is the impact? *Scientometrics, 105*(3), 1809–1831. https://doi.org/10.1007/s11192-015-1645-z

Gabor, D., & Brooks, S. (2017). The digital revolution in financial inclusion: International development in the Fintech era. *New Political Economy, 22*(4), 423–436.

Gai, K., Qiu, M., & Sun, X. (2018). A survey on Fintech. *Journal of Network and Computer Applications, 103*(March 2017), 262–273. https://doi.org/10.1016/j.jnca.2017.10.011

Gargano, A., & Rossi, A. G. (2024). Goal setting and-saving in the FinTech era. *Journal of Finance, 79*(3), 1931–1976. https://doi.org/10.1111/jofi.13339

Gomber, P., Kauffman, R. J., Parker, C., & Weber, B. W. (2018). On the Fintech revolution: Interpreting the forces of innovation, disruption, and transformation in financial services. *Journal of Management Information Systems, 35*(1), 220–265. https://doi.org/10.1080/07421222.2018.1440766

Hu, Y., Liu, C., & Peng, J. G. (2021). Financial inclusion and agricultural total factor productivity growth in China. *Economic Modelling, 96*, 68–82. https://doi.org/10.1016/j.econmod.2020.12.021

Jiao, Z. L., Shahid, M. S., Mirza, N., & Tan, Z. X. (2021). Should the fourth industrial revolution be widespread or confined geographically? A country-level analysis of Fintech economies. *Technological Forecasting and Social Change, 163*, 120442. https://doi.org/10.1016/j.techfore.2020.120442

Kim, D. W., Yu, J. S., & Hassan, M. K. (2018). Financial inclusion and economic growth in OIC countries. *Research in International Business and Finance, 43*, 1–14. https://doi.org/10.1016/j.ribaf.2017.07.178

Kim, D. W., Yu, J. S., & Hassan, M. K. (2020). The influence of religion and social inequality on financial inclusion. *Singapore Economic Review, 65*(1), 193–216. https://doi.org/10.1142/S0217590817460031

Kong, S. T., & Loubere, N. (2021). Digitally down to the countryside: Fintech and rural development in China. *The Journal of Development Studies, 57*, 1739. https://doi.org/10.1080/00220388.2021.1919631

Lagna, A., & Ravishankar, M. N. (2021). Making the world a better place with Fintech research. *Information Systems Journal*. https://doi.org/10.1111/isj.12333

Laidroo, L., Koroleva, E., Kliber, A., Rupeika-Apoga, R., & Grigaliuniene, Z. (2021). Business models of Fintech – Difference in similarity? *Electronic Commerce Research and Applications, 46*. https://doi.org/10.1016/j.elerap.2021.101034

Lashitew, A. A., van Tulder, R., & Liasse, Y. (2019). Mobile phones for financial inclusion: What explains the diffusion of mobile money innovations? *Research Policy, 48*(5), 1201–1215. https://doi.org/10.1016/j.respol.2018.12.010

Lee, I., & Shin, Y. J. (2018). Fintech: Ecosystem, business models, investment decisions, and challenges. *Business Horizons, 61*(1), 35–46. https://doi.org/10.1016/j.bushor.2017.09.003

Macchiavello, E., & Siri, M. (2020). Sustainable finance and Fintech: Can technology contribute to achieving environmental goals? A preliminary assessment of 'Green Fintech'. *SSRN Electronic Journal, 71*. https://doi.org/10.2139/ssrn.3672989

Moro-visconti, P. R. (2021, January). Do crowdfunded platforms improve sustainability and outreach do crowdfunded platforms improve sustainability and outreach in microfinance institutions?

Moro-Visconti, R., Rambaud, S. C., & Pascual, J. L. (2020). Sustainability in Fintechs: An explanation through business model scalability and market valuation. *Sustainability, 12*(24). https://doi.org/10.3390/su122410316

Newman, M. E. J., & Girvan, M. (2004). Finding and evaluating community structure in networks. *Physical Review E – Statistical, Nonlinear, and Soft Matter Physics, 69*(22), 1–15. https://doi.org/10.1103/PhysRevE.69.026113

Nikkel, B. (2020). Fintech forensics: Criminal investigation and digital evidence in financial technologies. *Forensic Science International-Digital*

Investigation, 33, 200908. https://doi.org/10.1016/j.fsidi.2020.200908

Pal, A., Herath, T., De, R., & Rao, H. R. (2021). Why do people use mobile payment technologies, and why would they continue? An examination and implications from India. *Research Policy, 50*(6), 104228. https://doi.org/10.1016/j.respol.2021.104228

Phan, D. H. B., Narayan, P. K., Rahman, R. E., & Hutabarat, A. R. (2020). Do financial technology firms influence bank performance? *Pacific-Basin Finance Journal, 62*, 101210. https://doi.org/10.1016/j.pacfin.2019.101210

Phuc Nguyen, C., Dinh Su, T., & Doytch, N. (2020). The drivers of financial development: Global evidence from internet and mobile usage. *Information Economics and Policy, 53*, 100892. https://doi.org/10.1016/j.infoecopol.2020.100892

Prabhakar, R. (2019). Financial inclusion: A tale of two literatures. *Social Policy and Society, 18*(1), 37–50. https://doi.org/10.1017/S1474746418000039

Pradhan, R. P., & Sahoo, P. P. (2021). Are there links between financial inclusion, mobile telephony, and economic growth? Evidence from Indian states. *Applied Economics Letters, 28*(4), 310–314. https://doi.org/10.1080/13504851.2020.1752893

Ryu, H. S., & Ko, K. S. (2020). Sustainable development of Fintech: Focused on uncertainty and perceived quality issues. *Sustainability, 12*(18). https://doi.org/10.3390/su12187669

Sarma, M., & Pais, J. (2011). Financial inclusion and development. *Journal of International Development, 23*(5), 613–628. https://doi.org/10.1002/jid.1698

Senyo, P. K., Osabutey, E. L. C., & Kan, K. A. S. (2020). Pathways to improving financial inclusion through mobile money: A fuzzy set qualitative comparative analysis. *Information Technology & People*. https://doi.org/10.1108/ITP-06-2020-0418

Shen, Y., Hueng, C. J., & Hu, W. X. (2021). Measurement and spillover effect of digital financial inclusion: A cross-country analysis. *Applied Economics Letters, 28*, 1738. https://doi.org/10.1080/13504851.2020.1853663

Shin, Y. J., & Choi, Y. (2019). Feasibility of the Fintech industry as an innovation platform for sustainable economic growth in Korea. *Sustainability, 11*(19). https://doi.org/10.3390/su11195351

UN (2015). Transforming Our World: The 2030 Agenda for Sustainable Development. Resolution Adopted by the General Assembly on 25 September 2015, A/RES/70/1. 1–35. https://www.un.org/en/development/desa/population/migration/generalassembly/docs/globalcompact/A_RES_70_1_E.pdf

Van, L. T. H., Vo, A. T., Nguyen, N. T., & Vo, D. H. (2021). Financial inclusion and economic growth: An international evidence. *Emerging Markets Finance and Trade, 57*(1), 239–263. https://doi.org/10.1080/1540496X.2019.1697672

World Bank. (2019). Europe and Central Asia Economic Update, Spring 2019: Financial Inclusion. Washington, DC: World Bank. © World Bank, Page 1–105. https://documents1.worldbank.org/curated/ar/377571554411400279/pdf/Financial-Inclusion.pdf

Yang, W., Sui, X. P., & Qi, Z. (2021). Can Fintech improve the efficiency of commercial banks?-An analysis based on big data. *Research in International Business and Finance, 55*. https://doi.org/10.1016/j.ribaf.2020.101338

Zetzsche, D. A., Buckley, R. P., & Arner, D. W. (2019). Fintech for financial inclusion: Driving sustainable growth. *Journal of Finance, First Edit (February)*, 1–26. https://papers.ssrn.com/sol3/papers.cfm?abstract_id=3245287

Part II

Conceptual Concerns

4 "Financial Inclusion Reduces Inequality and Poverty": Myth or Reality?

Asima Shirazi and Shehnaz Al-Helo

... financial inclusion is not an end, but rather a means to an end—and it is why inclusive finance is a major enabler for many of the UN Sustainable Development Goals (SDGs) and an important element for the 2030 Agenda. (UNSGSA, 2020)

Abstract

The Sustainable Development Goals (SDGs) have been developed with a view to enhancing the quality of life in both developed and developing nations. Financial inclusion (SDG 8.10) emphasizes the need for capacity building of domestic financial institutions to provide widespread access to financial services. This chapter investigates the impact of financial inclusion on poverty alleviation, income inequality and non-income dimensions of inequality, namely malnourishment, health, and education. Panel data regression analysis was carried out for 138 developed and developing countries spanning a period of 9 years from 2010 to 2018. Accordingly, changes within countries and across countries were captured over this period of time. Our results indicate that financial inclusion as measured by the Financial Inclusion Index has a strong negative association with poverty but not with the other SDGs that were tested. Therefore, it seems reasonable to conclude that this has ramifications for policy design and implementation. The main contribution of this chapter is that it confirms the significance of developing conducive financial services that, in turn, provide equal opportunities, irrespective of gender, religion, and ethnicity. However, it also highlights that financial inclusion, though necessary, may not be sufficient in the quest to achieve the SDGs.

Keywords

Sustainable Development Goals · Financial inclusion · Inequality · Panel data · Health · Education

The 17 Sustainable Development Goals (SDGs) are a compendium of factors that can raise the standard of living of households and enhance their quality of life. These SDGs are a prescription for sustainable development and are equally applicable to developed nations as they are to developing nations. They are as relevant for the United States as for Liberia (Parr, 2016).

A. Shirazi (✉) · S. Al-Helo
University of Wollongong in Dubai,
Dubai, United Arab Emirates
e-mail: AsimaShirazi@uowdubai.ac.ae

G. W. Muschert et al. (eds.), *Financial Inclusion*, Sustainable Development Goals Series,
https://doi.org/10.1007/978-3-031-68803-4_4

Financial inclusion (SDG 8.10[1]) aims to "strengthen the capacity of domestic financial institutions to encourage and expand access to banking, insurance and financial services for all." This chapter investigates the impact of financial inclusion on poverty alleviation, income inequality, and non-income dimensions of inequality, namely malnourishment, health, and education. Panel data regression analysis examines data from 138 countries spanning 9 years from 2010 to 2018. This approach captures changes within countries and across countries over this period. Our results indicate that financial inclusion, as measured by the Financial Inclusion Index, has a strong negative association with poverty but not with the other SDGs tested. This has implications for policy design and implementation.

4.1 Introduction

Our research aims to unravel the effect of financial inclusion (FI) on poverty, income inequality, and non-income inequality. The definition of financial inclusion could be unidimensional or multidimensional. Financial inclusion is a multifaceted phenomenon not adequately captured by a single indicator such as the number of bank branches or bank accounts. Consequently, we constructed a financial index to capture the multidimensional nature of financial inclusion.

Inequality and poverty can be alleviated if the poorest households in an economy have access to financial services such as loans and insurance. Financial inclusion can be an enabler to this end and contribute toward achieving other SDGs, such as poverty reduction and inequality (Klapper et al. 2016: UNSGSA, 2020). The highest level of financial inclusion should enable households to even out income and expenditure in the short and long run. Households should have access to the credit market to meet shortfalls in their income which can arise either due to endogenous factors such as ill health or exogenous reasons, for instance, job loss during a recession. However, there is a need to be cognizant of the negative impact of financial inclusion if households are over-leveraged and credit markets need to be more adequately regulated (Mader, 2018). Figure 4.1 below shows the benefits of financial inclusion.

Leyshon and Thrift (1995) define financial exclusion as the "processes that serve to prevent certain social groups and individuals from gaining access to the formal financial system." Barriers that prevent access to the formal financial system can be either voluntary or involuntary.

Figure 4.2 below sets out the barriers to financial services and products.

Policymakers need to reduce the impediments to financial inclusion both on the supply and demand sides. On the supply side, efficient and well-regulated financial markets and services should be provided, whereas on the demand side, the focus should be on education and financial literacy. New technologies should be explored in this age of digitalization to increase financial access for those currently excluded.

Financial inclusion can be an enabler for achieving other SDGs. For instance, increasing access to savings accounts or informal savings technologies can lead to an increase in consumption, productivity, income, and investment in preventive health and reduce vulnerability to illness and other unexpected events (Kunt & Klapper, 2013).

To understand the impact of financial inclusion on poverty alleviation, income inequality, and non-income dimensions of inequality, namely malnourishment, health, and education, panel data regression analysis examines 138 countries spanning 9 years from 2010 to 2018. A contribution of this chapter is to go beyond analyzing the effects of financial inclusion on poverty and income inequality and extend it to other developmental goals, including eradicating hunger, enhancing health and well-being, and improving education.

This chapter's organization is as follows: Section 4.2 presents the literature review. Section 4.3 outlines the methodology and the models. Section 4.4 sets out the results and their discussion, whereas Section 4.5 concludes.

[1]https://sdgs.un.org/goals/goal8

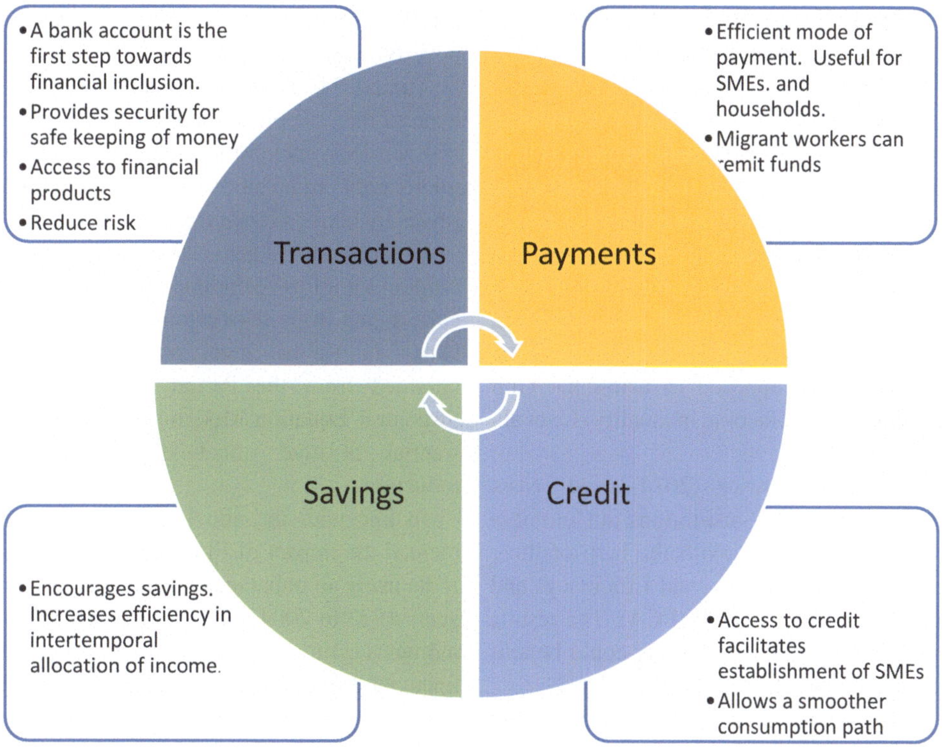

Fig. 4.1 Benefits of financial inclusion

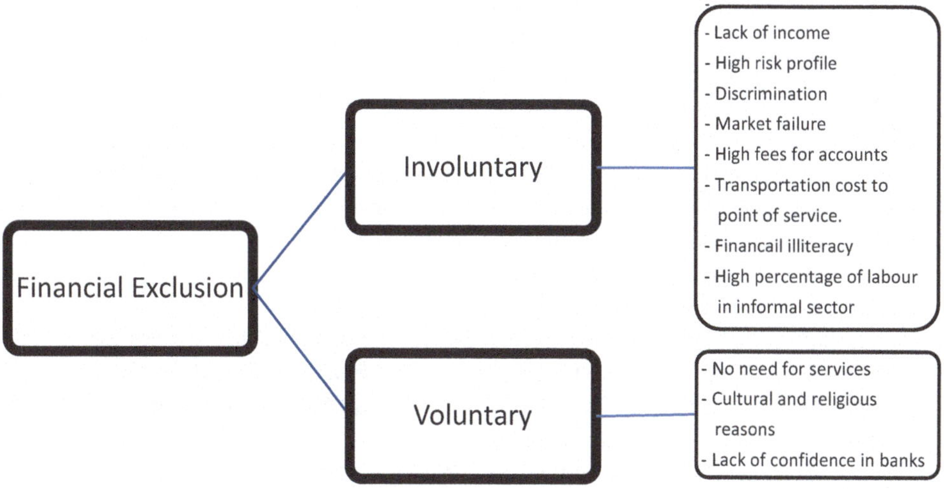

Fig. 4.2 Barriers to financial inclusion. (Source: Adapted from World Bank (2014 and 2016))

4.2 Literature Review

The literature review on the impact of financial inclusion (FI) falls into two categories. The first strand considers the effect of financial inclusion on poverty and income inequality, and the second strand extends the analysis to other SDGs such as hunger, health, and education.

4.2.1 Financial Inclusion: Income Inequality and Poverty

Park and Mercado (2015) studied 37 developing countries using an ordinary least squares (OLS) model. They confirmed that the Gini coefficient declines with the increased rule of law, the rate of completion in primary education, and growth in banks. Other empirical studies focused on using the ratio between agricultural and industrial value-added as a proportion of gross domestic product (GDP), as opposed to using the Gini coefficient to present income inequality (Le et al., 2019).

Meanwhile, Karpowicz (2014) used cross-sectional data from 942 institutions in Colombia to estimate the FI index, with the help of three dimensions (Access, Depth, and Efficiency) and principal component analysis (PCA). The results indicated that constrained workers could benefit once financial markets developed.

A study conducted in the United States by Beck et al. (2007) collected data on bank deregulation between the years 1976 to 2006 and applied the Weibul hazard model to determine whether bank branch expansion leads to a reduction in income inequality. The conclusion was that income disparity declined when banks were less regulated. The cross-sectional data collected by Mookerjee and Kalipioni (2010) based on the availability of financial services (the number of bank branches per 100,000 people) alludes to the improvements observed in terms of income inequality in developed and developing countries, with the reduction in obstacles that hinder access to financial services. With advancements in information access, Kim (2015) built on this work by expanding the sample to include 40 OECD and European countries and data covering the period from 2004 to 2011. Kim concluded that financial inclusion has a positive impact on economic growth, reducing income inequality.

Despite the evidence above, skeptics acknowledged that poverty reduction due to FI takes place until it reaches a particular threshold of economic development (Sehrawat & Giri, 2016). These findings are consistent with those of Lal (2018), who concluded that access to credit is imperative for disadvantaged individuals to break out of poverty.

Yunus (2007) ascertained that those in poverty express their skepticism concerning using financial services, including taking out loans, while microcredit organizations amass wealth for themselves by taking advantage of the poor.

Umaru and Chibuzo (2018) utilized a simple random sampling technique of a questionnaire to collect data from 384 customers of microfinance banks in Kebbi State, Nigeria. The results obtained from the Partial Least Square and Structural Equation Modeling illustrated a substantial positive effect of FI on poverty reduction.

In line with the above, Inoue (2019) documented the impact of FI on poverty reduction in India using an unbalanced panel dataset from the years 1973 to 2004 and a generalized method of moments estimation. The study presupposed that financial deepening along with FI has an inverse relationship to the poverty ratio for public sector banks. However, the same does not hold for private-sector banks.

In a related study, Omar and Inaba (2020) used a fixed effect estimation method to unequivocally reveal that income inequality, per capita income, ratio of internet users, inflation, and age dependency ratio have a consequential effect on the level of FI in emerging markets. Their paper further pointed to the fact that FI reduces poverty rates and income disparity in developing economies.

4.2.2 Financial Inclusion: Zero Hunger, Good Health, and Well-Being and Education

Wight et al. (2014) stated that improvements to financial services could empower individuals to save and monitor their spending habits, countering the risks of hunger.

In terms of health and financial inclusion, Ajefu et al. (2020) indicated that FI boosts health in two crucial ways: facilitating access to insurance facilities and providing peace of mind. The research found that having a degree of financial capability plays a significant role in mental and physical health (Nussbaum, 2009).

Aguila et al. (2016) panel data concentrated on those aged between 51 and 90 of Hispanic race living in the United States. Subjects of this particular age and cultural background were assessed due to the high likelihood of encountering hindrances in acquiring financial services. Their study concluded that older Hispanics had improved mental health, especially if they owned a bank account. Hence, there was a positive relationship.

A multivariate logistic regression and generalized Poisson regression by Gyasi et al. (2019) considered FI concerning older individuals' psychological, self-related, and health services in Ghana. In their study, they collected data through an FI questionnaire, which considered the following: recent withdrawal of money from an account, ownership of a mobile money service account, the use of automatic teller machines (ATMs), the level of savings undergone with a local bank, and access to loans. Their results revealed that bank account ownership fundamentally enhances mental and self-related health. However, they did not determine a definitive relationship with the use of health services.

A critical study by Bruhn and Love (2014) revealed that an individual (economic agent) who was previously financially excluded is more likely to invest in education once he has access to financial services.

There are several empirical studies on the impact of financial inclusion on poverty and income inequality. This chapter contributes to the existing literature by extending the evaluation of financial inclusion as a channel for attaining additional SDGs.

4.3 Methodology

We first constructed a financial inclusion index and then used this index in the five models as an independent variable together with control variables.

The regression analysis involved unbalanced annual panel data for 138 countries covering 2010 to 2018. The availability of data determined the choice of countries and years.

4.3.1 Financial Inclusion Index

The three aspects of financial inclusion, availability, usage, and access, are measured by the three variables financial institution, deposits, and loans. In addition to commercial banks and mobile phone agents, microfinance and credit union institutions are included in the index to reflect access to financial services amongst poorer households (Holden & Prokopenko, 2001).

Using Principal Component Analyses, a financial inclusion index (FIIND) was calculated. The variables included in the index reflect availability, usage, and access to financial services. The choice of variables and the construction of the index is based on the approaches adopted by other researchers (Omar & Inaba, 2020; Amidžić et al., 2014; Sarma, 2012; Honohan, 2007). Table 4.1 below lists the variables used in the construction of the index.

The first step was to normalize the variables used in the calculation of the index. The following equation was applied:

$$n\_X_{it} = (X_{it} - \min)/(\max - \min)$$

where

n_X_{it} = normalized value of X_{it}

X_{it} = original value

Min = minimum value of the variable

Max = maximum value of the variable

The values of the normalized variable lie between 0 and 1. Normalization is essential when variables are measured in different units and could have distinct average numerical values.

Principal Component Analysis (PCA) was carried out, and weights for the index were determined using the result of the PCA.[2] The Financial Inclusion Index (FIIND) was calculated as follows:

[2]The squared values of the eigenvector, which add to 1, are used as weights.

Table 4.1 Variables used in the construction of the financial inclusion index

Variable	Description	Source
Financial institutions (Finstit)	Aggregate the following financial institutions: Number of commercial bank branches per 1000 km^2 Number of ATMs per 1000 km^2 Number of registered mobile money agent outlets per 1000 km^2 Number of credit union and credit cooperative branches per 1000 km^2 Number of all microfinance institution branches per 1000 km^2	IMF Macroeconomic and Financial Data[a]
Deposits (Depfiins)	Aggregate of the following deposit accounts: Deposit accounts with commercial banks Deposit accounts with credit unions and credit cooperatives Deposit accounts with deposit-taking microfinance institutions	IMF Macroeconomic and Financial Data
Loans (Loanfin)	Aggregate of the following loan accounts: Loan accounts with commercial banks Loan accounts with credit unions and credit cooperatives Loan accounts with deposit-taking microfinance institutions	IMF Macroeconomic and Financial Data

[a]https://data.imf.org/?sk=E5DCAB7E-A5CA-4892-A6EA-598B5463A34C

$$\text{FIIND} = \left(\left(0.3511\,^\wedge 2\right) * \text{n\_finstit}\right) + \left(\left(0.6648\,^\wedge 2\right) * \text{n\_depfiins}\right) + \left(\left(0.6594\,^\wedge 2\right) * \text{n\_loanfin}\right) \tag{4.1}$$

The Financial Inclusion Index (FIIND) was used as an independent variable in the subsequent models to determine the impact of financial inclusion on other SDGs.

4.3.2 Models

The variables used in the estimation of the five models appear in Table 4.2 below:

Both fixed and random effect panel data regression models were estimated, and the appropriate one was selected based on the Hausman test. The general specification of the fixed effects (FE) and random effects (RE) models are given in Eqs. 4.2 and 4.3.

$$Y_{it} = \beta_1 X_{it} + \alpha_i + u_{it} \tag{4.2}$$

Where

α_i (i = 1….n) is the unknown intercept for each entity (n entity-specific intercepts).
Y_{it} is the dependent variable (DV) where i = entity and t = time.
X_{it} represents one independent variable (IV),
β_1 is the coefficient for that IV,
u_{it} is the error term.

In the random effects model, unlike the fixed effects model, the variation across entities is assumed to be random and uncorrelated with the independent variables included in the model:

$$Y_{it} = \beta_1 X_{it} + \alpha + u_{it} + \varepsilon_{it} \tag{4.3}$$

u_{it} = between entity error
ε_{it} = within entity error

4.3.2.1 Model 1 Poverty

Sustainable Development Goal 1 aims to eradicate poverty. The choice of variables was undertaken after a review of the literature. The dependent variable is the percentage of the population living below \$1.90 per day. The independent variables included internet and mobile use, secondary school gross enrollment, trade as a percentage of GDP, age dependency ratio, inflation, unemployment, log of GDP per capita, and the FIIND.

$$\begin{aligned} sdg1\_wpc_{it} &= \alpha_i + \beta_1 sdg9\_\text{intuse}_{it} + \beta_2 sdg9\_\text{mobuse}_{it} + \beta_3 \text{secscenl}_{it} + \\ &\beta_4 \text{tdegdp}_{it} + \beta_5 \text{agedep}_{it} + \beta_6 \text{inflation}_{it} + \beta_7 \text{unemp}_{it} + \\ &\beta_8 \text{gdpcap2}_{it} + \beta_9 \text{FIIND}_{it} + u_{it} \end{aligned} \tag{4.4}$$

Table 4.2 Variables used in the panel data models

Name	Description	Source
sdg1_wpc	Poverty headcount ratio at $1.90/day (%) It measures the percentage of the population living at or below $1.90 per day.	Sustainable Development Solutions Network (SDSN)
sdg2_undernsh	Prevalence of undernourishment (%)	SDSN
sdg3_u5mort	Mortality rate under 5 (per 1000 live births)	SDSN
sdg9_intuse	Population using the internet (%)	SDSN
sdg9_mobuse	Mobile broadband subscriptions (per 100 population)	SDSN
GDPpcap	Per capita GDP	World Development Indicators (WDI)
GDPpcap2	Log of per capita GDP	WDI
GDPpcapgr	Per capita GDP growth rate	WDI
Tde%GDP	Trade as a percentage of GDP	
Agedep	Age dependency ratio. The age dependency ratio is the ratio of dependents—people younger than 15 or older than 64—to the working-age population—those ages 15–64. Data appear as the proportion of dependents per 100 working-age population.	WDI
Inflation	Inflation rate based on CPI.	WDI
SecScenl	Gross Secondary School Enrollment	WDI
Gini	Gini Coefficient, a measure of income inequality	WDI
ghthgdp	Government Expenditure on health as a percentage of GDP	WDI
gedugdp	Government Expenditure on education as a percentage of GDP	WDI
unemp	Unemployment rate	WDI

[a]https://www.sdgindex.org/

4.3.2.2 Model 2 Inequality

The Gini index is a measure of inequality in the distribution of income. A higher value of the Gini index is associated with greater income distribution inequality. SDG 10 focuses on the reduction of inequality within and among countries.

$$\begin{aligned}\text{Gini}_{it} &= \alpha_i + \beta_1 sdg9\_\text{intuse}_{it} + \\ &\beta_2 sdg9\_\text{mobuse}_{it} + \beta_3 \text{secscenl}_{it} + \\ &\beta_4 \text{agedep}_{it} + \beta_5 \text{inflation}_{it} + \\ &\beta_6 \text{unemp}_{it} + \beta_7 \text{gdpcap2}_{it} + \\ &\beta_8 \text{FIIND}_{it} + \text{tdegdp}_{it} + u_{it}\end{aligned} \tag{4.5}$$

4.3.2.3 Model 3 Hunger

The eradication of hunger is the goal of SDG 2. Several indicators contribute to the fulfillment of this goal. For our analysis, the dependent variable is the prevalence of undernourishment.[3]

$$\begin{aligned}sdg2\_\text{undernsh}_{it} &= \alpha_i + \beta_1 sdg9\_\text{intuse}_{it} + \\ &\beta_2 sdg9\_\text{mobuse}_{it} + \beta_3 \text{secscenl}_{it} + \\ &\beta_4 sdg1\_wpc_{it} + \\ &\beta_5 \text{agedep}_{it} + \beta_6 \text{inflation}_{it} + \\ &\beta_7 \text{unemp}_{it} + \beta_8 \text{gdpcap2}_{it} + \beta_9 \text{FIIND}_{it} + u_{it}\end{aligned} \tag{4.6}$$

4.3.2.4 Model 4 Infant Mortality

Infant mortality under 5 years is one of the indicators of SDG 3 health and well-being. In this model, infant mortality is the dependent variable.

[3]The prevalence of undernourishment is an estimate of the proportion of the population whose habitual food consumption is insufficient to provide the dietary energy levels that are required to maintain a normally active and healthy life. It is expressed as a percentage. http://www.fao.org/sustainable-development-goals/indicators/211/en/

$$sdg3\_\text{u5mort}_{it} = \alpha_i + \beta_1 sdg9\_\text{intuse}_{it} + \beta_2\text{agedep}_{it} + \beta_3\text{unemp}_{it} + \beta_4\text{gdpcap2}_{it} + \beta_5\text{FIIND}_{it} + \beta_6 + \text{secscenl}_{it} + \beta_7 sdg1\_wpc_{it} + \beta_8\text{ghthgdp}_{it} + u_{it} \quad (4.7)$$

4.3.2.5 Model 5 School Enrollment

SDG 4 aims to provide inclusive and equitable quality education and promote lifelong learning opportunities for all.[4] The dependent variable is gross secondary school enrollment which is the ratio of total enrollment, regardless of age, to the population of the age group. Secondary education completes the provision of basic education and enables lifelong learning.

$$\text{secscenl}_{it} = \alpha_i + \beta_1 sdg9\_\text{mobuse}_{it} + \beta_2\text{agedep}_{it} + \beta_3\text{unemp}_{it} + \beta_4\text{gdpcap2}_{it} + \beta_5\text{FIIND}_{it} + \beta_6\text{gedugdp}_{it} + \beta_7 sdg9\_\text{intuse}_{it} + \beta_8 sdg1\_wpc_{it} + \beta_9\text{inflation}_{it} + u_{it} \quad (4.8)$$

4.4 Discussion of Results

The results of the panel data regression models appear in Table 4.3 below.

It is evident from the estimations that financial inclusion had an influential negative association with headcount poverty but not for any other dependent variable. As financial inclusion increases, the poverty level reduces. Several studies have reported similar results (Burgess & Pande, 2005; Jalilian & Kirkpatrick, 2005; Brune et al., 2011; Swamy, 2014; Park & Mercado, 2015, 2018a, b; Rewilak, 2017; Omar & Inaba, 2020; Koomson et al., 2020; Nsiah et al., 2021). An increase in the Financial Inclusion Index (FIIND) leads to economic growth, accompanied by job creation and higher incomes, which can pull people out of poverty. When financial inclusion policies can serve the needs of the poor and marginalized segments of the population, they can have direct impacts on reducing poverty. Other control variables that significantly affect poverty are the age dependency ratio and per capita GDP. These results are consistent with our expectations. If the age dependency ratio increases due to an aging population or a large number of children in the family, income per capita will decline to increase poverty. If this ratio increases due to an aging population, it indicates a decline in the working-age population. A reduction in the child dependency ratio leads to an increase in gross domestic product per capita growth and a poverty reduction (Cruz & Ahmed, 2018). There is a negative association between poverty and per capita GDP, which means that as income levels increase, poverty will decline. The other two variables that significantly affect poverty are inflation (negative) and secondary school enrollment. A negative association between inflation and poverty implies that as inflation increases, poverty decreases when an economy expands due to rising aggregate demand, output, and the price level increase. In most cases, mild inflation is associated with economic growth. An increase in the gross secondary school enrollment rate will increase productivity and household incomes and will accompany a lower poverty level.

The second model's dependent variable measuring income inequality is the Gini coefficient. Several empirical studies conclude a negative and significant correlation between Financial inclusion and inequality (Park & Mercado, 2015, 2018a, b; Mookerjee & Kalipioni, 2010; Beck et al., 2007; Hamori & Hashiguchi, 2012; Karpowicz, 2014; Kim, 2015; Fouejieu et al., 2020; Cihak & Sahay, 2020; Omar & Inaba, 2020). The independent variable Financial Inclusion Index (FIIND), is not significant.

Kling et al. (2020) report a mixed outcome. Income inequality worsens if households rely on formal or informal loans, whereas access to bank accounts improves households' prospects in future income distribution.

Financial inclusion has little bearing on inequality in our model since the sample of countries includes both developed and developing nations. The impact of financial inclusion on

[4] https://sdg4education2030.org/the-goal

Table 4.3 Results of model estimations

	Model 1	Model 2	Model 3	Model 4	Model 5
Independent variables	DV = sdg1_wpc	DV = Gini Index	DV = sdg2_ undernsh	DV = sdg3_ u5mort	DV = secscenl
Constant	82.9*** (8.19)	−30.37** (−2.10)	114.9*** (7.38)	69.18*** (3.06)	67.558 (0.95)
sdg9_intuse	−0.007 (− 0.90)	−0.045*** (−3.94)	0.003 (0.24)	−0.111*** (−6.44)	0.013 (0.25)
sdg9_mobuse	0.006* (1.86)	−0.015*** (−3.59)	0.014*** (2.65)	–	0.106*** (5.06)
tdegdp	−0.005 (−1.16)	0.017*** (3.04)	–	–	–
agedep	0.256*** (9.49)	0.169*** (4.90)	0.084** (2.08)	0.591*** (9.54)	0.453*** (2.79)
inflation	−0.03** (2.36)	−0.029* (−1.81)	−0.027 (1.55)	–	0.007 (0.11)
unemp	−0.026 (−0.77)	0.223*** (5.79)	−0.194*** (−3.86)	0.049 (0.59)	0.307 (1.42)
gdpcap2	−9.65*** (−8.74)	6.089*** (4.04)	−12.109*** (−7.06)	−7.414*** (−2.89)	−0.519 (−0.07)
FIIND	−53.1*** (−8.44)	−5.007 (−0.35)	23.26** (2.37)	−19.015 (−1.20)	19.95 (0.41)
secscenl	−0.021** (−2.55)	0.005 (0.63)	−0.013 (−2.55)	−0.114*** (−5.78)	–
sdg1_wpc	–	–	−0.316*** (−5.74)	0.55*** (7.17)	−0.853*** (−3.73)
ghthgdp	–	–	–	−0.407 (−1.42)	–
gedugdp	–	–	–	–	−0.02 (−0.05)
F statistic	73.35	10.55	8.44	94.92	15.62
P-Value	0.000	0.000	0.000	0.000	0.000
R squared:					
within	0.48	0.18	0.09	0.49	0.23
between	0.57	0.05	0.36	0.75	0.53
overall	0.55	0.11	0.38	0.76	0.53
Estimation Method	Fixed effects	Fixed effects	Fixed effects	Fixed effects	Fixed effects
Hausman Test	Chi2 = 77.34 Prob>Chi2 = 0.00	Chi2 = 26.96 Prob>Chi2 = 0.000	Chi2 = 126.73 Prob>Chi2 = 0.00	Chi2 = 25.50 Prob>Chi2 = 0.00	Chi2 = 31.55 Prob>Chi2 = 0.00

Notes: *DV* Dependent Variable. *T*-ratios are in parenthesis. (*), (**), and (***) denote significance at the 10%, 5%, and 1% levels, respectively. Only gdpcap2 (GDP per capita) appears in log form

income inequality is also challenging to predict since the relationship may be non-linear. Kuznets (1955) posited that the path of income inequality might be an inverted U-shaped curve, also known as Kuznet's curve. During the early stages of development, internal migration from rural to urban areas will escalate inequality. However, the next generation of rural migrants will see an increase in income as they benefit from urban opportunities (Greenwood & Jovanovic, 1990). The speed at which the migrant population catches up is dependent on, among other things, access to good quality education, health facilities, employment opportunities, and access to financial markets. Financial inclusion could be an enabler of more significant opportunities. This reasoning also applies to displaced persons who migrate to host countries. Several studies have confirmed evidence of an inverted U-curve relationship between financial development and income inequality (Nikoloski, 2012; Jauch & Watzka, 2016; Greenwood & Jovanovic, 1990). The relationship between economic development (including financial inclusion) and inequality has political underpinnings. Development accompanied by increased inequality can lead to political instability and "democratization," or there can be nondemocratic paths, namely an "autocratic disaster" with high inequality and low output or an "East Asian Miracle" with low inequality and high output (Acemoglu & Robinson, 2002).

The control variables that significantly affect income inequality are internet and mobile phone use. These are indicators of technological development. They are both negative and significant since an increase in their usage is associated with lower income inequality. Trade to GDP ratio has a positive relationship with the Gini coefficient indicating an increase in inequality associated with openness. The age dependency ratio and unemployment are both positively related and significant indicating that an increase in these variables will lead to greater inequality. There is a positive relationship between inequality and per capita GDP. This confirms that an increase in per capita GDP can be associated with greater inequality.

Models 3, 4, and 5 estimate the association between Financial Inclusion and undernourishment, infant mortality, and secondary school enrollment, respectively. The relationship between the Financial Inclusion Index and undernourishment is significant at 5% and positive, indicating that an increase in Financial Inclusion will be associated with undernourishment. This is possible if FI is associated with greater income inequality. The control variables include age dependency, which is positive, indicating that an increase is associated with higher malnutrition. The per capita GDP is significant and negative indicating that an increase in per capita GDP will lead to a decrease in undernourishment. Unemployment and poverty explanatory variables are significant and negative, suggesting that a decrease in these two variables will accompany an increase in undernourishment. Low unemployment may be associated with higher inflation leading to increased food prices and undernourishment. We must be cognizant that malnutrition results from several factors, including awareness of healthy eating habits, availability, and access to food at affordable prices (Masters, 2016, p. 46).

For health and education (under-five mortality and gross secondary school enrollment), FI is not a significant explanatory variable. In the health model, the control variables internet use, GDP per capita, and secondary school enrollment have a negative and highly significant association with under-5 mortality. An increase in these variables will lead to a reduction in under-5 mortality. Age dependency and poverty have a positive relationship indicating that an increase in these variables is associated with an increase in infant mortality. In the education model, mobile use and age dependency positively affect school enrollment, whereas poverty has a negative relationship. Age dependency can be high due to an increase in the child dependency ratio leading to more child enrollment in schools. An increase in poverty will lead to lower school enrollment. In both models, government expenditure as a percentage of GDP on health and education does not significantly influence the dependent variables. It is evident in this context that the outcomes in the spheres of health and education are influenced not just by the level of spending but also by the efficiency and efficacy of public expenditure (Baldacci et al., 2003).

4.5 Conclusion and Recommendations

Empirical evidence on the link between financial inclusion and poverty alleviation is positive, and the result of our model confirms it. However, the contribution of financial inclusion to income inequality needs to be more conclusive. In our study, the effect is insignificant. There needs to be more empirical data on the association between financial inclusion and the attainment of other SDGs. We did not find any significant contribution of financial inclusion to infant mortality and secondary school enrollment. Undernourishment increases with financial inclusion, possibly due to increased inequality associated with financial development. Policymakers should consider these findings to develop financial services that provide equal opportunities to all, irrespective of gender, religion, and ethnicity. Additionally, they should ensure that financial inclusion channels intersect positively with country-specific policy initiatives designed to achieve the SDGs.

The results of the present study have shown that the age dependency ratio and per capita GDP influence the dependent variables in our models. An aging population and child dependency influence the age-dependency ratio.

Although economic growth and prosperity accompany financial inclusion, it is possible that even in developed countries, people will need more access to financial markets and products. In developing countries, the barriers to financial inclusion could be more widespread, and a large segment of the population may need access to formal financial services.

There is an overarching need to remove impediments to financial inclusion. Both demand and supply-side barriers need to be overcome. Financial products and services must develop to meet the needs of end users in both urban and rural areas. Sustainable development can only take place if there is a nexus between economic growth, social inclusion, and environmental protection.

References

Acemoglu, D., & Robinson, J. A. (2002). The political economy of the Kuznets curve. *Review of Development Economics, 6*(2), 183–203.

Aguila, E., Angrisani, M., & Blanco, L. (2016). Ownership of a bank account and health of older Hispanics. *Economics Letters, 144*, 41–44.

Ajefu, J. B., Demir, A., & Haghpanahan, H. (2020). The impact of financial inclusion on mental health. *SSM - Population Health, 11*, 100630–100630.

Amidžić, G., Massara, A., & Mialou, A. (2014). 'Assessing countries' Financial Inclusion Standing—A New Composite Index. *International Monetary Fund Working Paper No. 14/36*. Washington, DC.

Baldacci, E., Guin-Siu, M. T., & de Mello, L. (2003). More on the effectiveness of public spending on health care and education: A covariance structure model. *Journal of International Development, 15*(6), 709–725.

Barajas, A., Beck, T., Belhaj, M., & Naceur, S. B. (2020). Financial inclusion: What have we learned so far? What do we have to learn?, *IMF Working Paper WP/20/157.*

Beck, T., Demirgüç-Kunt, A., & Levine, R. (2007). Finance, inequality and the poor. *Journal of Economic Growth, 12*(1), 27–49.

Bruhn, M., & Love, I. (2014). The real impact of improved access to finance: Evidence from Mexico. *Journal of Finance, 69*(3), 1347–1376.

Brune, L., Giné, X., Goldberg, J., & Yang, D. (2011). Commitmentstosave:AfieldexperimentinruralMalawi. *World Bank Policy Research Working Paper 5748*. https://openknowledge.worldbank.org/bitstream/handle/10986/3510/WPS5748.pdf?sequence=1&isAllowed=y. [Accessed 18 July 2021].

Burgess, R., & Pande, R. (2005). Do rural banks matter? Evidence from the Indian social banking experiment. *American Economic Review, 95*(3), 780–795.

Cihak, M., & Sahay, R. (2020). *Finance and inequality. IMF Staff Discussion Notes 2020/001*. International Monetary Fund.

Cruz, M., & Ahmed, S. A. (2018). On the impact of demographic change on economic growth and poverty. *World Development, 105*, 95–106.

Fouejieu, A., Sahay, R., Cihak, M., & Chen, S. (2020). Financial inclusion and inequality: A cross-country analysis. *The Journal of International Trade & Economic Development, 29*(8), 1018–1048.

Greenwood, J., & Jovanovic, B. (1990). Financial development, growth, and the distribution of income. *Journal of Political Economy, 98*(5), 1076–1107.

Gyasi, R. M., Adam, A. M., & Phillips, D. R. (2019). Financial inclusion, health-seeking behavior, and health outcomes among older adults in Ghana. *Research on Aging, 41*(8), 794–820.

Gyasi, R. M., Phillips, D. R., & Adam, A. M. (2021). How far is inclusivity of financial services associated with food insecurity in later life? Implications for health policy and sustainable development goals. *Journal of Applied Gerontology, 40*(2), 189–200.

Hamori, S., & Hashiguchi, Y. (2012). The effect of financial deepening on inequality: Some international evidence. *Journal of Asian Economics, 23*, 353–359.

Holden, P., & Prokopenko, V. (2001). Financial development and poverty alleviation: Issues and policy implications for developing and transition countries, *IMF Working Paper Reference No. WP/01/160.*

Honohan, P. (2007). Cross-country variation in household access to financial services, *World Bank paper conference on "access to finance"*. Washington, DC.

Inoue, T. (2019). Financial inclusion and poverty reduction in India. *Journal of Financial Economic Policy, 11*(1), 21–33.

Jalilian, H., & Kirkpatrick, C. (2005). Does financial development contribute to poverty reduction? *The Journal of Development Studies, 41*(4), 636–656.

Jauch, S., & Watzka, S. (2016). Financial development and income inequality: A panel data approach. *Empirical Economics, 51*, 291–314.

Karpowicz, I. (2014). Financial inclusion, growth, and inequality: A model application to Colombia. *IMF Working Papers, 14*(166), 1–32.

Kim, J. H. (2015). A study on the effect of financial inclusion on the relationship between income inequality and economic growth. *Emerging Markets Finance and Trade, 52*, 498–512. https://doi.org/10.1080/1540496X.2016.1110467 [Accessed on 21 July 2021].

Klapper, L., El-Zoghbi, M., & Hess, J. (2016). *Achieving the sustainable development goals: The role of financial inclusion*. CGAP.

Kling, G., Pesque-Cela, V., Tian, L., & Luo, D. (2020). A theory of financial inclusion and income inequality. *The European Journal of Finance*, 1–21. https://www.tandfonline.com/doi/full/10.1080/1351847X.2020.1792960. [Accessed on 1 August 2021].

Koomson, I., Villano, R. A., & Hadley, D. (2020). Effect of financial inclusion on poverty and vulnerability to poverty: Evidence using a multidimensional measure of financial inclusion. *Social Indicators Research, 149*, 613–639.

Kunt, A. D., & Klapper, L. (2013). *Measuring financial inclusion: Explaining variation and use of financial services across and within countries* (Brookings Papers on Economic Activity). World Bank.

Kuznets, S. (1955). Economic growth and income inequality. *American Economic Review, 45*, 1–28.

Lal, T. (2018). Impact of financial inclusion on poverty alleviation through cooperative banks. *International Journal of Social Economics, 45*(5), 808–828.

Le, Q. H., Ho, H. L., & Mai, N. C. (2019). The impact of financial inclusion on income inequality in transition economies. *Management Science Letters, 9*(5), 661–672.

Leyshon, A., & Thrift, N. (1995). Geographies of financial exclusion: Financial abandonment in Britain and the United States. *Transactions of the Institute of British Geographers, New Series, 20*(3), 312–341.

Mader, P. (2018). Contesting financial inclusion. *Development Change, 49*, 461–483.

Masters, W. A. (2016). The economic consequences of malnutrition. In M. Eggersdorfer et al. (Eds.), *Good nutrition: Perspectives for the 21st century* (pp. 46–57). Sight & Life.

Mookerjee, R. & Kalipioni, P. (2010). Availability of financial services and income inequality: Evidence from many countries. *Emerging Markets Review 11*, 404–408.

Nikoloski, Z. (2012). Financial sector development and inequality. Is there a financial Kuznets curve? *Journal of International Development, 25*(7), 897–911.

Nsiah, A. Y., Yusif, H., Tweneboah, G., Agyei, K., & Baidoo, S. T. (2021). The effect of financial inclusion on poverty reduction in Sub-Sahara Africa: Does threshold matter? *Cogent Social Sciences, 7*(1).

Nussbaum, M. C. (2009). Creating capabilities: The human development approach and its implementation. *Hypatia, 24*(3), 211–215.

Omar, M. A., & Inaba, K. (2020). Does financial inclusion reduce poverty and income inequality in developing countries? A panel data analysis. *Journal of Economic Structures, 9*(37), 1–25.

Park, C. Y., & Mercado, R. V. (2015). *Financial inclusion, poverty, and income inequality in developing Asia* (ADB Economics Working Paper Series No. 426). Asian Development Bank. https://www.adb.org/sites/default/fles/publication/153143/ewp-426.pdf. [Accessed 17 July 2021].

Park, C. Y., & Mercado, R. V. (2018a). Financial inclusion, poverty, and income inequality. *Singapore Economic Review, 63*(1), 185–206. https://searchebscohostcom.ezproxy.uow.edu.au/login.aspx?direct=true&db=bth&AN=127867819&site=ehost-live. [Accessed on 21 July 2021].

Park, C. Y., & Mercado, R. V. (2018b). Financial Inclusion: New Measurement and Cross-Country Impact Assessment, *ADB Economics Working Paper Series No. 539*. https://www.adb.org/sites/default/files/publication/408621/ewp-539-financial-inclusion.pdf. [Accessed 17 July 2021].

Parr, S. K. (2016). From the millennium development goals to the sustainable development goals: Shifts in purpose, concept, and politics of global goal setting for development. *Gender and Development, 24*(1), 43–52.

Rewilak, J. (2017). The role of financial development in poverty reduction. *Review of Development Finance, 7*, 169–176.

Sarma, M. (2012). Index of financial inclusion—a measure of financial sector inclusiveness, *Berlin Working Papers on Money, Finance, Trade and Development*, 07/2012. https://finance-and-trade.htw berlin.de/fileadmin/HTW/Forschung/Money_Finance_Trade_Development/working_paper_series/wp_07_2012_Sarma_Index-of-Financial-Inclusion.pdf. [Accessed on 10 July 2021].

Sehrawat, M., & Giri, A. K. (2016). Financial development and poverty reduction: Panel data analysis of south Asian countries. *International Journal of Social Economics, 43*(4), 400–416.

Swamy, V. (2014). Financial inclusion, gender dimension, and economic impact on poor households. *World Development, 56*, 1–15.

Umaru, H., & Chibuzo, I. C. (2018). The effects of financial inclusion on poverty reduction: The moderating effects of microfinance. *International Journal of Multidisciplinary Research and Development, 5*(12), 188–198.

UNSGSA. (2020). Financial Inclusion Beyond Access and Usage to Quality, *UNSGSA Annual Report*. https://www.unsgsa.org/sites/default/files/resources-files/2020%2010/UNSGSA_2020_Annual_Report_Digital.pdf. [Accessed on 20 July 2021].

Wight, V., Kaushal, N., Waldfogel, J., & Garfinkel, I. (2014). Understanding the link between poverty and food insecurity among children: Does the definition of poverty matter? *Journal of Children and Poverty, 20*(1), 1–20.

World Bank. (2014). Financial Inclusion. In *Global Financial Development Report 2014*. World Bank.

World Bank. (2016). Payment aspects of financial inclusion committee on payments and Market Infrastructures World Bank Group.

Yunus, M. (2007). Banker to the poor: Micro-lending and the battle against world poverty, *Public Affairs*, New York.

Zhang, Q., & Posso, A. (2019). Thinking inside the box: A closer look at financial inclusion and household income. *The Journal of Development Studies, 55*(7), 1616–1631.

5 Driving Financial Inclusion Through Frugal Innovation to Achieve UN SDG Target 8.10

Kerry Slade

Abstract

Traditional banking solutions have not succeeded in creating financial inclusion for the unbanked. Mobile money services, through which telecommunications companies offer financial services through a mobile phone number, increase the accessibility of financial services. For that reason, the proportion of adults with mobile money accounts is one indicator of success for Sustainable Development Goal Target 8.10, which seeks to "strengthen the capacity of domestic financial institutions to encourage and expand access to banking, insurance and financial services for all." M-Pesa is the world's first mobile money service, which was born of frugal innovation and increased financial inclusion in East Africa.

Through a case study M-Pesa in Kenya, this chapter demonstrates how approaching financial inclusion from the perspective of frugal innovation generates high-value solutions to help meet Sustainable Development Goals (SDG) 8 and Target 8.10. Frugal innovation, an innovation management mindset that reduces complexity and resource use, subverts the traditional model and trades highly structured and expensive research and development for flexible and inclusive adaptation. As efforts to develop more inclusive financial systems increase, the discipline can be transformational for both developing and developed markets.

Keywords

Financial inclusion · Frugal innovation · Mobile money · M-Pesa · Target 8.10 · Financial services · Kenya · Africa · International development · Unbanked

5.1 Introduction

To achieve the financial inclusion goals set in Target 8.10, financial products and services must be made more accessible and relevant to those currently excluded. Target 8.10 seeks to "strengthen the capacity of domestic financial institutions to encourage and expand access to banking, insurance, and financial services for all." In addition to measures of ATMs, commercial bank branches, and traditional bank accounts, the proportion of adults with mobile money accounts is one indicator of success for Target 8.10. This chapter explores the case of M-Pesa, the world's first mobile money service born of frugal innovation and increased financial inclusion in East Africa (Shapshak, 2020).

K. Slade (✉)
Fox School of Business, Temple University, Philadelphia, PA, USA

G. W. Muschert et al. (eds.), *Financial Inclusion*, Sustainable Development Goals Series,
https://doi.org/10.1007/978-3-031-68803-4_5

The purpose of Sustainable Development Goals (SDG) 8 is to promote economic growth that is inclusive and sustainable. Financial inclusion, measured in part by Target 8.10, is integral to SDG 8 because economic growth requires access to financial services that enable saving, borrowing, and risk-hedging. Relying on traditional banking solutions to provide financial inclusiveness ignores that they thus far have yet to succeed in including the unbanked. Mobile money services, where telecommunications companies offer financial services through a mobile phone number, increase the accessibility of financial services. Data suggest a positive correlation exists between having an account of any kind and increased financial security (Demirgüc-Kunt et al., 2018).

Through a case study of M-Pesa in Kenya, this chapter demonstrates how approaching financial inclusion from the perspective of frugal innovation can generate high-value solutions to help meet SDG 8 and Target 8.10 (Leliveld & Knorringa, 2017; Shapshak, 2020). Frugal innovation, an innovation management mindset that reduces complexity and resource use, subverts the traditional model and trades highly structured and expensive research and development for flexible and inclusive adaptation. As efforts to develop more inclusive financial systems increase, the discipline can be transformational for developing and developed markets (Kahn, 2016).

5.2 About Mobile Money

A mobile money account is a mobile-phone-based money service usually operated by a telecommunications network carrier that is not linked to a traditional bank or financial institution. These accounts allow people to send and receive payments and save money through an account linked to their phone number (Demirgüc-Kunt et al., 2018). In contrast with mobile banking, mobile money does not require a user to have a bank account and allows the transfer, deposit, and withdrawal of money without one. Mobile money, therefore, is considered outside of the formal banking system. Many mobile money services do not require smartphone capabilities or internet access. Cash transactions can occur at one of the thousands of agent locations dispersed throughout the country, even near small villages in rural areas.

There were 1.21 billion registered mobile money accounts worldwide in 2020, an increase of 12.7% from the prior year, double the projected growth rate. The global value of transactions in 2020 exceeded $2 billion daily (Andersson-Manjang & Naghavi, 2021). As of 2018, 83% of adults in developing countries have a mobile phone, and this percentage likely has increased since then (Klapper, 2019). Women, the less educated, and lower-income persons own disproportionately fewer phones (Demirgüc-Kunt et al., 2018).

In addition to the M-Pesa example from East Africa discussed in this chapter, mobile money services include Jazz in Pakistan, Zain in Iraq and Jordan, Tigo in Bolivia and Guatemala, Grameenphone in Bangladesh, and Orange in the Middle East, North Africa, and sub-Saharan Africa (Andersson-Manjang & Naghavi, 2021). Mobile money seems to be a phenomenon of the "developing" world—the Global Findex Database reports that as of 2017, only three high-income countries had these services available: Chile, Singapore, and the United Arab Emirates (Demirgüc-Kunt et al., 2018).

Mobile money is already increasing financial inclusion worldwide, as explored in this case study. The overall decreased transaction costs, transportation costs, and time involved in mobile money enable more people to access it. Once a member, one of four mobile money providers offers savings accounts and credit, either directly or indirectly. Customer data from mobile money transactions are used to assess creditworthiness for potential loans. As of June 2020, customers had saved an average of $39 each, and mobile money providers had loaned $423 million. Further, mobile money platforms serve as a conduit for customers to access traditional banks. The value of transactions between mobile money platforms and banks grew from $15 billion in 2015 to $68 billion in 2020 (Andersson-Manjang & Naghavi, 2021).

By using mobile money infrastructure for payments, over-the-top services have flourished. A new "mobile-enabled insurance" category allows microinsurance to cover millions of uninsured, primarily with life and health insurance. Forty-three million policies were written in 2020. The agricultural sector is another beneficiary of mobile money services, with 34% of providers reporting partnerships with agricultural companies (Andersson-Manjang & Naghavi, 2021). Mobile money is a robust technology that increases access to financial services for marginalized populations, enabling the developing world to leapfrog existing financial systems.

5.3 Introduction to M-Pesa

M-Pesa is a mobile money service born in Kenya and operates in seven sub-Saharan African countries ("pesa" is Swahili for money). The service has nearly 50 million customers and almost a million agents that process transactions for $20 billion a month (Vodaphone, 2021; Gilbert, 2020b). Mobile-phone-based money services in sub-Saharan Africa are widespread; the region accounted for 52% of the world's mobile money services by 2015 (Suri, 2017).

Mobile phone systems in Kenya work on a prepaid basis, where users purchase "airtime" credits and load them onto their accounts using numerical codes. M-Pesa launched in 2007 after executives at Safaricom, one of Kenya's leading mobile phone network providers, became aware that phone users were transferring network credits for calling and texting to each other as a proxy for money. Kenyans rapidly adopted the service; by 2014, 97% of households in Kenya had an M-Pesa account.

M-Pesa has become a vital part of the Kenyan economy. As of the first quarter of 2020, 29.1 million Kenyans had mobile money accounts (99% market share), and 173,000 M-Pesa agents serviced customers across the country (Gilbert, 2020a). The COVID-19 pandemic propelled growing mobile money use throughout the continent, and in 2020, M-Pesa users in Kenya made 5 billion transactions, with a total value of $82 billion (Gilbert, 2020b).

5.4 Who Are the Unbanked

According to the World Bank's Global Findex Database 2017, 69% of adults worldwide have an account at some financial institution (72% of men and 65% of women). However, the remaining 31% represent 1.7 billion people without an account at a bank or mobile money provider, and the vast majority live in developing countries. In 2017, just 63% of adults in developing countries had any account. In addition, more than 20% of bank accounts in developing countries have been inactive for at least a year (Demirgüc-Kunt et al., 2018).

Women are 56% of the unbanked globally. Lower-income households are overrepresented in the unbanked but do not exclusively represent it—50% of the unbanked are in the country's top 60% of income (Demirgüc-Kunt et al., 2018). This data imply other barriers—physical distance and lack of transportation, bureaucracy, lack of education, excessive fees, or lack of dignity—in accessing traditional financial institutions.

Indeed, the 2017 Global Findex Database survey asked adults without a bank account why they did not have one. Two-thirds said they did not have enough money; 26% said accounts were too expensive; 26% also said that a family member already had an account, including many women. Twenty-two percent said financial institutions were too far away. See Fig. 5.1 (Demirgüc-Kunt et al., 2018).

Mobile money addresses several reasons why the unbanked do not use an account. The account becomes more affordable with a zero minimum balance and lower fees. There is no need to travel farther than the local agent, who is from the customer's community and, therefore, is trusted. Further, two-thirds of these unbanked adults, and 76% of adults with inactive bank accounts, already have a mobile phone, so conversion to mobile money is simplified. The World Bank reports that "mobile phone ownership offers large

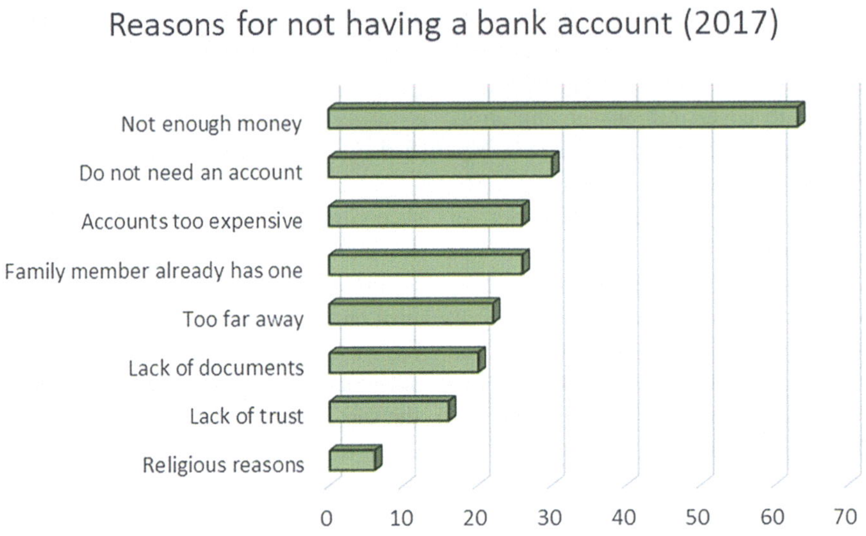

Fig. 5.1 Global Findex Database (Demirgüc-Kunt et al., 2018)

opportunities among the unbanked." (Demirgüc-Kunt et al., 2018).

Below, we explore the definition of frugal innovation most likely to further financial inclusion and then examine how the popular Kenyan mobile money service provider M-Pesa exemplifies frugal principles. We then present evidence of increased financial inclusion directly and indirectly from the platform. The case of M-Pesa explored below is instructive because it examines the oldest, most robust example of mobile money services and demonstrates their capability to transform economies in pursuit of SDG 8 and its targets.

5.5 A Definition of Frugal Innovation That Furthers Inclusion

A values-based approach to frugal innovation can inspire and guide the creation of future products and services to further financial inclusion and SDG 8.10. Because various definitions of frugal innovation have appeared over the years, it is more complex than advising that the practice be adopted. Instead, we must briefly review some of the literature surrounding the definition to make deliberate choices about pursuing the goal of financial inclusion.

Many credit Renault-Nissan's Carlos Ghosn with coining the term "frugal engineering" in 2006, referring to technical innovation under severe resource constraints (Pisoni et al., 2018). In 2010, the Economist published an introduction to its special issue on innovation in emerging markets in which it advocated that companies become more familiar with frugal innovation. It equated the practice to the Indian tradition of jugaad; in Hindi, jugaad means a flexible and creative attitude that allows one to adapt to circumstances that may be uncertain or less than ideal (Economist, 2010; Khan, 2016; Leliveld & Knorringa, 2017).

5.6 Jugaad Innovation

The book *Jugaad Innovation* was published in 2012 and proposed that if readers "think frugal, [and] be flexible," they would "generate breakthrough growth." (Radjou et al., 2012). The jugaad mindset of flexibility and experimentation contrasted with the West's structured and linear innovation processes. The six principles of jugaad align well with SDG 8 and are incorporated heavily in this chapter.

1. Seek opportunity in adversity. Jugaad innovators drive to find ways to turn problems into advantages. They are "driven to take charge of events and steer them in a direction of their own choosing. For them, adversity exists largely in the mind" (pg. 32).
2. Do more with less. Scarcity can take various forms: capital, infrastructure, natural resources, or talent. "Rather than doing everything themselves, [jugaad innovators] rely extensively on partners to perform various operations, thus saving time and energy" (p. 61). This principle can also be thought of as "use what we already have." It increases the sustainability of solutions by using resources efficiently and involving a support network.
3. Think and act flexibly. Jugaad innovators steer away from structured processes and constantly experiment and adapt as obstacles arise. By its definition, inclusion requires the flexibility to meet the differing needs of the world's varied populations.
4. Keep it simple. Creative simplicity is jugaad's fundamental principle. Simple products are helpful for three reasons: they are cheaper to make and more affordable for customers, are easier to install and maintain, and can satisfy a wider audience (designed for the most fundamental skill level).
5. Include the margin. Jugaad innovators engage low-income communities as active value co-creators in inclusive business models. A sense of fairness drives them. Solutions developed in conjunction with potential customers based on understanding their needs. Radjou et al. also cite a starker reason for including the margin: "[a]n economy rife with [] inequality and downward mobility—the so-called 'plutonomy'—is simply not sustainable" (p. 145).
6. Follow your heart. Jugaad innovators empathize with marginalized populations and tap into their passion and purpose to innovate to improve lives. "[E]mpathy forms the cornerstone of jugaad innovators' practice of an altruistic form of capitalism that is shaped by enlightened self-interest" (p. 161).

In jugaad, frugal innovation depends on a mindset of flexibility, ingenuity, and inclusiveness. This mindset and values—rather than any technical characteristics or a product or service—guide innovators.

5.7 Literature Reviews

Other scholars have proposed more specific definitions of frugal innovation with mixed success. One rigorous literature review combined with management interviews proposed three criteria for a frugal innovation: substantial cost reduction, focus on core functionality, and optimized performance. The authors concluded that substantial cost reduction must be from a customer perspective so that the innovation becomes "affordable" (Wayrauch & Herstatt, 2016, p. 7). They determine, without method, that the cost reduction must be at least one-third from "comparable products." (p. 8). Although price is equivalent to frugality in some contexts, comparing existing solutions presumes that solutions may be comparable in nature. Further, it needs to pay attention that government and nongovernmental groups are often significant customers for frugal solutions, not individuals. This chapter does not focus on cost reduction as a necessary criterion. The second criterion, concentration on core functionalities, is similar to the jugaad values of "do more with less" and "keep it simple." The third criterion of optimized performance level requires not too much nor too little performance capability and seems more akin to a variety of concentrating on core functionalities. The article does not mention "bottom-up" or inclusive values as part of the definition (Wayrauch & Herstatt, 2016). Thus, the three-pronged definition is minimally helpful in addressing financial inclusion and SDGs.

Pisoni et al. also conducted a literature review and determined that there have been three generations of discussion in the scholarship surrounding frugal innovation. Now, the third layer of literature dissects the various iterations of the definition yet further, in some cases, to examine whether and how frugal innovation can be applied in the context of more resource-rich countries, given the common misconception that it is only applicable in emerging markets. The "first generation" of frugal innovation concepts focused on the product or service itself. This definition focuses, among other things, on the minimal use of materials and resources and ease of use. The "second generation" of concepts is split between market applications, focusing on developing products to sell to base of the pyramid (BOP) populations, and frugal innovation as a process, which focuses on serving the needs of the BOP. This distinction is the topic of the next section.[1] (Pisoni et al., 2018). Based on this review, Pisoni et al. consider frugal "as an approach related to the whole innovation process or product/service characteristics that applies across different typologies of innovation." (Pisoni et al., 2018, p. 112). A flexible definition such as this one accords with this chapter's recommendation that frugal solutions are a generally applicable innovation management approach.

5.8 The Dangers of a Corporate Co-Opt of Frugal Innovation

In an unsurprising co-opt by corporate forces, frugal innovation often appears as a market approach that enables multinational corporations to design products and services to profit from low-income consumers (Kahn, 2016). One famous example is Unilever's small single-use sachets of washing powder and other consumer products, which are more affordable and easier to store in tiny houses (Mahajan, 2016). Another market-based theory is that engaging with the BOP populations as equals—entrepreneurs, producers, suppliers, and customers—allows companies to profit and lifts communities out of poverty, a "win-win" that should rebut any questions about the sincerity of corporate intentions (Kahn, 2016; Prahalad & Hammond, 2002). Even jugaad innovation sometimes trends toward a market incentive, noting that "the choice to include the margin promises many potentially lucrative opportunities to build entirely new businesses." (Radjou et al., 2012, p. 133).

Development scholars have roundly rejected and criticized the couching of frugal innovation as a business strategy. The likely reality is that most of those practicing frugal innovation principles are not multinational corporations but individuals, families, and communities who do so to survive (Leliveld & Knorringa, 2017). There is also evidence that the free-market approach that allows the sale of all products in local economies harms the poor. In the UN Department of Social and Economic Affairs working paper, *The Bottom of the Pyramid Strategy for Reducing Poverty: A Failed Promise*, data indicate low-income persons may spend less on essential nutritional items as consumer goods become readily available (Karnani, 2009). Further, this approach creates the risk that multinational corporations will commandeer frugal innovations conceived by BOP populations for profit (Meagher, 2017). At its worst, "frugal innovation involves wealthier corporate partners drawing on the organizational resources of more marginalized economic sectors for their own purposes in ways that reinforce rather than reduce inequality" (Meagher, 2017, p. 22).

In a systemic criticism, Pansera points out that innovation is unconsciously positioned as apolitical. Resource imbalances result from choices, privilege, and externalization by many of these multinationals we look to as exemplars of frugal innovation. Pansera criticizes management scholars as complicit with the idea that BOP populations need material goods rather than social justice and opportunity. He suggests combining frugal innovation studies with disciplines concerned with power in society to "re-politicize the links between poverty, inequality, and innovation" (2018, p. 9). Further, a market approach to poverty alleviation absolves governments of accountability (Karnani, 2009).

[1]"Reverse innovation" involves solutions originating in emerging markets that adapted from those fitting developed markets (Khan,2016).

5.9 Frugal Innovation as a Values-Based Innovation Management Approach

Frugal innovation principles and practices are essential in designing financial products, services, and systems to advance financial inclusion. However, the frugal innovation perspective that will help us achieve Goal 8 and Target 8.10 is a values-based innovation management approach, not a market strategy. This vein of frugal innovation aligns closely with jugaad innovation, as it is not only about creating more accessible products and services. It also involves a mindset focused on inclusion and social value. This moral orientation means collaborating with local customers, conserving resources, and designing for simplicity. Accessibility can be created by the product or service itself or through the business model and channels surrounding it. Rooting ourselves in a values-based definition of frugal innovation prevents the process from becoming an unmoored agent of capitalism.

The success story of M-Pesa is highly instructive when it comes specifically to frugal innovations in financial products and services. Here, we focus on its impact and influence in Kenya, its home country, allowing the most robust data.

5.10 M-Pesa as a Frugal Innovation

The origin of M-Pesa needs to receive more attention, especially since it illustrates one of the basic principles of frugal innovation. That is, that marginalized people themselves are experts concerning their challenges, having had to live their entire lives according to such constraints. The idea for mobile money did not emerge from a boardroom, computer lab, or at the business level—it arose organically from Kenyan users of Safaricom's network. Safaricom and many other mobile networks in sub-Saharan Africa use a prepayment system. Customers purchase small paper cards for credit amounts called "airtime" from agent-franchised brick-and-mortar kiosks throughout the country. Like a lottery ticket, customers scratch off the wax on the back to reveal a code to type into their phone to activate the credits (Peša, 2017; Meagher, 2017).

One of the original features of the Safaricom service was the ability to transfer credits to others on the Safaricom network. Incidentally, meaning one by one by one, users began to use this ability to transfer airtime credits over the network as a substitute for money (Peša, 2017; Meagher, 2017; Van Hov & Dubus, 2019). Safaricom corporate learned that its customers were using airtime credits in this way. In a joint venture with Vodafone, it then set out to build a mobile money program funded partly by a grant from the UK government. Since M-Pesa's launch in 2007, it has ballooned in popularity. The service operates across seven countries with 50 million customers: Kenya, Tanzania, the Democratic Republic of Congo, Egypt, Ghana, Lesotho, and Mozambique.

To use M-Pesa, one must first have a Safaricom mobile account, opened by complying with "know your customer" rules, and requires government identification. This process is much faster than opening a bank account. There is no minimum balance, and the account stays open even if the balance is $0. Once cash is deposited through an agent, the customer can make payments using their phone number to anyone with an M-Pesa account. Transactions can originate via an app or SMS texting alone. Notably, this means that one does not need a smartphone to use the service; one source reports that only 25% of the 40 million M-Pesa users have a smartphone, but this grows by about 10% yearly (Shapshak, 2020). Such a customer texts a code to M-Pesa, which replies with a numbered menu of options, such as "send money," "withdraw cash," and other functions. Using the numbers to respond, users can navigate the menu to make the transfer or withdrawal (Suri, 2017; Onamu, 2020). Safaricom updated its menus for non-smartphone users for the first time in 2020, which was part of an effort described by CEO Ndegwa to "simplify our range of products and services for our customers" (Onamu, 2020).

With 97% of households in Kenya using M-Pesa, the customers are varied. However, domestic remittances, money sent from a relative or friend in the country, are high in sub-Saharan Africa, where 45% of people report receiving money from a relative or friend in another part of the country in the past 12 months (in some countries as high as 70%) (Demirgüc-Kunt et al., 2018). Kenya has one of the region's highest rates of domestic remittances, and 89% of these occur using some account, with M-Pesa being the overwhelming favorite (Demirgüc-Kunt et al., 2018). Foreign remittances are another source of funds; M-Pesa is the medium for 60% of remittances into Kenya (M-Pesa, 2024). Users can also pay bills, receive or pay wages, or pay for goods or services on the system.

Agents serve customers seeking to deposit or withdraw cash into their accounts. They must ensure they maintain enough "float" to service cash withdrawal requests and often sell other products or have businesses of which M-Pesa income is a small part. As of 2015, there were over 65,000 M-Pesa agents in Kenya (Suri, 2017).

Using an existing resource instead of building a new one is one reason M-Pesa was adopted so quickly. When M-Pesa launched, it took advantage of its existing network of agents throughout the country, selling airtime for calling and texting. Further, because agents exist throughout local communities, they know their customers and sometimes assist them with transactions. Agents may even enter the customer's PIN when a customer needs help using the phone (Wellen & van Dijk, 2019). Thus, the flexibility needed to serve the particular local population is embedded within the business model (Table 5.1).

Table 5.1 Frugal innovation elements of M-Pesa

Elements	Principles exemplified
Based on an organic innovation by users	Ground-up innovation; think and act flexibly
Designed for use via text on non-smartphones	Keep it simple, include the margin, and do more with less
Relied on distribution network (agents) already in place	Do more with less; use what you already have
Local agents are empowered to help customers based on needs	Include the margin, think and act flexibly
Spanned time and geographic constraints of existing financial systems	Seek opportunity in adversity

5.11 M-Pesa Contributes Directly and Indirectly to Financial Inclusion

M-Pesa's rapid penetration of marginalized populations significantly increased financial inclusion in Kenya, in furtherance of the financial inclusion goals set in Target 8.10, as well as the broader aims of SDG 8 for greater economic empowerment and participation.

First, M-Pesa gave access to financial services to those to previously lacked it. By 2011, 4 years after its founding, 75% of Kenyans outside Nairobi without a bank account used M-Pesa. Even among the poorest of the poor, those living on less than $1.25 a day, 72% used M-Pesa by 2011 (Suri & Jack, 2012). Over time, M-Pesa has expanded the financial services offered to its customers and businesses with incremental frugal innovations, such as continuous overdraft protection, called Fuliza, which automatically covers a shortfall in funds up to a specific limit that can grow over time. Customers pay a 1% access fee and a varied daily maintenance fee on any outstanding balance (Safaricom, 2021).

Mobile money may also give women an advantage over cash. M-Pesa gave 186,000 women the ability to transition from agriculture to business and retail as an occupation (Suri, 2017). Mobile money may lead to greater empowerment for women, who lag financially behind men in nearly every country due to the increased privacy and control digital platforms provide (Klapper, 2019).

Second, M-Pesa increased absolute income to households and made them more resistant to disruptions, facilitating overarching Goal 8 of inclusive economic growth. Impressively, one study attributes a 2% decline in the poverty rate in Kenya by 2014 as a result of the platform (nearly 200,000 households) (Suri, 2017). The data show that M-Pesa users in both urban and rural environments gain increased income from remittances, making the households better able to cope with financial risks and other shocks (Morawczynski, 2009; Suri & Jack, 2012). Persons with health issues could spend more on medical care without sacrificing spending in other vital areas, such as food and education. "[B]y keeping [households vulnerable to poverty] away from the precipice, M-Pesa could conceivably have long-term impacts on poverty rates among otherwise vulnerable households" (Suri & Jack, 2012).

Third, M-Pesa is a bridge to formal banking, saving, and investing. Commercial Bank of Africa (CBA) partnered with Safaricom to release M-Shwari ("calm" in Swahili), a savings product that enables customers to open a bank account connected with M-Pesa. Customers can also request a loan from the bank, and CBA uses the customer's data on mobile money transactions to assess the credit risk when deciding whether to issue the loan (Suri, 2017; M-PESA, 2024). Beginning in 2017, the Kenyan government began to offer M-Akiba ("savings"), a 10% infrastructure bond available to be purchased with an account linked to M-Pesa. To facilitate this, the legislature lowered the minimum legal investment in treasury bonds from more than $1000 to about $30 and allowed the opening of a treasury account with a mobile phone (Suri, 2017). This is an example of the government's support for the mobile money industry.

M-Pesa has also enabled small businesses which fall under Target 8.10 and SDG 8 more broadly. Lipa na M-Pesa emerged as a retail payment platform with lower fees that the retailer can pay or pass onto the consumer. Transaction fees are 0.5% or 0% if under a certain amount. M-Pesa also offers a portal for businesses to manage sales and expenses, send bills, place orders, and access credit (M-PESA, 2024). The platform can integrate with business websites, whereby customers enter their phone numbers and confirm via their phone to complete payments. Research has shown that mobile money is safer for entrepreneurs to do business with than cash (Klapper, 2019).

There also has been a proliferation of over-the-top services (OTT), which rely on the M-Pesa platform for payments. These are frugal innovations that take advantage of already-existing systems and streamline aspects of traditional business models. For example, insurance for health, agriculture, funerals, and businesses is available from M-Pesa partners, with payment automatically sent via the platform (M-PESA, 2024; Kilimo Salama). Two solar panel companies, M-Kopa and Angaza, use M-Pesa to collect small payments from customers over time to pay to equip houses with solar panels, thus enabling asset ownership for these households (Suri, 2017). The M-Farm service, available through SMS text or an app, allows farmers to get prices, connect with buyers, and get advice on when to plant. Revenue comes in through M-Pesa and is easy to manage. It also has a group selling tool that allows multiple small farmers to aggregate their produce to achieve a higher volume (Altamirano & van Beers, 2017). For farmers, M-Pesa payments facilitate increased inputs per acre, which in one study paid off with significantly higher yield for banana farmers ($3435 annual mean income for those with M-Pesa accounts vs. $1854 for those without) (Kikulwe et al., 2014).

Today, M-Pesa customers can access premier worldwide financial networks. In 2018, M-Pesa and Safaricom announced a collaboration with Western Union to enable payments worldwide. People with an M-Pesa account can send money to and from over 200 countries (Western Union, 2018). That same year, PayPal partnered with linked accounts to increase e-commerce (Dahir, 2018). In 2020, Safaricom partnered with Visa to expand options for digital payments. It offers a prepaid M-Pesa Visa Safari card for withdrawing cash in any currency from one's connected M-Pesa account at Visa ATMs worldwide and making purchases with a Visa card (Muli, 2019; Miriri, 2020). For customers in Kenya, mobile money has enabled them to move closer to the reality envisioned by Target 8.10 and SDG 8.[2]

[2] Only some agree that M-Pesa has been a resounding success in creating inclusion. Meagher rightly points out that Safaricom, "having benefited from extensive informal innovation and the use of a low-cost, informal business infrastructure, [] proceeded to centralize control over the management of the system and the distribution of gains."

5.12 Mobile Money Can Help Achieve SDG 8 and Target 8.10

SDG 8 paints a vision of economic growth for all, and Target 8.10 recognizes that inclusion in financial systems is vital to achieving this economic growth. M-Pesa's simplicity and broad applicability have opened up an entirely new range of financial products and services to those formerly marginalized, thus creating a ripple effect of inclusion for its 50 million customers and their families. The case study demonstrates that mobile money can catalyze the sustainable economic growth and financial inclusion envisioned by SDG 8 and Target 8.10.

Mobile money services directly reduce poverty and increase economic opportunity and financial inclusion, outcomes directly aligned with SDG 8 and Target 8.10. The poverty rate in Kenya decreased by 2% due to M-Pesa. For the poorest households, although mobile money may not immediately lift them out of poverty, it does provide increased stability over the long term. Household incomes increase as relatives and friends can more easily send small amounts to smooth disruptions. Both parties save the time, transportation money, and fees it would take to complete the transaction at a bank or other in-person locations. This smoothing effect elevates the long-term economic prospects of the family as aspired to by SDG 8.

Mobile money contributes to financial inclusion, addressed explicitly by Target 8.10 by providing a first entry point into banking and more traditional services, such as loans and savings. Worldwide, 58% of account holders saved in 2017 versus 28% of unbanked persons. After opening a mobile money account, customers increase savings and shift from informal to formal savings. Kenya had one of the highest savings rates of developing economies in 2017, including one of the highest standard savings rates (Demirgüc-Kunt et al., 2018). Beyond local savings, data from mobile financial transactions can be used to develop credit ratings and significantly expand access to traditional services. There also is the potential for mobile-phone-based point-of-sale systems to leapfrog credit and debit card payments directly to mobile payments. In short, mobile money enables citizens in developing economies to experience levels of financial inclusion closer to those of wealthier countries (*See* Demirgüc-Kunt et al., 2018).

The potential benefits of mobile money are especially salient in specific contexts. Rural persons are significantly cut off from traditional banks and, therefore, are prime customers for mobile services. In agricultural economies, managing risk is essential due to exposure to poor weather or disease. Mobile money helps smooth these risks and gives agricultural workers access to payments, insurance, and information. In the case of corrupt economies and governments, or conflict areas, digital payments can reduce impropriety, bribery, and theft opportunities. Payments become more secure, and transparency increases (Demirgüc-Kunt et al., 2018). The privacy of mobile money may, for the first time, help many women gain control over their income and expenditures.

Government support is necessary to create a friendly legislative environment. Regulation that legitimizes mobile transactions protects consumers, allows flexible documentation, and provides other functions that can encourage development. Governments can also take the lead by digitizing their wages and benefits payments and by creating digital portals to receive payments. Data show that a substantial proportion of current account holders opened the account to allow them to receive a digital payment. Similarly, encouraging the private sector to digitize its payments would also help convert the unbanked, as most unbanked adults who receive private wages in cash have a mobile phone (Demirgüc-Kunt et al., 2018).

(2017, p. 28). One way it did so was to consolidate its dealers from hundreds to approximately 10, each supervised at least 2000 sub-agents. Sub-agents received lower commissions, while wealth became more concentrated at the top (Meagher, 2017). The underpayment of valuable agents in this manner, who provide a crucial component of continuity and flexibility for BOP customers, is different from the values-based approach to frugal innovation advanced in this chapter.

Mobile money is more than an alternative payment system accessible to marginalized populations. It is a technology that will enable developing countries to leapfrog existing financial systems and fully realize the vision of SDG 8 and Target 8.10. To increase equity of access in developing areas, efforts must promote mobile money and include more individuals in these burgeoning systems. Using frugal innovation principles as a guide will ensure that those at the margins are genuinely included.

References

A special report on innovation in emerging markets: First break all the rules. (2010, April 17). *The Economist*.

Andersson-Manjang, S. K., & Naghavi, N. (2021). *State of the industry report on mobile money 2021*. Group Speciale Mobile Association.

Altamirano, M., and van Beers, C. (2017). Frugal Innovations in Technological and Institutional Infrastructure: Impact of Mobile Phone Technology on Productivity, Public Service Provision and Inclusiveness. *European Journal of Research Development. 30*(1), 1–24.

Dahir, A. L. (2018). M-Pesa and PayPal are partnering to boost e-commerce in Kenya. *Quartz Africa*. https://qz.com/africa/1247879/safaricom-m-pesa-and-paypal-deal-to-boost-kenya-e-commerce/. Accessed 29 Jan 2022.

Demirgüc-Kunt, A., et al. (2018). *The Global Findex Database 2017*. World Bank Group.

Gilbert, P. (2020a, July 7). M-Pesa has almost 99% market share in Kenya. *Connecting Africa*. http://www.connectingafrica.com/author.asp?section_id=761&doc_id=762180. Accessed 2 Aug 2021.

Gilbert, P. (2020b, November 25). Pandemic propels mobile money growth. *Connecting Africa*. http://www.connectingafrica.com/author.asp?section_id=761&doc_id=765686. Accessed 15 Aug 2021.

Karnani, A. (2009). *The bottom of the pyramid strategy for reducing poverty: A failed promise*. United Nations Department of Economics and Social Affairs.

Khan, R. (2016). How Frugal Innovation Promotes Social Sustainability. *Sustainability. 8*(10), 1034.

Kikulwe, E. M., Fischer, E., & Qaim, M. (2014). Mobile money, smallholder farmers, and household welfare in Kenya. *PLoS One, 9*, e109804.

Klapper, L. (2019). Mobile phones are key to economic development. Are women missing out? *Brookings.edu*. https://www.brookings.edu/blog/future-development/2019/04/10/mobile-phones-are-key-to-economic-development-are-women-missing-out/. Accessed 21 Jan 2022.

Leliveld, A., & Knorringa, P. (2017). Frugal innovation and development research. *European Journal of Development Research, 30*(1), 1–16.

Mahajan, V. (2016). How Unilever reaches rural customers in emerging markets. *Harvard Business Review*.

Meagher, K. (2017). Cannibalizing the informal economy: Frugal innovation and economic inclusion in Africa. *European Journal of Development Research, 30*(1), 17–33.

Miriri, D. (2020). Visa gives Kenya's Safaricom 'missing link' for global payments. *Reuters.com*. https://www.reuters.com/article/us-kenya-safaricom-visa/visa-gives-kenyas-safaricom-missing-link-for-global-payments-idUSKBN22C1M3. (Accessed 21 Jan 2022).

Morawczynski, O. (2009). Exploring the usage and impact of 'transformational' mobile phone financial services: The case of M-Pesa in Kenya. *Journal of East African Studies, 3*, 509–525.

Muli, F. (2019). How to get and use the M-Pesa prepay visa safari card. *Kahawatungu.com*. https://www.kahawatungu.com/how-to-get-use-m-pesa-prepay-visa-safari-card/ (Accessed 21 Jan 2022).

M-PESA Website. (2024). https://www.vodafone.com/about-vodafone/what-we-do/m-pesa.

Onamu, A. (2020). How to use M-Pesa: A guide for non-smartphone users. *Gadgets Africa*. https://gadgets-africa.com/2020/09/10/m-pesa-is-now-accessible-via-ussd-for-non-smartphone-users/. Accessed 14 July 2021.

Pansera, M. (2018). Frugal or fair? The unfulfilled promises of frugal innovation. *Technology Innovation Management Review, 8*(4), 6–13.

Peša, I. (2017). The developmental potential of frugal innovation among Mobile Money Agents in Kitwe, Zambia. *European Journal of Development Research, 30*, 49–65.

Pisoni, A., Michelini, L., & Martignoni, G. (2018). Frugal approach to innovation: State of the art and future perspectives. *Journal of Cleaner Production, 171*, 107–126.

Prahalad, C. K., & Hammond, A. (2002, September). Serving the world's poor, profitably. *Harvard Business Review*.

Radjou, N., Prabhu, J., & Ahuja, S. (2012). *Jugaad Innovation*. Jossey-Bass.

Safaricom. *Fuliza FAQ*. https://www.safaricom.co.ke/faqs/faq/512. Accessed 8 Aug 2021.

Shapshak, T. (2020). Vodacom and Safaricom acquire M-Pesa to accelerator mobile money services in Africa. *Forbes*. https://www.forbes.com/sites/tobyshapshak/2020/04/06/vodacom-and-safaricom-acquire-m-pesa-to-accelerate-mobile-money-services-in-africa/?sh=766e03e2a392. Accessed 21 Jan 2022.

Suri, T. (2017). Mobile Money. *Annual Review of Economics, 9*, 497–520.

Suri, T., & Jack, B. (2012). Reaching the poor: Mobile banking and financial inclusion. *Slate.com*. https://slate.com/technology/2012/02/m-pesa-ict4d-and-mobile-banking-for-the-poor.html. Accessed 19 July 2021.

Van Hove, L., and Dubus, A. (2019). M-PESA and Financial Inclusion in Kenya: Of Paying Comes Saving? *Sustainability. 11(3)*, 568.

Vodaphone. *What is M-Pesa?* https://www.vodafone.com/about-vodafone/what-we-do/consumer-products-and-services/m-pesa. Accessed 10 July 2021.

Wayrauch, T., & Herstatt, C. (2016). What is frugal innovation? Three defining criteria. *Journal of Frugal Innovation, 2*, 1. https://doi.org/10.1186/s40669-016-0005-y

Wellen, L., & van Dijk, M. P. (2019). Frugal financial innovations for inclusive finance: The experience with customer care at M-Pesa in Kenya. *Enterprise Development and Microfinance, 29*, 262–275.

Western Union Powers Safaricom's M-PESA Mobile Wallet to Send Money Globally. (2018). Westernunion.com. https://ir.westernunion.com/news/archived-press-releases/press-release-details/2018/Western-Union-Powers-Safaricoms-M-PESA-Mobile-Wallet-to-Send-Money-Globally/default.aspx#:~:text=M%2DPESA%20wallet%20holders%20already,pay%2Dout%20across%20the%20world. Accessed 29 Jan 2022.

The Benefits and Costs of Fintech

6

Imad A. Moosa

Abstract

Fintech refers to the use of the state-of-the-art technology to provide financial services. The development of fintech has been enhanced by massive data generation, advances in computer algorithms, and growth in processing power, which have been facilitated by broadband internet, cloud computing, and artificial intelligence. It is typically claimed that digital finance is boosting financial inclusion and that digital financial inclusion is conducive to economic growth. While most of the views expressed on the desirability of fintech are optimistic, it is plausible to suggest that the introduction of digital finance should be judged in terms of costs and benefits.

Keywords

Fintech · Financial Literacy · Financial Inclusion · Digital Financial Services · Operational Risk · Bitcoin · War on Cash

6.1 Introduction

Financial technology (fintech) is the "technology-enabled innovation in financial services that could result in new business models, applications, processes, or products with an associated material effect on the provision of financial services" (Sahay et al., 2020, p. ix). It is also defined as technology and innovation that aims to compete with traditional financial methods in delivering financial services (Van Loo, 2018). These are not the only two definitions of fintech, which simply refers to the computerization and digitization of financial services and transactions.

The advent of the COVID-19 pandemic has made it desirable to conduct digital financial transactions to avoid contact in the spirit of social distancing and lockdowns. During the pandemic, technology created new opportunities for digital financial services amid social distancing and containment measures. Sahay et al. (2020) claim that the advent of the pandemic has also enhanced financial inclusion, even though it is unclear how the acceleration in the use of technology to settle financial transactions and provide financial services is necessarily associated with a greater tendency for people to go that way. In other words, there is a mix-up between fintech's supply and demand sides.

Three fundamental changes have influenced the development of fintech: massive data generation, computer algorithm advances, and

I. A. Moosa (✉)
Kuwait University, Kuwait City, Kuwait

G. W. Muschert et al. (eds.), *Financial Inclusion*, Sustainable Development Goals Series,
https://doi.org/10.1007/978-3-031-68803-4_6

processing power growth. High-speed broadband internet, cloud computing, and artificial intelligence have allowed these changes by enabling big data analytics, blockchain technology, and biometric identification.

Sahay et al. (2020) assert that at least two positive outcomes have come from fintech. The first is that digital finance boosts financial inclusion, even where traditional financial inclusion is declining. The second is that digital financial inclusion propels and is conducive to economic growth. However, they also recognize some side effects and state some caveats. The first is the potential negative effect on financial stability without proper financial regulation, mainly when financial literacy is low. They also recognize that digital finance can create new risks to financial inclusion. However, two aspects of the fintech revolution are noteworthy. The first is that while digital finance may be conducive to financial inclusion and economic growth, it may also lead to a rising level of operational risk, primarily the risk of fraud and technology-related risks. The second is that no one seems to talk about how digital finance is (by design or otherwise) a component of an elaborate plan to convince or force people not to use cash for benefit the banking system. The war on cash can be won by encouraging migration to digital finance, related to the disastrous policy of ultra-low or negative interest rates and the bail-in legislation introduced in many countries to allow banks to confiscate deposits when they are in trouble.

The effect of fintech on financial inclusion is particularly relevant to international efforts to expand financial inclusion. The Alliance for Financial Inclusion (AFI), a global network of policymakers, sets out its primary objectives in the Maya Declaration of 2011. In the 2010 Seoul Summit, the G20 leaders discussed financial inclusion, endorsing a Financial Inclusion Action Plan (FIAP) and creating the Global Partnership for Financial Inclusion (GPFI). In 2015, the United Nations adopted the Sustainable Development Goals (SDGs) for 2030, which feature prominent financial inclusion. In 2016, the AFI and GPFI identified technology as a core aspect of financial inclusion, creating a new work stream called "Fintech for Financial Inclusion." In the 2018 annual International Monetary Fund (IMF) and World Bank meetings, the two institutions launched the Bali Fintech Agenda, which lays out the broad principles for the safe development of fintech to support financial inclusion.

This chapter aims to point out that scholars should judge the introduction and spread of digital finance in terms of costs and benefits. Most views expressed on the desirability of fintech are relatively sanguine because the potential adverse outcomes are ignored, inadvertently or intentionally. For this purpose, two largely ignored issues are the topics of discussion: fintech as a source of operational risk, particularly the risk of fraud and technology-related risks, and the role of fintech in the war on cash.

6.2 Some Basic Concepts

Fintech has been defined and described in more than one way. Schueffel (2017, p. 32) defines it as a "new financial industry that applies technology to improve financial activities." Sanicola (2017, n.p.) suggests the definition as "the new applications, processes, products, or business models in the financial services industry, composed of one or more complementary financial services and provided as an end-to-end process via the internet." In general, fintech covers innovative ideas that improve the provision of financial services by proposing technology solutions to various business situations. A related concept is digital finance, which includes digital payments and lending. Digital payments occur using mobile phones or online payment systems, excluding debit and credit cards. Digital lending is the extension of funds through digital means, including peer-to-peer lending, e-commerce lending, online lending, and mobile lending.

Examples of fintech, in general, include using smartphones for mobile banking, investing, borrowing services, and cryptocurrencies—these are technologies intended to make financial services more accessible to the general public. The technologies used in the financial services industry include artificial intelligence, big data, robotic process automation, and blockchain. Artificial intelligence uses computer algorithms to predict changes in the stock market and the economy. Big data can predict client investments and market changes, create new strategies and portfolios, and analyze customer spending habits. Financial institutions utilize big data to create segmented marketing strategies and optimize operations. Robotic process automation is an artificial intelligence technology that automates specific repetitive tasks to process financial information such as accounts payable and receivable. Blockchain is the technology associated with, but not limited to, cryptocurrencies, which enables the decentralization of financial services such that no need arises to trust a third party to execute transactions.

Fintech may enhance financial inclusion, which is the access to and use of formal financial services, including transactions, savings, credit, and insurance (Sahay et al., 2015b). Digital financial inclusion, or fintech-enabled financial inclusion, is digital access to and usage of formal financial services through mobile phones and computers. This concept includes the services provided by fintech companies and financial institutions.

Fintech is changing the delivery of financial services to small businesses and low-income households. Banks, nonbank financial institutions, and informal systems (such as relying on relatives, microlending clubs, and money lenders) have traditionally provided financial services. The traditional system predominantly relies on cash transactions and face-to-face interactions with financial service providers. Interactions between providers and users are used to monitor creditworthiness and are sources of financial literacy. With the development of digital finance tools accessible from mobile phones or computers, there is less need for face-to-face interactions. The mobility restrictions introduced to contain the COVID-19 pandemic have amplified the benefits of expanding digital financial services.

Facing a deluge of customer service calls, many retail banks rely more than ever on technology to boost call center operations (for example, by using conversational AI agents that can engage directly with customers for rapid and real-time banking transactions and resolutions). This trend emerged long before the current economic and societal factors came into play with the Covid-19 pandemic. The need for better-digitized customer experiences has only crystallized and accelerated, with consumers staying home or, at the very least, away from physical bank branches.

6.3 The Effects of Fintech and Digital Finance

The availability of cross-country data has propelled the growth of research on fintech and financial inclusion. The early literature relied primarily on survey work on individual countries or single measures of financial inclusion, such as the number of bank branches, ATMs, and bank accounts per capita (Beck et al., 2007; Honohan, 2008). The launch of databases, such as the IMF's Financial Access Survey (FAS) and the World Bank's Global Findex database, has made it possible to develop more multidimensional, composite measures of financial inclusion, taking into account various aspects of access and usage by household and firms (Amidžić et al., 2014; Dabla-Norris et al., 2015a; Camara & Tuesta, 2017). The macroeconomic impact has been examined by Sahay et al. (2015a, b), Svirydzenka (2016), Dabla-Norris et al. (2015b), and Loukoianova and Yang (2018). Rojas-Suárez (2016), Rojas-Suárez and Amado (2014), and Deléchat et al. (2018) have all looked into the factors that contribute to financial inclusion.

The empirical literature on digital financial inclusion has been carried out mainly on a country or regional basis. Two interview-based studies of fintech providers and regulators explore the role of fintech in financial inclusion. Patwardhan et al. (2018) highlight the importance of mobile and other person-to-person payment methods and the development of new ways of complying with customer due diligence. Likewise, the study of Ghose et al. (2020) emphasizes the role of identity, mobile money, platform-based services, and microcredit. Some studies examine the role of fintech in supporting access to credit, an essential dimension of financial inclusion (Ayyagari et al., 2017). Some studies focus on the determinants of digital lending, underscoring the importance of regulation, financial development, digital infrastructure, or market structure (Rau, 2019; Claessens et al., 2018). Fintech lenders are faster at processing mortgage applications than traditional lenders (Fuster et al., 2018), offering lower-cost credit than traditional lenders (Jagtiani & Lemieux, 2017; de Roure et al., 2018). There are not many studies that look at the possible stability and inclusion risks that come with fintech, the effects on financial stability of how closely fintech companies and banks are becoming connected, and how well people are aware of the risk that fintech could lead to financial exclusion (Adrian & Mancini-Griffoli, 2019; Brunnermeier et al., 2019). Ozili (2018) argues that digital finance and financial inclusion benefit financial services users, digital finance providers, governments, and the economy. However, he acknowledges several issues that, if addressed, can make digital finance work better for individuals, businesses, and governments.

The most comprehensive study is that of Sahay et al. (2020, p. xi), who find that digital finance is conducive to financial inclusion and economic growth. Accordingly, they suggest that "digital financial inclusion could play an important role in mitigating the economic and social impact of the ongoing COVID-19 crisis." Unfortunately, Sahay et al. (2020) overstate the case. Using a mobile phone to pay for a hair dryer will not help the effects of the pandemic-related economic crisis, which combines aggregate demand and supply shocks. They further argue that "[b]roadening the financial access of low-income households and small businesses could also support a more inclusive recovery" (Sahay et al., 2020, p. xi). That is right. However, the more inclusive recovery comes from something other than fintech, which is drastic fiscal actions. Imposing a one-off pandemic-related wealth tax on a billionaire will contribute much more to the recovery than convincing a poor or middle-income person to shop online. However, they admit that these potentials "cannot be taken for granted, as the pandemic could accelerate pre-existing risks of financial exclusion and give rise to new risks to the fintech sector itself" (Sahay et al., 2020, p. xi).

It was stated in the introduction argued that expanding digital financial services will only sometimes lead to adoption by the general public, hence financial inclusion. Sahay et al. (2020) argue that this would be the case because digital financial services are faster, more efficient, and typically cheaper than traditional financial services, which means they are increasingly reaching lower-income households and small- and medium-sized enterprises. The problem here is the implicit assumption that most people have a high degree of computer literacy, which is not the case. People think more than twice before clicking "pay" on an internet transaction. A typical person would instead explore the options of renewing a deposit by talking to someone face-to-face rather than doing that by using Internet banking. More importantly, people are typically worried about technology-enabled fraud and prefer cash transactions. A large segment of the population in developed countries, particularly older people, still needs to get credit cards. Sahay et al. (2020) warn of cybersecurity risks or inappropriate lending practices by unregulated institutions, which could jeopardize trust.

In their comprehensive study, Sahay et al. (2020) reached several specific findings based on data covering 52 countries. The first is that digital finance boosts financial inclusion, even where traditional financial inclusion is declining. They explain this outcome by arguing that traditional financial service delivery tends to leave a gap in the market for digital financial inclusion. For example, the absence of bank branches in rural areas can be overcome by doing Internet banking. Naturally, this can only happen with a good internet connection, which is less prevalent in developing countries.

The second finding is that digital financial inclusion boosts economic growth, claiming that fintech could be essential in mitigating the economic impact of the COVID-19 pandemic and supporting economic recovery. Their explanations for the cause-and-effect process need to be more convincing. They argue that countries with higher digital financial inclusion (i) find it relatively easier to ensure continued access to financial services while keeping people safe; (ii) deliver government support effectively and securely; and (iii) support consumption, innovation, and hence productivity through digital economy developments. The problem is that the pandemic has already aggravated inequality, particularly in the distribution of wealth. It is unlikely that a government support check arriving by post is less effective in alleviating misery than a transfer to the electronic wallet on the phone. One can only wonder how the homeless and beggars who depend on cash donations will benefit from fintech.

The rest of their findings make the picture less rosy. They suggest that "the safe development of digital financial inclusion rests on factors such as regulation and financial literacy" (Sahay et al., 2020, p. 7). They also argue that "digital finance can create new risks to financial inclusion" (Sahay et al., 2020, p. 7), stemming from unequal access to digital infrastructure, financial and digital literacy constraints, and potential biases amplified by new data sources and data analytics.

6.4 Fintech as a Source of Operational Risk

By its very nature, fintech aggravates exposure to operational risk, which may lead to significant operational losses. The definition of "operational risk" first appeared in a joint report by Robert Morris Associates and other organizations in 1999: "Operational risk is the direct or indirect loss resulting from inadequate or failed internal processes, people, and systems, or from external events." The Basel Committee on Banking Supervision (BCBS) adopted this definition as it was initially, but subsequently eliminated the reference to indirect losses (including those resulting from the loss of reputation). The BCBS (2004) defines operational risk as "the risk arising from inadequate or failed internal processes, people, and systems, or from external events." Based on the underlying causes of operational risk, this definition was used to develop the concept by considering the distinction between causes, events, and the resulting operational losses (consequences or effects).

Mestchian (2003) suggests the decomposition of the definition of the BCBS into four components. Inefficiencies or ineffectiveness in the various business processes within the company are what cause process risks. Human-centric risks arise from employee error, misdeeds, unavailability, inadequate employee development, and recruitment. System failures, problems with data quality and integrity, a lack of capacity, and poor project management are the leading causes of technology (or system) risks. External risks, which resemble the iceberg hazard for the Titanic, result from both macroeconomic and socioeconomic events as well as the actions of external parties (such as competitor behavior, external fraud, and regulatory changes).

The BCBS (2004) classifies operational loss events into seven categories, as shown in Table 6.1. Out of the seven events, fraud (both internal and external) is highly relevant to fin-

Table 6.1 BCBS classification of operational loss events

Event	BCBS Definition
Internal fraud	Losses due to acts of fraud involving at least one internal party.
External fraud	Same as internal fraud, except that an external party carries it out.
Employment practices and workplace safety	Losses arising from violations of employment and health and safety laws.
Clients, products, and business practices	Losses arising from failure to meet client obligations or the design of a product.
Damage to physical assets	Losses arising from damage inflicted on physical assets by a natural disaster or another event.
Business disruption and system failures	Losses arising from disruptions to or failures in systems, telecommunication, and utilities.
Execution, delivery, and process management	Losses arising from failed transaction processing with counterparties such as vendors

Source: BCBS (2004)

tech. Some loss events have no connection, like damage to physical assets by fire or natural disasters. However, it is arguable that installing expensive IT systems makes this type of loss event more costly. A direct connection is evident between fintech and system failures. In Table 6.2, five of the seven categories are relevant to fintech. For example, external fraud often occurs via computer hacking, which becomes more frequent when transactions are computerized. Problems with hardware, software, utility outages, and disruptions become more severe as computerization intensifies. The same goes for errors in data entry. In Table 6.3, we can see fintech's relevance to operational risk sources. Technology risk is particularly relevant. For a detailed discussion of these classifications, see Moosa (2007, 2008).

Aguayo and Slusarczyk (2020) refer explicitly to operational risk arising from the digitization of banking services. They use the concept of "digital operational risk," which they attribute to internal sources (defects or failures) and external sources (fraud risk, technological risk, and cyber risk). They recognize electronic banking services to include (i) account statements for a customer; (ii) information on banking products (deposits, loans, and securities); (iii) applications for opening deposits and obtaining loans and bank cards; (iv) internal transfers to bank accounts; (v) transfers to accounts in other banks; and (vi) currency conversion. On the other hand, they recognize the sources of operational risk as (i) defects and failures; (ii) loss of data integrity and unauthorized access to customer data; (iii) violation of a technical system in information space; (iv) cyberattacks; and (v) data integrity and unauthorized access to customer data.

Let us concentrate on digital finance fraud, often called "digital fraud." As financial institutions and businesses adopt the latest systems to make digital payments more secure, the nature of digital fraud continues to evolve. The new-age computer hacker is using more sophisticated ways to obtain valuable customer information and login credentials to hack into accounts. As a result, it has become relatively common to lose security authentication data, download malicious

Table 6.3 Sources of operational risk relevant to Fintech

Risk	Examples
People Risk	Misuse of important information Embezzlement Unauthorized fund transfers Unauthorized trading
Process Risk	Employee error Inadequate quality control Inadequate security Inadequate supervision Inadequate account reconciliation Inadequate transaction completion Inadequate transaction execution Inadequate transaction settlement Lack of proper due diligence Loss of critical information
Technology Risk	New technology failure Technology-related operational errors System failure Outdated hardware Computer virus Data security Hacking Inadequate testing System failure Incompatible software Inadequate systems System maintenance

Source: BCBS (2004)

Table 6.2 Loss events relevant to Fintech with examples

Event	Examples
Internal fraud	Account takeover and impersonation Credit fraud
External fraud	Computer hacking Theft of information Forgery
Clients, products, and business practices	Account churning Breach of privacy Misuse of confidential information Customer discloser violations
Business disruption and system failures	Hardware Software Telecommunications Utility outage Utility Disruption
Execution, delivery, and process management	Incorrect client records Data entry, maintenance, or loading error Miscommunication

Source: BCBS (2004)

apps, and be the victim of credit card fraud. Chris (2020) refers to the "fraud pandemic," which is associated with the "accelerated shift to online banking and mobile transactions," which has pushed financial institutions to re-evaluate the resilience of their fraud management strategies.

Common types of fraud in digital transactions include malware, phishing, account takeover, and card not present (CNP). Malware is a computer program that can collect information on financial transactions when installed on a device. Phising can be used for gathering personally identifiable information via deceptive emails and fake websites to access customer accounts. Account takeover occurs when a hacker, posing as a genuine customer, takes control of an account and makes unauthorized transactions. Card not present or remote purchase fraud involves the fraudulent use of card details obtained through skimming, digital attacks, and unsolicited e-mails or calls.

The failure of computer systems can cause huge losses without fraud—the best example is algorithmic trading gone wrong. Algorithmic trading, where a computer automatically executes trades based on preprogrammed instructions, has been around for a long time and is now a significant factor in generating daily stock market volatility. Computer algorithms quickly discern short-lived profit opportunities over units of time as small as microseconds and determine optimal order submission strategies. The algorithms are helpful to monitor prices and liquidity across different markets. Croft (2020) explains how algorithmic trading programs can worsen bad stock market days.

High-frequency algorithmic trading has been the subject of intense public debate since the May 6, 2010, "flash crash" (Mehta & Kisling, 2010; Bowley, 2010; Spicer & Younglai, 2010). The Dow Jones Industrial Average, which measures the US market, experienced its most significant intraday point loss ever during the flash crash. However, it quickly recovered most of the loss (for a brief period, $1 trillion in market value vanished). Waddell & Reed Financial, a little-known mutual fund, initiated the crash. Portfolio managers at Waddell were increasingly worried that fears of a European financial crisis would spread to the United States, where the stock market was already bearish. As a result, they decided to make a trade that would limit their losses if the market continued to decline, instructing their computers to start an automated program of selling financial contracts linked to the DJIA. This action triggered a wave of selling that overwhelmed the market as the Dow fell by 600 points in 15 minutes.

6.5 The Bitcoin "Scam"

One of the technologies used in fintech is blockchain, an invention that led to the development of cryptocurrencies, including Bitcoin, which was the first and most successful. The expression "Bitcoin scam" is used in two different ways. The first pertains to the risk of fraud causing the loss of one's Bitcoin balance. The second argument is more general: Bitcoin is a scam because its creators, the "Satoshis," are market manipulators who promote it despite its lack of intrinsic value.

Let us look at the first possibility, which has hurt many ordinary people who put their life savings in Bitcoin. Liebkind (2020, n.p.) talks about five different Bitcoin-related scams. He refers to "one of the downsides of new investors entering the market," which is "the increase in the number of scams, frauds, and stories of retail investors who lose their coins to shady ventures." Scammers use various means to swindle holders of Bitcoin. The first is hardware wallet theft, where the wallet is a physical device that stores the owner's private keys. The scam may entail selling hardware wallets to users with a pre-configured seed phrase hidden under a scratch card. This situation creates a backdoor that allows hackers to drain funds once a wallet is activated. Exchange scams occur because cryptocurrency exchanges are not regulated; some are traps. Number three is fake initial currency offerings (ICOs), whereby a scammer creates fake websites that resemble an ICO and instructs users to deposit coins into a compromised wallet. Number four is cloud mining schemes, where mining is the extraction of new coins, an incredibly resource-intensive activity. Potential miners are scammed by paying some server space rent to mine coins at a set rate. Some scammers operate Ponzi schemes that can lead to massive losses. Lastly, multilevel marketing schemes offer naïve investors "excellent opportunities" for progressively more significant sums of Bitcoin. Butler (2020, n.p.) mentions an estimated $100 million that "has been wheedled out of people around the world in a cryptocurrency scam run out of a Kiev call centre." He interviews four victims who tell horror stories.

For some people, however, the whole concept of Bitcoin, and cryptocurrencies in general, is a scam. For example, Byrne (2017, n.p.) refers to "a class of fraud that uses technology, rather than a scheme operator, to mediate the interactions between an investment scam's beneficiaries and its dupes." He also describes what he calls the "Nakamoto Scheme" as "an automated hybrid of a Ponzi scheme and a pyramid scheme which has, from the perspective of operating a criminal enterprise, the strengths of both and (currently) the weaknesses of neither." Gandal et al. (2018) identify the impact of suspicious trading activity on the Mt. Gox Bitcoin currency exchange, in which approximately 600,000 Bitcoins valued at $188 million were acquired fraudulently. They find that the price of Bitcoin rose by an average of 4% on days when suspicious trades took place, compared to a slight decline on days without suspicious activity. They reached the plausible conclusion that the suspicious trading activity likely caused the unprecedented spike in the price of Bitcoin in late 2013, when the price jumped from around $150 to more than $1000 in two months (Moosa, 2020a).

Figures 6.1 and 6.2, respectively, display the dollar price of Bitcoin and the trading turnover (in billion dollars) from August 2014 to December 2020. We can see that the price rose dramatically in 2017 before sharply declining in a typical formation of a bubble followed by a crash (Moosa, 2020a). Another spectacular rise started in September 2020 amid the COVID-19 pandemic. Figure 6.2 shows spectacular growth in trading turnover (in billion dollars), showing how much Bitcoin attracts individuals and corporations

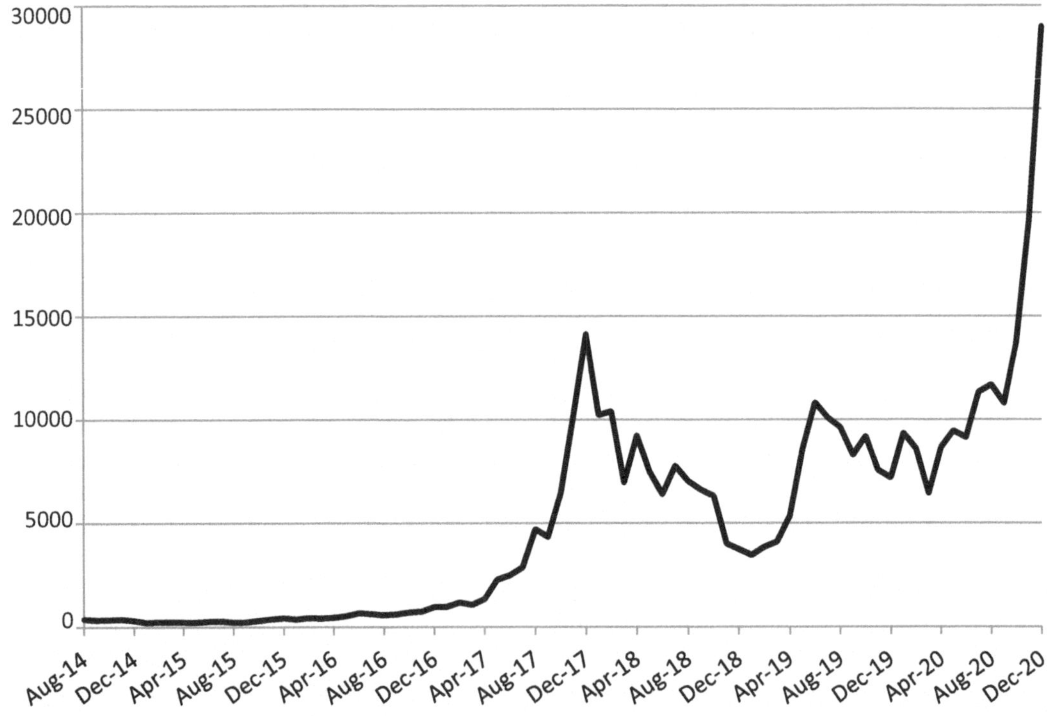

Fig. 6.1 The dollar price of Bitcoin. (Source of data: Yahoo Finance)

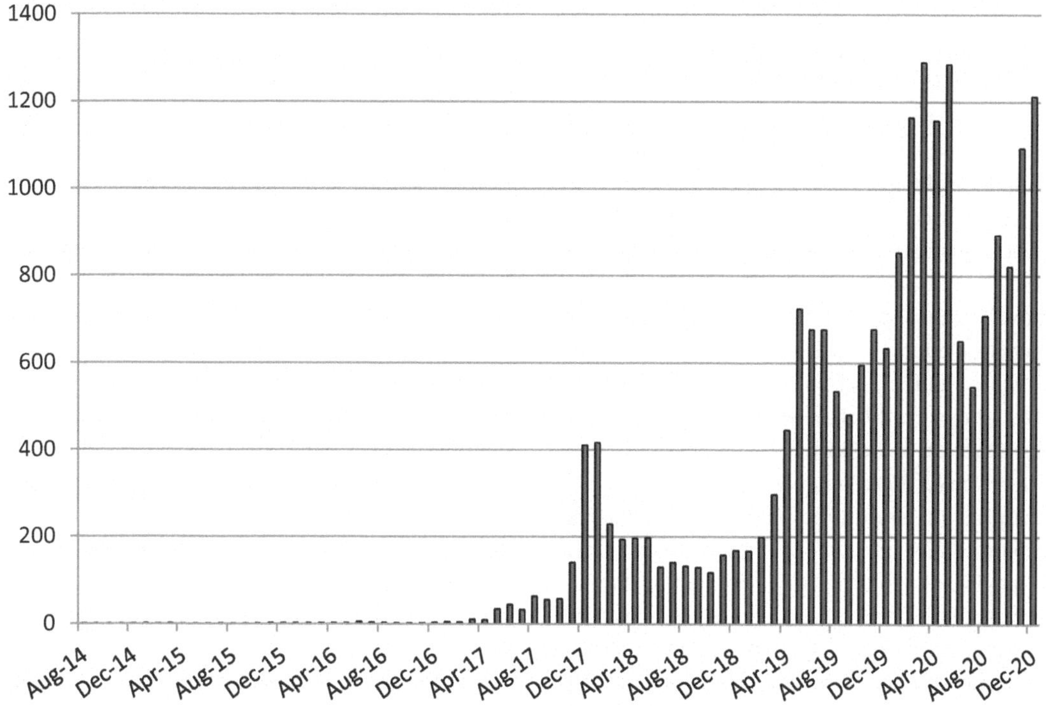

Fig. 6.2 The dollar value of Bitcoin turnover ($ billion). (Source of data: Yahoo Finance)

seeking higher returns in a low or negative return environment.

Let us try to discover why people scramble to hold a mysterious "asset" without an intrinsic value. Bitcoin is not even a currency, but people buy it on the assumption that it will keep on appreciating. Bitcoin's proponents assert that its scarcity makes it as valuable as gold, but gold is not a product of human labor, and God or nature determines its scarcity. On the other hand, Bitcoin is scarce only because the establishment has told us that the Bitcoin stock will stay within a predetermined level. This prediction, however, is not a law of nature but a "promise" of the establishment. It relates to a scene in Arnold Schwarzenegger's movie where he tells one of the bad guys, "Do you remember when I promised to kill you last—well, I lied?" Furthermore, gold has commercial use in jewelry and industrial use in various applications.

The rise in the price relies on several factors, the first of which is rising actual and expected inflation arising from the disastrous policy of massive monetary expansion. Even though governments lie about the actual inflation figures, people can feel it and try to circumvent the consequences (Moosa, 2020b). Another factor is increased adoption by payment applications like PayPal. However, another factor is that publicly traded companies have been acquiring Bitcoin to make a "quick buck." Conway (2020) refers to Bitcoin's historical trend of closely following its halving stock-to-flow model, which creates a bullish outlook. Conway, however, fails to mention an essential factor: ultra-low and negative interest rates, which push people to bear more risk to preserve the actual value of capital.

As Moosa discussed (2020a, p. 96), the surge in late 2020 marks the second wave of the cryptocurrency bubble, following the initial wave in 2017. During that period, analysts attributed the bubble to fervent speculative activity. For instance, Turanova (2017, n.p.) drew parallels between the Bitcoin market and historical speculative events such as the tulip fever in seventeenth-century Europe, the late 1980s Japanese stock market rally, and the internet mania in 1999. Janet Yellen, the former Fed chair, characterized Bitcoin as a "highly speculative asset" based on conventional wisdom from traditional finance.

Thompson (2017) likened Bitcoin's ascent to an "astronomical trajectory," drawing comparisons to collectible crazes like Beanie Babies in the late 1990s. Detrixhe (2017) also cited Robert Shiller, who identified Bitcoin as "the best example [of a bubble] right now."

The Bitcoin phenomenon, whether it is the first or second wave, resembles that of dotcom: both are associated with the development of novel technologies (the internet in the case of the dot.com bubble and blockchain in the case of Bitcoin) and supported by media frenzy, which encourages herd behavior. Implying similarity to the situation in the tulip bulb market of the 1630s, Morrison (2017) reminds us that in January 1637, the price of a rare tulip bulb in Amsterdam was ten times the annual wage for a skilled worker, as a single bulb sold for 1000 pounds of cheese at the height of the mania. There are similarities between the Bitcoin bubble and the bubble in mortgage-backed securities in 2007 (Moosa, 2020a). There is a natural and quickly growing threat that Bitcoin and other cryptocurrencies could behave like mortgage-backed securities and be used as the foundation for hazardous and complicated financial instruments, such as credit default swaps that were traded in massive amounts without any limits based on the number of mortgages (Morrison, 2017).

Irrespective of whether or not this is a second wave of the bubble, technology has enabled what seems to be a scam. Naive investors have lost their life savings because of greed or a desire to preserve capital value. At the same time, a minority of shrewd people used technology enormously at the expense of the majority. Fintech is to blame for this mishap.

6.6 Fintech as a Weapon in the War on Cash

Fintech enables the global war on cash because it offers alternative payment methods. The alleged reason for pursuing the war on cash is to fight the underground economy. In reality, however, the war on cash is a component of an elaborate scheme for the benefit of the banking oligarchy. In addition to the war on cash, this scheme involves ultra-low or negative interest rates and bail-in legislation whereby deposits can be confiscated to recapitalize failed banks and pay their CEOs bonuses and golden parachutes (Moosa, 2021).

Let us start with ultra-low/negative interest rates that we observed worldwide before the resurgence of inflation. Ultra-low interest rates are supposed to boost the economy when it is contracting. Growth happens because companies borrow to expand their productive capacity and employ people. After all, households borrow to buy consumer durables because savings on mortgage payments go to goods and services. This nonsense was made very clear during the COVID-19 pandemic. Policymakers forget that taking the horse to the water is possible but not forcing it to drink.

On the other hand, low interest rates encourage debt accumulation and create or maintain bubbles in the financial and property markets. Even worse, low interest rates discourage fiscal discipline by making it more tantalizing for governments to accumulate debt. The monetary authorities in "Western" nations have never conducted a more reckless and dangerous monetary experiment than negative interest rates. They destroy the entire risk-return system that has been the basis of investing for the last several centuries. The biggest beneficiaries of negative interest rates are banks, which would tax depositors. Banks can also make more money by offering mortgages at a −1% rate while charging borrowers 4% in commissions and administrative fees. Believing banks could offer customers negative interest-rate mortgages is a triumph of wishful thinking over reason.

Negative interest rates are bound to encourage people to keep money under the bed or in a hole in the ground rather than with banks. This works against the feasibility of bail-in legislation, whereby failed banks are allowed to confiscate depositors' money. Therefore, how do governments retaliate against those "irresponsible" depositors who withdraw their money from banks? For one thing, governments are waging war on cash transactions, not to combat the underground economy (as they claim) but to force people to keep money at banks so that the money will be confiscated eventually. People are threatened with confinement behind bars if they use cash to settle transactions over a particular value (so much for human rights). However, this is not all, as economists acting as "hired guns" have devised innovative proposals to force people to keep money at banks. One idea is to cancel, by a random selection of serial numbers, 10% of the banknotes in circulation. Those who keep money at home will face an expected rate of return of −10% on their cash holdings, in which case they will be better off (or less at risk) by keeping their money with banks at the rate of −2%, hoping that a bail-in will not materialize (Mankiw, 2009).

Dowd (2017) has written a devastating critique of the war on cash, which was pursued to combat the underground economy, tax evasion, and the coronavirus. The following is how he describes the situation:

> The abolition of cash threatens to destroy what is left of our privacy and our freedom; we wouldn't be able to buy a stick of gum without the government knowing about it and giving its approval. The cash abolitionists want total control over your money and what you can do with it. Besides making us all entirely dependent on the whim of the state, banning cash also threatens to cause widespread economic damage and have a devastating impact on the most vulnerable in our society. Quite simply, the government's war against cash is the state's war against us (p. 1).

He links the war on cash to the policy of ultra-low or negative interest rates, identifying the adverse consequences of such a policy as negative impacts on saving and long-term capital accumulation, the losses inflicted on savers, pension funds, and the retired, and the enormous and ongoing damage caused by boom-bust cycles in markets across the world, as well as the costs of asset mispricing and associated capital misallocations produced by the conduit to ultra-low/negative interest rates: quantitative easing. In his reply to the Mankiw proposal, he writes the following:

> Well, Greg, I can't exactly see people being "delighted" at having to lend at minus 3 percent when the alternative imposed on them is to hold cash under the mattress and expect to lose 10 percent. Instead, I can see them holding the minimum possible amounts of working cash and putting their money elsewhere where their wealth is safe—into foreign bank accounts, into gold, silver, commodities, shares, cryptocurrencies or whatever. I can also see considerable distress when people find that their banknotes are invalidated through a lottery and they lose their money through no fault of their own. I can see this distress falling especially on the poor, who are more dependent on banknotes and less able to bear the losses. And I can see considerable public anger at the injustice and the needless hassle of everyone being regularly forced to check the serial numbers on their banknotes to see if they are still worth anything (Dowd, 2017, p. 12).

In what is an argument against abolishing cash transactions in the name of fintech, Dowd (2017) lists the benefits of cash as follows: (i) it is a very efficient way of handling small transactions; (ii) it is costless and easy to use; (iii) cash transactions are immediate and flexible; (iv) cash is highly anonymous, and traditionally, the anonymity of cash is one of its most significant benefits; (v) cash does not need a password; (vi) unlike a bank account, it cannot be hacked; (vii) it involves state-of-the-art in anti-counterfeiting technology, which makes it more challenging to replicate or corrupt than digital currency; and (viii) the usefulness of cash is not dependent on sophisticated technology that might break down.

Similar arguments emerged against a bill debated at the Australian Senate to restrict the amount of cash used to settle transactions to $10,000. For example, there is no clear rationale for the $10,000 limit, which will have a limited impact on money laundering and gang finances. Nevertheless, it "will remove a right we currently have to use cash as we please, thus narrowing the definition of legal tender" (Digital Finance Analytics, 2020). Eventually, the bill was withdrawn (for example, Khadem, 2020). These and similar arguments presented by others are worth considering when someone tends to praise fintech for reducing cash transactions, thus protecting us from the underground economy and the coronavirus.

6.7 Concluding Remarks

The advent of COVID-19 has already reinforced two related trends: fintech and the tendency to use less cash. This is because measures of lockdown and social distancing make it more compelling to engage in contactless financial transactions. Enthusiasm for fintech and digital finance is so rampant that most observers look at the benefits while overlooking the costs. Fintech is conducive to financial inclusion and supports economic growth. Even though it is unclear how that would happen, it would aid in the recovery from the pandemic-induced economic recession.

Those who think this way seem to forget the embarrassment of being unable to pay for dinner in a restaurant because the credit card is not working or the system is down. They also forget that forcing people to pay electronically intrudes on privacy and civil liberties. Accordingly, it follows that eliminating cash is a good idea, which means a government-imposed ban on cash is justifiable.

Fintech and digital finance costs can be enormous for both financial institutions and their customers. Fintech is a weapon in the war on cash, a component of an elaborate scheme combining cash bans with negative interest rates and bail-in legislation for the benefit of the banking establishment. Fintech boosts both the frequency and severity of operational loss events by enhancing the opportunity to commit fraud and the possibility of system failure. Algorithmic trading of financial assets causes enormous volatility and may lead to significant losses. Blockchain technology has created Bitcoin, which allows scams and may be a scam. Following the invention of Bitcoin, some 2000 cryptocurrencies have been created to emulate the success of the Bitcoin scam. While technology is lovely, we must be aware of the hazards of "technologizing" the financial sector beyond recognition.

References

Adrian, T., & Mancini-Griffoli, T. (2019). The rise of digital money. *IMF Fintech Notes, No 19/01.*

Aguayo, F. Z., & Slusarczyk, B. (2020). Risks of banking services' digitalization: The practice of diversification and sustainable development goals. *Sustainability* (published online, 14 May).

Amidžić, G., Massara, A., & Mialou, A. (2014). Assessing countries' financial inclusion standing—A new composite index. *IMF Working Papers, No. 14/36.*

Ayyagari, M., Demirgüç-Kunt, A., & Maksimovic, V. (2017). SME finance. *World Bank Policy Research Working Papers, No. 8241.*

BCBS. (2004). *Basel II: International convergence of capital measurement and capital standards: A revised framework.* Bank for International Settlements, June.

Beck, T., Demirgüç-Kunt, A., & Peria, M. S. M. (2007). Reaching out: Access to and use of banking service across countries. *Journal of Financial Economics, 85,* 234–266.

Bowley, G. (2010, October 2). Lone $4.1 billion sale led to 'flash crash' in May. *New York Times.* http://www.nytimes.com/2010/10/02/business/02flash.html?_r=2&scp=1&sq=flash+crash&st=nyt&

Brunnermeier, M., Harold, J., & Landau, J. P. (2019). The digitalization of money. *NBER Working Papers, No. 26300.*

Butler, B. (2020, March 2). Swindled with Bitcoin: Australian victims count cost of online finance scam. *The Guardian.*

Byrne, P. (2017, December 8). *The problem with calling bitcoin a "Ponzi Scheme".* https://prestonbyrne.com/2017/12/08/Bitcoin_ponzi/

Camara, N., & Tuesta, D. (2017, July 14). Measuring financial inclusion: A multidimensional index, seminar at the ISI world statistics congress on financial inclusion. *Marrakech.*

Chris, C. (2020, November 12). *Preparing for the fraud pandemic.* https://bluenotes.anz.com/posts/2020/11/gbgplc-financial-institutions-fraud-scams-digital-banking

Claessens, S., Frost, J., Turner, G., & Zhu, F. (2018). Fintech credit markets around the world: Size, drivers and policy issues. *BIS Quarterly Review,* September, 29–49.

Conway, L. (2020, November 5). Why is Bitcoin's Price rising? *Investopedia.*

Croft, A. (2020, March 11). How high-frequency algorithmic trading programs can make bad stock market days even worse. *Fortune.*

Dabla-Norris, E., Deng, Y., Ivanova, A., Karpowicz, I., Unsal, F., VanLeemput, E., & Wong, J (2015a). Financial Inclusion: Zooming in on Latin America. *IMF Working Papers, No. 15/206.*

Dabla-Norris, E., Ji, Y., Townsend, R., & Unsal, D. F. (2015b). Distinguishing constraints on financial inclusion and their impact on GDP, TFP, and inequality. *NBER Working Papers, No. 20821.*

De Roure, C., Pelizzon, L., & Thakor, A. V. (2018). P2P lenders versus banks: Creak skimming or bottom fishing? *SAFE Working Papers, No. 206.*

Deléchat, C., Newiak, M., Xu, R., Yang, F., & Aslan, G. (2018). What is driving Women's financial inclusion across countries? *IMF Working Papers, No. 18/38.*

Detrixhe, J. (2017, September 5). *Irrational exuberance.* https://qz.com/1067557/robert-shiller-wrote-the-book-on-bubbles-he-says-the-best-example-right-now-is-Bitcoin/

Digital Finance Analytics. (2020, January 19). *The Cash Countdown*. Downloaded from https://digital--finance-analytics-dfa-blog-11.castos.com/episodes/the-cash-countdown-podcast

Dowd, K. (2017, April). Killing the Cash Cow: Why Andy Haldane Is Wrong about Demonetisation. *Adam Smith Institute Briefing Papers*. Downloaded from https://www.adamsmith.org/research/killing-the-cash-cow-why-andy-haldane-is-wrong-about-demonetisation

Fuster, A., Plosser, M., Schnabl, P., & Vickery, J. (2018). The role of technology in mortgage lending. *NBER Working Papers, No. 24500.*

Gandal, N., Hamrick, J. T., Moore, T., & Oberman, T. (2018). Price manipulation in the bitcoin ecosystem. *Journal of Monetary Economics, 95*, 86.

Ghose, R., Bajaj, R., Khallouf, M., Shirvaikar, A., Annibale, R., & Nava, J.R. (2020). *Banking the next billion: Digital financial inclusion in action*. https://www.citivelocity.com/citigps/banking-next-billion/

Honohan, P. (2008). Cross-country variation in household access to financial services. *Journal of Banking and Finance, 32*, 2493–2500.

Jagtiani, J., & Lemieux, C. (2017). Fintech lending: Financial inclusion, risk pricing, and alternative information. *Federal Reserve Bank of Philadelphia Working Papers, No. 17–17.*

Khadem, N. (2020, December 7). Law to ban cash for purchases of $10,000 or more abandoned, but don't be surprised if it's revived. *ABC News*.

Liebkind, J. (2020, September 24). *Beware of these 5 bitcoin scams*. https://www.investopedia.com/articles/forex/042315/beware-these-five-Bitcoin-scams.asp

Loukoianova, E., & Yang, Y. (2018). Financial inclusion in Asia-Pacific. *IMF Asia and Pacific Departmental Papers, No. 18/17.*

Mankiw, N. G. (2009, April 18). It may be time for the fed to go negative. *New York Times*.

Mehta, N., & Kisling, W. (2010, October 2). Futures sale spurred 6 May panic as traders lost faith in data. *Bloomberg*. http://www.bloomberg.com/news/2010-10-01/automatic-trade-of-futures-drove-may-6-stock-crash-report-says.html

Mestchian, P. (2003). Operational risk management: The solution is in the problem. In *Advances in operational risk: Firm-wide issues for financial institutions*. Risk Books.

Moosa, I. A. (2007). *Operational risk management*. Palgrave.

Moosa, I. A. (2008). *Quantification of operational risk under Basel II: The good, bad, and ugly*. Palgrave.

Moosa, I. A. (2020a). The bitcoin: A sparkling bubble or price discovery? *Journal of Industrial and Business Economics, 47*, 93–113.

Moosa, I. A. (2020b). *Controversies in economics and finance: Puzzles and myths*. Edward Elgar.

Moosa, I. A. (2021). *The Washington consensus: A critique of the principles and implications for economic development*. World Scientific (in press).

Morrison, R. (2017, December 20). *The bitcoin bubble*. https://www.globalresearch.ca/the-Bitcoin-bubble/5623349

Ozili, P. Z. (2018). Impact of digital finance on financial inclusion and stability. *Borsa Istanbul Review, 18*, 329–340.

Patwardhan, A., Singleton, K., & Schmitz, K. (2018). *Financial inclusion in the digital age*. International Finance Corporation.

Rau, R. P. (2019). *Law, trust, and the development of crowdfunding*. Cambridge Judge Business School, Cambridge University.

Robert Morris Associates, British Bankers' Association, and International Swaps and Derivatives Association. (1999). *Operational risk: The next frontier*. RMA.

Rojas-Suárez, L. (2016). Financial inclusion in Latin America: Facts, obstacles and central banks' policy issues. *IDB Discussion Papers, No. DP-464.*

Rojas-Suárez, L., & Amado, M. A. (2014). Understanding Latin America's financial inclusion gap. *CGD Working Papers, No. 367.*

Sahay, R., Čihák, M., N'Diaye, P., Barajas, A., Mitra, S., Kyobe, A., Mooi, Y. N., & Yousefi, S. R. (2015a). Financial inclusion: Can it meet multiple macroeconomic goals?. *IMF Staff Discussion Notes, No. 15/17.*

Sahay, R., Čihák, M., N'Diaye, P., Barajas, A., Bi, R., Ayala, D., Gao, Y., Kyobe, A., Nguyen, L., Saborowski, C., Svirydzenka, K., & Yousefi, S. R. (2015b). Rethinking financial deepening: Stability and growth in emerging markets. *IMF Staff Discussion Notes, No. 15/08.*

Sahay, R., von Allmen, E.U., Lahreche, A., Khera, P., Ogawa, S., Bazarbash, M., & Beaton, K. (2020). The promise of Fintech financial inclusion in the post-COVID-19 era. *International Monetary Fund, Monetary and Capital Markets Department Papers, No. 20/09.*

Sanicola, L. (2017, August 20). What is FinTech? *Huffington Post.*

Schueffel, P. (2017). Taming the beast: A scientific definition of Fintech. *Journal of Innovation Management, 4*, 32–54.

Spicer, J., & Younglai, R. (2010, October 1). 4-single U.S. trade helped spark May's flash crash. *Reuters*. http://www.reuters.com/article/2010/10/01/financial-regulation-flashcrash-idUKN0114164220101001

Svirydzenka, K. (2016). Introducing a new broad-based index of financial development. *IMF Working Papers, No. 16/5.*

Thompson, D. (2017, December 9). *Is bitcoin the most obvious bubble ever?*. https://www.theatlantic.com/business/archive/2017/12/Bitcoin-bubble/547952/

Turanova, G. (2017, December 29). *Is bitcoin a bubble?*. http://www.nasdaq.com/article/is-Bitcoin-a-bubble-cm898150

Van Loo, R. (2018). Making innovation more competitive: The case of Fintech. *UCLA Law Review*. https://scholarship.law.bu.edu/faculty_scholarship/50

Digital Finance, Agriculture, and Inclusive Development

7

Haruna Issahaku

Abstract

This chapter explores how the deployment of financial digitalization can foster agriculture and inclusive development. It provides critical insights into recent developments in digital finance, how digital finance can drive agriculture and inclusive development, and the challenges associated with financial digitalization. The chapter recommends options for harnessing financial digitalization to foster agriculture and inclusive development.

Keywords

Financial digitalization · Digital finance · Agriculture development · Inclusive development

7.1 Introduction

Digital finance is a relatively new concept. It refers to accessing, using, or providing financial services and products through digital channels. Digital finance encompasses mobile phones, computers, or cards used over point-of-sale (POS) devices connecting individuals and businesses to a digitized national payments infrastructure, enabling seamless transactions across all parties. It also covers all types of financial services, such as payments, savings accounts, credit, insurance, and other financial products; all types of users, including individuals at all income levels, businesses of all sizes, and government entities at all levels; and all types of providers of financial services, including banks, payment providers, other financial institutions, telecommunications companies, financial technology (fintech) start-ups, retailers, and other businesses. This chapter aims to explore the state-of-the-art trends in financial digitalization and discusses how we can effectively deploy digital finance to foster agriculture and inclusive development.

An assessment of the link between financial digitalization and agriculture, as well as inclusive development, is an essential contribution to knowledge and practice for several reasons. In the first place, evidence of the effect of digital finance on agriculture and inclusive development is mainly anecdotal and needs rigorous demonstration. "We have yet to see significant, robust, causative evidence that links the digitization of financial activities to reduced hunger, higher incomes, greater resilience, or reduced poverty for smallholder farmers" (Miller, 2019, p. 1). Second, financial digitalization is a dynamic and fluid concept and needs more discussion in academic and development circles to fully deploy it

H. Issahaku (✉)
Department of Finance, School of Business,
University for Development Studies, Tamale, Ghana
e-mail: iharuna@uds.edu.gh

G. W. Muschert et al. (eds.), *Financial Inclusion*, Sustainable Development Goals Series,
https://doi.org/10.1007/978-3-031-68803-4_7

for the good of society while minimizing potential adverse effects on the vulnerable.

The third reason we should explore the digital innovation–agriculture–inclusive development nexus is that digital finance presents a vital opportunity to uplift the majority of people experiencing poverty, most of whom are farmers, from poverty by addressing the intractable problem of financial exclusion (Abor et al., 2018). According to the United Nations Development Programme and Oxford Poverty and Human Development Initiative (2020), over 1.3 billion people (22% of the population) in developing countries live in multidimensional poverty. Interestingly, 65% of the world's working poor rely on agriculture to make a living (World Bank, 2020). If agriculture could be made more productive through digital innovation, millions of the world's poor would rise out of poverty.

Lastly, the coronavirus pandemic presents another reason to investigate and harness the potential of financial digitization. Financial digitization will help improve the economic well-being and reduce the spread of the pandemic by minimizing, if not eliminating, the need for face-to-face contact in financial transactions. Digital financial inclusion will enable farm households to make or receive payments wherever they are without the need to visit a financial institution in person and risk contracting or spreading the virus. Thus, digital finance is COVID-19 compliant, and this feature helps protect the lives of the vulnerable from the pandemic.

7.2 Recent Developments in Digital Finance

The digital revolution is on the move, and every aspect of life may be impacted. In the business world, robots team up with humans for firefighting, vehicle manufacturing and assembly, delivery of health services, hotel and restaurant services, and legal services, among others. Similarly, finance has been revolutionized by digitization, moving finance from brick-and-motor banking to mobile banking, automation, and cashless banking (Deloitte, 2018). Some trends drive this new normal of financial digitization. Some of these trends feature in Tmob (2021), Capital Bank Solutions and IBS Intelligence (2021), and Deloitte (2018), which is summarized below.

Trend 1: Rapid Adoption and Diffusion of New-Age Technologies

The finance arena has seen rapid adoption and spread of new-age impactful technologies such as Artificial Intelligence (AI), the Internet of Things (IoT), Robotic Process Automation (RPA), Big data, and blockchain. Though some discoveries have been made, the financial industry is awash with trials and experiments with these new-age technologies to find ways of optimizing them. The financial industry is already deriving good value from new-age technologies. For instance, JP Morgan introduced an AI program called COIN (contract intelligence) to reduce loan processing mistakes and optimize time use. COIN is a utility for reviewing and interpreting loan contracts. COIN can cut down 360,000 h of human work by attorneys and loan officers (Capital Bank Solutions and IBS Intelligence, 2021). An estimate is that AI and machine learning will save the financial industry over $1 trillion over time (Sastry, 2020).

Big data is taking shape in the financial industry and is driving financial innovation. Financial sector players are using big data to understand their customers better and to design products and services that delight the customer. Industry actors also realize the potential of using big data to reduce work hours done by Information Technology (IT) officers.

Like other new-age technologies, RPA is a game changer in the financial industry. RPA is helping the financial industry to save costs, increase earnings, and boost productivity. RPA can increase estimated return on investment between 30% and 200% in the first year of adoption (Capital Bank Solutions and IBS Intelligence, 2021). Mixon (2021) reports that about 80% of leaders in the finance industry have either adopted or are planning to roll out RPA.

IoT is already receiving significant applications in the finance space. It holds much promise for the finance industry, including enhanced banking experience, delightful service beyond banking, efficient branch banking, enhanced credit and debit card experience, and many more (Pareek, 2020). The installation of beacons in some branches of Citibank to allow customers to access ATMs through Bluetooth technology on their equipment is a significant breakthrough (Capital Bank Solutions and IBS Intelligence, 2021).

Trend 2: Mobile Phone-Driven Banking Is on the Ascendancy

The emergence of mobile telephony has changed the course of banking for good. During the peak of the coronavirus pandemic, mobile money became a lifesaver for many as it offered remote financing solutions amid lockdowns, social distancing, and other restrictions. Mobile money accounts grew by 12.7% in 2020 to reach 1.21 billion accounts from about 1.1 billion in 2019 (Andersson-Manjang & Naghavi, 2021). The demand for mobile phone-based banking is driven by the availability of less expensive mobile phone brands, interoperability, the desire by customers for faster and more convenient banking, and the circumstances imposed by the pandemic.

Trend 3: Upsurge of Fintechs

It has now become clear that fintechs are here to stay, dictating trends in the emerging technological innovations in the financial sector. Fintechs have led the way in mobile money services, digital wallets, peer-to-peer lending, small ticket loans, alternative credit scoring, and payment gateways, among other innovations. Globally, the value of FinTech investments ballooned from $9 billion in 2010 to $135.7 billion in 2019 (see Fig. 7.1).

Trend 4: Advent of Open Banking

Open banking is changing the way banking is done. Open banking permits the networking of customer accounts and data across banks and nonbank institutions via application programming interfaces (APIs) so that this data is harnessed for use by customers, the financial system, and other third-party users (Investopedia, 2020). Open banking can raise revenue by 20%, customer experience by 78.3%, and about 83% of banks view private API as imperative (Capital Bank Solutions and IBS Intelligence, 2021). The advent of open banking has increased collaborations between banks and fintechs. This has led to banks providing their services based on fintech services while fintechs consider themselves the new-era banks (Tmob, 2021). The potentials of open banking are limitless.

Trend 5: Emergence of New Interactive Modes of Banking

Globally, bank visits are dwindling by the day as new interactive banking channels take over the finance space. It has been projected that customer visits to bank branches will fall 36% between 2017 and 2022, whereas mobile transactions will skyrocket 121% over the same period (see Pilcher, 2021). Banks are reinventing to cope with the phasing out of branch banking by introducing mobile kiosks, cardless and card banking, and other innovations, all driven by digital technology. In the age of wearables, financial institutions are investing in apps that enable access to and usage of financial services via smartwatches.

Trend 6: Enhanced Customer Experience

Customers have developed a taste for speed, convenience, cost-effectiveness, and trendiness in delivering services across the board. Acquaintances with social media such as Facebook, Twitter, WhatsApp, and Instagram, among others, and encounters with technology firms such as Apple, Google, and Amazon are changing consumers' experiences. This trend has put pressure on the financial industry to find ways of improving customer experiences in ways that tech companies do. Banks are investing in biometric technology to make banking easy. Self-service and contactless banking services are increasingly rolled out to delight the customer.

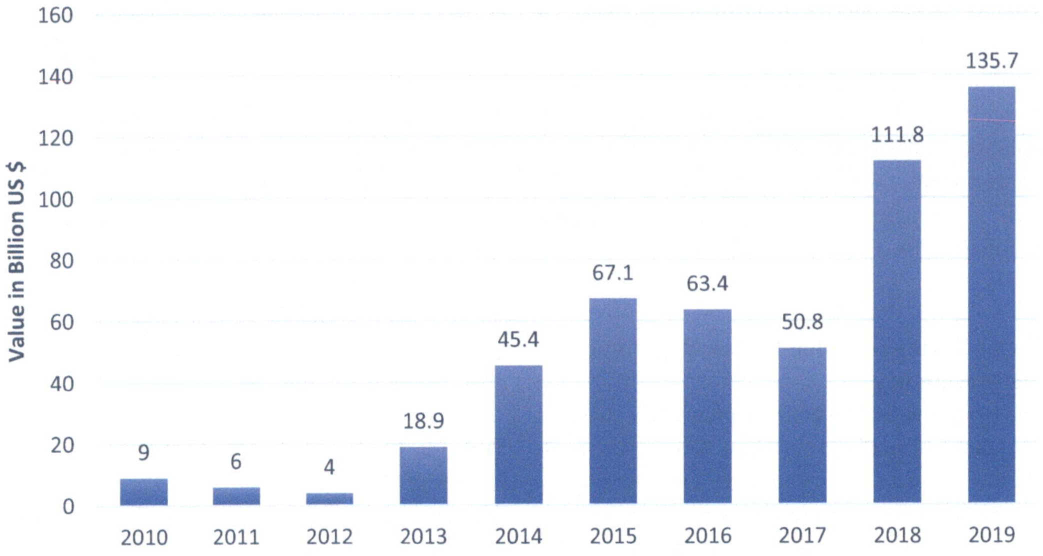

Fig. 7.1 Trends in fintech investments (US $, Billions): 2010–2019. (Source: Tmob, 2021)

Other Trends

Other trends underlining the digital finance revolution include increased optimization of back-end processes, increased relevance of data security, financial system architecture overhaul, the transformation of organizational structures of financial institutions, and rapid evolution of new financial technologies.

7.3 Driving Agriculture Development Through Digital Finance

Financial digitalization portends much promise for transforming financial services rendered to the agriculture sector. Digitalization affects the operating models, profitability, product design, competition, costs, and distributional strategies of financial services and, by so doing, leads to the development of financial services and products with enhanced accessibility and affordability and customized to meet the peculiar needs of the agriculture sector (International Finance Corporation [IFC], 2018).

7.3.1 Critical Factors Defining the Success of Digital Financial Innovations in Agriculture

Critical factors defining the success of digital financial innovations in agriculture include complexity-related factors, design factors, operational factors, growth and innovation factors, and connectivity (Miller, 2019). Agriculture is associated with complex problems, and these complex problems require multidimensional solutions. The constraints faced by the agriculture sector are varied and include seasonality, perishability, weak linkages in the value chain, lack of political will, lack of access to markets and inputs, inadequate supply of extension workers, susceptibility to weather variability, lack of access to improved technology, and the list goes on. Only some of these problems can be solved by digitalization. In rolling out digital financial innovations, care must be taken to ensure they are complemented with other solutions that comprehensively address the agriculture sector's challenges.

When designing digital financial solutions for farmers, the agriculture sector's unique circumstances must be considered. Financial services for farmers must mimic their production cycle in order to be successful. For instance, crop farmers have irregular income flows due to the seasonal nature of such crops. Thus, in designing a credit package, repayments should occur after harvest when the farmer is in an excellent position to honor their obligations. Monthly instalment payments will not suit a farmer face with seasonality; bullet repayments may be better. A study in Uganda showed that farmers were more willing to take up loans to procure modern equipment when repayments were scheduled to occur after harvest (Matsumoto et al., 2013). Likewise, a Kenyan study found that when insurance premium payments are delayed after harvest, farmer participation increases from 5% to 72% (Casaburi & Wills, 2018).

Operational factors play a critical role in the success of any intervention. When discussing digital innovations, there is a tendency to take out the human interface or to think that digitalization completely replaces human effort. However, this is only partially the case. The field force's role is critical in operationalizing digital interventions to farmers and should be factored into the roll-out plan. These field forces may entail extension officers, field officers, sales agents, and mobile money agents. These agents have various roles to play in the success of a digitalization intervention, including helping farmers with registration processes, supply of digital equipment, repair of equipment, taking farmers through digital procedures, helping with deposits and withdrawals, resolving challenges encountered by farmers, convincing farmers to adopt digital solutions, among others. Thus, the role of various human actors must be acknowledged and factored into the operationalization of digital financial intervention.

There is a general belief that digital innovations pick up very quickly. This is usually different in agriculture. Adopting digital technology in agriculture usually picks up slowly, so patience is needed. Knowing this, one should take more time to scale up digital interventions in the sector. Donor funding is critical in rolling out digital financial services to agriculture. Subsidies are needed at the experimentation stage to develop suitable digital innovations for farmers. Similarly, there may be a need to subsidize the innovations during the implementation phase to attract farmers to the program. When they realize the benefits, they will continue using those innovations.

Connectivity factors are critical, especially in rural areas where the digital infrastructure is lacking. This requires governments to collaborate with the private sector, mainly the telecommunication companies, to ensure robust connectivity across the board. With a strong communication network and internet connectivity, the potential of digital finance for agriculture can be fully unlocked.

7.3.2 Increasing Access of Farmers to Financial Services Through Digitalization

Digital financial services will be appealing to farmers because they reduce the cost of traveling over long distances with cash or to make or receive cash payments, save time and effort, are more convenient, ensure greater transparency and accountability, reduce the risk of theft and fraud associated with handling cash, and provide an entry point for limitless financial and nonfinancial services (IFC, 2018). Digital technology can be harnessed to develop products and services that will engender help better lending policies, risk management, and restructuring agricultural value chains (McIntosh & Mansini, 2018). Briefly discussed below are some digital financial services and products that can improve access to and uptake of financial services in agriculture.

Digital Payments and Receipts

If agricultural commercialization is to be promoted, digital payments and receipts are essential. Fintechs and collaborations between Fintechs and banks have driven the arena of digital payments and receipts. The World Bank Global Findex 2017 shows that 59% of the 235 million adults globally who do not have bank accounts receive cash payments through sales of agricultural produce and have access to a mobile phone. This means mobile phone-driven financial digitalization can lift millions of farmers from financial exclusion. Digital transactions can enable farmers to purchase inputs remotely and sell their products without traveling long distances amid all the associated risks.

Digital Credit

Credit is one of the most critical determinants of returns to investments in the agriculture sector and should be given serious attention in proposing innovative solutions. Microfinance can be retooled to meet the lending needs of agriculture by employing digitalization. For instance, they can use digital channels to collect data on prospective clients to target beneficiaries appropriately. Again, disbursements, repayments, payments, credit information, and business advisory services can be done through mobile phones, which are more accessible to farmers.

Digital Savings

One of the criticisms often leveled against smallholder farmers is that they do not save. A significant reason for this is a need for more suitable savings channels. Digital finance enables farmers to save anywhere and everywhere without the need to visit a formal financial institution. Savings through mobile money is becoming more popular with farmers. Additionally, some ATMs now receive cash, saving the farmer the ordeal of dealing with an unwelcoming bank teller. Encouraging more farmers to save through digital means will require proper regulatory oversight from central banks to ensure interest payments on savings are attractive while withdrawal costs are low.

Digital Insurance

If any enterprise needs insurance, it is agriculture enterprises, especially rain-fed agriculture. Unfortunately, even the gold standard in insurance, index insurance, has yet to be met due to low demand. It is hard to find any wide-ranging success story on farm insurance. The low demand for index insurance is attributable to several reasons, including high premiums, poor credibility and timeliness of insurance payouts, lack of comprehensive insurance coverage, and "ambiguity aversion" (McIntosh & Mansini, 2018). Ambiguity aversion refers to the situation where a person does not like paying for a product when they do not fully understand how shocks and payouts are distributed (Bryan, 2010). Digital innovations can improve the demand for agriculture insurance in several ways. First, by employing big data, cloud computing, and artificial intelligence, the data required to construct indexes for insurance and assess risk can be obtained timeously, affordably, and with high precision. Second, the low credibility and timeliness in the insurance sector can be minimized by using digital channels to pay premiums and payouts. Digital platforms can also promote agriculture insurance products and general financial literacy. Digital insurance solutions for agriculture must comprehensively address all aspects of the farmer's life within and outside agriculture in order to appeal to farmers. Thus, digital finance must come with a bundle of solutions beyond insurance, addressing the needs of farmers holistically (IFC, 2018).

E-commerce

Electronic commerce (e-commerce) entails selling and purchasing items or services or transmitting funds and data over a digital platform. There are many e-commerce models, including business-to-business (B2B), business-to-customer (B2C), consumer-to-business (C2B), consumer-to-consumer (C2C), and business-to-administration (B2A), among others. Some examples of e-commerce market platforms include Amazon, Alibaba, eBay, Jumia, Wayfair, Overstock, Tonatom, and Jiji. The widespread adoption and use of e-commerce in agriculture will impact the sector immensely. The potential benefits of e-commerce in agriculture include cost efficiency, speedy purchases, limitless access to the market, limitless online store space, quick response to customers, and high payment flexibility for customers (Amazon, 2021).

The B2C concept can link small-scale farmers to input and output markets, eliminating or reducing the need for intermediaries. For large agribusiness firms, the B2C can also be used to reach out to customers anywhere they are worldwide. Agribusiness companies can harness the B2A approach in dealing with government ministries, departments, and agencies.

Cross-Cutting Digital Solutions

The above-discussed digital solutions are only indicative as digital products and services are emerging daily. Crowdfunding and crowdlending are gaining ground and are being used to fund several projects, including those about agriculture. Most stock exchanges have gone digital, allowing trading to occur seamlessly over a 24-h cycle. Other uses of digitalization entail precision agriculture, where artificial intelligence ensures the tailored application of fertilizer, water, and other treatments to plants. Complete digitization of agriculture value chains is another offering deployed to strengthen agriculture value chains. Delivery of extension advice and weather information through digital media has also been the focus of experimentation.

7.3.3 Empirical Evidence on Impact of Digitalization on Agriculture Development

Though the findings are sometimes mixed, the empirical literature weighs strongly in favor of a positive impact of digitalization on agriculture development. The work of Nzie et al. (2017) assessed the impact of mobile telephony on transaction costs and prices of vegetable farmers in Cameroon. The findings revealed that mobile phone use increased transaction costs, hence the price at which vegetables were sold. The authors attributed the positive effect of digitalization on transaction costs to the high cost of phone calls. The finding of Nzie et al. (2017) is contradicted by Abor et al. (2018), who revealed that new technologies reduced information costs and improve market efficiency in Ghana.

Issahaku et al. (2018) found a more favorable impact of a mobile telephone on agriculture development. This study investigated the impact of mobile phone ownership and use on the productivity of maize farmers in Ghana. Using propensity score matching, Issahaku et al. (2018) revealed that mobile phone use increased the productivity of maize farmers by a minimum of 261.20 kg/ha per farming season. The authors identified that mobile telephony impacted agriculture productivity by boosting extension services, technology adoption, and market participation. According to the authors, by strengthening farmer resilience, fintech makes agriculture climate-smart-compliant and transforms agriculture in line with critical sustainable development goals, particularly 1, 2, 10, 12, 13, and 15. Digital crop insurance can be harnessed to increase the resilience of smallholder farmers against the vagaries of climate change (Hinson et al., 2019).

Miller (2019) has discussed the impact of digital finance on agriculture development based on product type. Starting the discussion with digital transaction-based products, Suri and Jack (2016) and Kikulwe et al. (2014) have shown that mobile payments and transactions improved the livelihood of farmers mainly through the remittance channel. By facilitating the receipt of remittances, digital payments increased farm households' investment and improved households' resilience against shocks. In terms of digital savings, the impact of digital savings on agriculture has been buttressed by randomized control trials (RCT). For example, Brune et al. (2016) found that farmers whose harvests were paid digitally into bank accounts tended to save leading up to the following season and invested more in inputs in that particular season. Subsequently, such farmers had a higher output, sales and profits, and increased expenditure above the control group. An RCT study on the impact of a digital credit bundle comprising fertilizer, training, and seed in the Western part of Kenya revealed that the output and profit of beneficiary maize and beans farmers increased by 34% and 20%, respectively (see Miller, 2019).

7.4 Driving Inclusive Development Through Digital Finance

The role of digital finance goes beyond agriculture to include poverty eradication, inequality reduction, and spurring entrepreneurship, women empowerment, health, employment, and financial inclusion. In Abor et al. (2018), we learned how mobile phone usage improves consumption, reduces poverty, and improves financial inclusion by increasing users' access to various financial services and products, reducing the cost of information, and opening up limitless opportunities. Bongomin et al. (2018) have emphasized how mobile money has increased the access of people with low incomes to financial services. The authors revealed that social networks amplify the impact of mobile money on financial inclusion. The impact of digital finance on access to finance goes beyond agriculture or rural settings. Employing enterprise survey data, Gosavi (2017) publicized that small and medium firms that used mobile money had more access to credit and were more productive than their counterparts who did not. Kirui et al. (2013) have provided evidence to show that mobile money significantly impacts rural poverty by correcting market failures and increasing access to finance.

The literature also discusses the role of digitalization in women's empowerment. Asongu (2015) and Ondiege (2013) have discussed how digital financial services have improved women's access to finance, thereby reducing the gender finance gap. In providing corroborating evidence, Ojo et al. (2012) revealed that mobile telephony improved the welfare of women via three main channels: stimulating regulations that enhance access of marginalized groups to financial services and products, facilitating community mobilization, and creating and strengthening linkages between rural folk and development agencies.

The concern in the literature has been whether or not the impact of digitalization on inclusive development is long-lived. Suri and Jack (2016) have answered this question by examining the influence of M-PESA (mobile money services) on consumption and poverty in Kenya. The findings showed that mobile money had significant poverty reduction and consumption-boosting effects in both the long and short runs. Furthermore, they found that these impacts are more pronounced in households headed by females, confirming the women empowerment abilities of digital finance.

7.5 Challenges of Digital Finance for Inclusive Development

Digital finance holds numerous opportunities for improving agriculture and inclusive development. Some of these opportunities include opportunities to reduce the cost of providing financial services, opportunities to innovate and provide customized financial services, opportunities to improve and expand access of firms and individuals to finance, opportunities to increase formalization and financial inclusion, opportunities for more efficient management of collateral, resulting in more firms accessing credit, and opportunities to increase capital market efficiency, freeing up intermediaries' capital for lending. These potentials notwithstanding, some challenges hinder the creation and widespread use of digital financial services and products, especially among small-scale farmers and small and medium enterprises. These challenges include security, regulatory concerns, lack of access to good-quality and affordable digital connectivity (broadband access), disruption of existing business models, and limited product offerings.

Security concerns: Some digital finance users are concerned about their fraud and password theft exposure. Fraudsters are cloning e-commerce platforms, and hijacking emails, social media platforms, and mobile money lines. Again, some Ponzi schemes have duped many unsuspecting customers of their hard-earned income using digital channels. These developments deter aspiring users of digital finance and raise doubts in the minds of users about the safety of digital finance. The onus lies on law enforcement agencies and regulatory authorities to clamp down on digital fraudsters and for innovators to enhance the security features of digital financial products and services.

Regulatory concerns: Regulation usually needs to catch up to digital innovation. Regulation cuts both ways. If regulations are too stringent, they may discourage innovation; if they are too lenient also, they may encourage fraud and inefficiencies. What is being advocated is risk-proportionate regulation—regulation that promotes innovation in a manner that balances the need for new products with the hazard associated with innovations. Collaboration between regulators and innovators will help regulatory bodies keep pace with innovations to prevent regulations from lagging too far behind innovation.

Lack of access to good-quality and affordable digital connectivity: The infrastructure to undergird digital finance is often fragile and, in most cases, absent in rural areas. Telecommunications networks, broadband access, and other digital infrastructure required to drive financial digitization are mostly in short supply in developing countries. Where they are present, quality and affordability are not assured. Solving the digital infrastructure deficit would require huge commitments and investments from governments and development partners in hard-to-reach areas.

Disruption of existing business models: Digital finance is often loathed because of its disruptive nature. Digitalization requires adjustment to business models and processes. This disruptive nature of digitalization can lead to the decline and collapse of some hard-to-adjust businesses. The disruption also means some old job avenues are closed while new ones are opened. The disruptive nature of digitalization is evident in the banking sector, where brick-and-mortar banking is fast giving way to branchless banking.

Limited product offerings: For smallholder farmers, only few digital products have been customized to meet their peculiar needs.

7.6 Conclusion

This chapter discussed how financial digitalization could be harnessed to support agriculture and inclusive development. This comes on the back of the need to activate the potential of the financial system to help achieve the sustainable development goals (SDGs). The upsurge in digital finance has been driven by trends such as the emergence of new-age technology (such as AI, IoT, RPA, Big Data, and blockchain), mobile money service, fintech, open banking, interactive modes of banking, and changing taste of customers. These trends have led to the developing of digital financial products and services for agriculture. These include digital transactions products, digital savings, credit and insurance, crowdfunding, and e-commerce. The discussion revealed that though there have been some mixed results, the evidence supports the positive impact of digital finance on agriculture and inclusive development. Digital finance has been shown to have a positive influence on output, productivity, consumption, and financial inclusion while having a dampening impact on poverty and inequality. The potential for digital finance to reduce the gender finance gap and empower women has also been established.

Even with the impact of digital finance, challenges such as security concerns, regulatory fears, lack of access to good-quality and affordable digital connectivity (broadband access), disruption of existing business models, and limited product offerings prevent widespread adoption and use. Addressing security fears customers raise regarding exposure to fraud will go a long way to help push the digital agenda among farmers. The burden is on law enforcement agencies and regulatory authorities to clamp down on digital fraudsters and for innovators to enhance the security features of digital financial products and services. Inculcation of change management in agribusiness and small- and medium-sized enterprises (SMEs) will help them navigate the disruptive nature of digital finance and technology. Smallholder farmers are still searching for appropriate digital products that suit their seasonal needs. This is a challenge that innovators and the financial sector must overcome. However, in doing so, innovators must be aware of the critical factors that decide the success or failure of digital innovations in agriculture. These success factors include complexity-related factors, design factors, operational factors, growth and innovation factors, and connectivity.

References

Abor, J. Y., Amidu, M., & Issahaku, H. (2018). Mobile telephony, financial inclusion, and inclusive growth. *Journal of African Business, 19*(3), 430–453.

Amazon. (2021). *A complete guide on the advantages of e-commerce to business*. Retrieved September 16, 20201, from https://sell.amazon.in/seller-blog/advantages-of-ecommerce

Andersson-Manjang, S. K., & Naghavi, N. (2021). *State of the industry report on mobile money 2021*. GSMA.

Asongu, S. (2015). The impact of mobile phone penetration on African inequality. *International Journal of Social Economics, 42*(8), 706–716.

Bongomin, G. O. C., Ntayi, J. M., Munene, J. C., & Malinga, C. A. (2018). Mobile money and financial inclusion in sub-Saharan Africa: The moderating role of social networks. *Journal of African Business*. https://doi.org/10.1080/15228916.2017.1416214

Brune, L., Giné, X., Goldberg, J., & Yang, D. (2016). Facilitating Savings for Agriculture: Field experimental evidence from Malawi. *Economic Development and Cultural Change, 64*(2), 187–220. (University of Chicago Press).

Bryan, G. (2010). *Ambiguity and insurance*. Unpublished manuscript.

Capital Bank Solutions and IBS Intelligence. (2021). *Emerging trends in digital banking*. Retrieved September 16, 2021, from https//www.capital--banking.com.

Casaburi, L., & Wills, J. (2018). Time vs. state in insurance: Experimental evidence from contract farming in Kenya. *American Economic Review, 108*(12), 3778–3813.

Deloitte. (2018). *Crunch time V finance 2025*. Deloitte Development LLC.

Gosavi, A. (2017). Can Mobile money help firms mitigate the problem of access to finance in Eastern sub-Saharan Africa? *Journal of African Business*. https://doi.org/10.1080/15228916.2017.1396791

Hinson, R., Lensink, R., & Mueller, A. (2019). Transforming agribusiness in developing countries: SDGs and the role of FinTech. *Current Opinion in Environmental Sustainability, 41*, 1–9.

IFC. (2018). *Digital financial services for agriculture*. International Finance Corporation, World Bank Group.

Investopedia. (2020). *Open banking*. Retrieved September 16, 2021, from https://www.investopedia.com/terms/o/open-banking.asp

Issahaku, H., Abu, B. M., & Nkegbe, P. K. (2018). Does the use of mobile phones by smallholder maize farmers affect productivity in Ghana? *Journal of African Business, 19*(3), 302–322.

Kikulwe, E. M., Fischer, E., & Qaim, M. (2014). Mobile money, smallholder farmers, and household welfare in Kenya. *PLoS One, 9*(10), e109804. https://doi.org/10.1371/journal.pone.0109804

Kirui, O. K., Okello, J. J., Nyikal, R. A., & Njiraini, G. W. (2013). Impact of Mobile phone-based money transfer Services in Agriculture: Evidence from Kenya. *Quarterly Journal of International Agriculture, 52*(2), 141–162.

Matsumoto, T., Yamano, T., & Sserunkuuma, D. (2013). *Technology adoption in agriculture: Evidence from experimental intervention in maize production in Uganda*. https://doi.org/10.1007/978-94-007-5760-8_12.

McIntosh, C., & Mansini, C. S. (2018). *The use of financial technology in the agriculture sector* (ADBI working paper, no. 872). Asian Development Bank Institute (ADBI).

Miller, H. (2019). What works for digital financial services in agricultural development?. USAID.

Mixon, J. (2021). *The role of robotic process automation (RPA) in finance automation*. Gartner Advisory. Retrieved September 16, 2021, from: https://www.gartner.com/en/finance/insights/robotics-in-finance

Nzie, J. R. M., Bidogeza, J. C., & Ngum, N. A. (2017). Mobile phone use, transaction costs, AND price: Evidence FROM rural vegetable farmers IN Cameroon. *Journal of African Business*. https://doi.org/10.1080/15228916.2017.1405704

Ojo, A., Janowski, T., & Awotwi, J. (2012). Enabling development through governance and mobile technology. *Government Information Quarterly, 30*(2013), S32–S45.

Ondiege, P. (2013). *Fostering financial inclusion with mobile banking*. African Development Bank. Retrieved September 16, 2021, from http://www.proparco.fr/webdav/site/proparco/shared/PORTAILS/Secteur_prive_developpement/PDF/SPD16/SPD16_Peter_Ondiege_UK.pdf

Pareek A. (2020). *Internet of Things (IoT) in Banking: Examples of IoT solutions in finance*. Retrieved September 16, 2021, from https://customerthink.com/internet-of-things-iot-in-banking-examples-of-iot-solutions-in-finance/

Pilcher, J. (2021). *Branches in decline: last one out, turn off the lights*. Retrieved September 16, 2021 from https://thefinancialbrand.com/66228/bank-credit-union-branch-traffic/

Sastry, V. V. L. N. (2020). *Artificial intelligence in financial services and banking industry*. Idea Publishing.

Suri, T., & Jack, W. (2016). The long-run poverty and gender impacts of mobile money. *Science, 354*(6317), 1288–1292.

Tmob (2021). *Financial services: 7 emerging technology trends that will accelerate your digital transformation.* Retrieved September 16, 2021, from: https://thinksmobility.com/7-emerging-technology-trends-that-will-accelerate-your-digital-transformation/

United Nations Development Programme and Oxford Poverty and Human Development Initiative. (2020). *Global MPI 2020 – Charting pathways out of multidimensional poverty: Achieving the SDGs.* United Nations Development Programme and Oxford Poverty and Human Development Initiative. Retrieved September 16, 2021, from https://ophi.org.uk/global-mpi-report-2020/

World Bank. (2020). *Agriculture and food.* Retrieved September 4, 2021, from https://www.worldbank.org/en/topic/agriculture/overview

8 Financial Literacy, Economic Inequality, and Trust in Government

Alina R. Oxendine

Abstract

United Nations (UN) Sustainable Development Goals associated with economic empowerment, financial inclusion, and reducing inequalities are fundamental and timely. This chapter analyzes the political implications of attitudes related to financial literacy, especially as they connect to the global challenge of growing economic inequality. Existing public opinion research suggests that exposure to high inequality influences people's attitudes toward politics. This discussion analyzes the relationship between inequality and trust in government and how financial literacy shapes these political beliefs.

Data were collected from approximately 1200 adults living in the United States using Amazon Mechanical Turk (MTurk) during January and February 2021. The survey instrument measures people's perceptions of their finances, financial literacy, trust in government, political efficacy, views about economic inequality, and demographic backgrounds. Results suggest that financial literacy strongly and positively connects to attitudes widely considered healthy in democratic contexts, including confidence in government and political efficacy. Financial literacy also moderates economic inequality's influence on people's trust in government. In addition to evaluating connections between inequality, trust, and financial attitudes, this discussion explores the potential long-term consequences of sustainability initiatives and other public policies that address economic injustices.

Keywords

Financial literacy · Economic inequality · Political psychology · Trust in government

The United Nations Sustainable Development sub-goal 8.10 seeks economic empowerment by increasing "…access to banking, insurance, and financial services for all." Even if these economic tools become more readily available, people must have confidence in these financial opportunities and be literate enough to sort through beneficial versus exploitative prospects. Vital to financial inclusion, this chapter analyzes the political implications of financial literacy in the contemporary context of rising income inequality and distrust of major political and economic institu-

This research is supported by a 2020–21 Research Grant from the American Political Science Association's Centennial Center for Public Affairs and a 2021–2 Hanna Faculty Development Grant from the College of Liberal Arts at Hamline University.

A. R. Oxendine (✉)
Department of Political Science, Hamline University, St. Paul, MN, USA
e-mail: aoxendine01@hamline.edu

G. W. Muschert et al. (eds.), *Financial Inclusion*, Sustainable Development Goals Series,
https://doi.org/10.1007/978-3-031-68803-4_8

tions. To what extent do personal attitudes about economic empowerment matter, especially regarding confidence in government and political participation? Do attitudes about money, particularly financial literacy, shape political beliefs and have political implications? Exploring connections between inequality, trust, and financial attitudes helps us understand intersections between different but related sustainability initiatives and public policies that address economic injustices. Through analysis of survey data, this project explores the global and national-level dynamics of these broader issues while also focusing on data specific to the United States, which has exceptionally high levels of economic inequality, low levels of political trust, and also struggled to democratize financial literacy (Council for Economic Education, 2020; Pew Research Center, 2017).

Just as many countries struggle with widening financial literacy and inclusion, many are also experiencing unprecedented levels of economic inequality. Research finds that income inequality creates and perpetuates several societal challenges, including weakened democracy, sharp political polarization, low social capital, and stunted economic growth (Dabla-Norris et al., 2015; McCarty et al., 2006; Muller & Seligson, 1994).

Concerns about economic justice led to United Nations Sustainable Development Goal 10, which seeks to reduce inequality within and between countries. Like financial inclusion, addressing the challenge of widening economic inequality is imperative (United Nations, 2020).

Scholars must also evaluate financial literacy as another fundamental challenge of rising economic inequality. Familiarity and aptitude tend to be higher among the wealthy, so when economic inequality is high, disparities pose a formidable obstacle in democratizing financial knowledge (Council for Economic Education, 2020; OECD, 2014). Moreover, if inequality erodes people's confidence in major political and economic institutions, this distrust could influence their willingness to integrate basic strategies, like saving and investing, into their financial decision-making.

8.1 The Political Relevance of Financial Literacy

An effective way to conceptualize financial literacy is to evaluate whether an individual (1) understands how to make effective financial decisions and (2) applies that knowledge to their own life. In short, financial literacy should capture "how well an individual can understand and use personal finance-related information" (Huston, 2010, p. 307). Levels of financial literacy worldwide vary tremendously, and Nordic countries lead, with over 70% of their populace adept in handling their finances. The United States tends to fall in the middle of international rankings, with about 57% of adults exhibiting literacy (Klapper et al., 2016). Even though these levels are reasonably high, financial literacy is unevenly distributed in the United States, with education, knowledge, and aptitude higher for affluent Americans. For instance, for schools with over 75% of students eligible for free/reduced lunch, only about 3% of students are required to take a personal finance course. In contrast, the rate is three times higher in schools with less than 25% eligible for free/reduced lunch (Council for Economic Education Report, 2020).

To what extent might these personal views about money management be related to political attitudes about government? Political scientists have studied many fascinating connections between economic and political perceptions. Scholars have found that, compared to one's economic situation, sociotropic concerns are even more important in shaping political attitudes and voter behavior (Kinder & Kiewiet, 1981; Mutz & Mondak, 1997). Researchers have also explored how people's perceptions of their relative economic standing affect necessary attitudes, like anxiety about money, willingness to support social welfare spending, and political engagement (Condon & Wichowsky, 2020; Kraus et al., 2017). Despite this progress in analyzing economic and political intersections, our understanding of financial literacy and its political relevance still needs to be improved. Financial literacy could provide confidence and empowerment relevant to how people view large-scale economic and political forces, particularly with heightened economic uncertainty associated with US free market capitalism.

Survey research suggests that Americans have lost confidence in major institutions, particularly the public sector, and their ability to address societal problems. They are incredibly conflicted about government spending when faced with taxation, partisanship, and politics (Bartels, 2008). Citizens disagree considerably about whether public or private entities should lead the way in solving many collective problems (Pew Research Center, 2015). Recent surveys reveal that only 20% of Americans "trust the government in Washington to do what is right" almost always or most of the time. These numbers were closer to three-quarters of those surveyed during the 1960s and 1970s (Pew Research Center, 2017). This project analyzes trust in government (also termed political trust) and a similar concept of political "internal" efficacy, which is citizens' confidence in their abilities to understand and influence public affairs. Collective political internal efficacy has been more consistent over time than political trust. Most scholars acknowledge that having a citizenry high in both trust in government and political efficacy is helpful for democracy, but these two concepts are not necessarily strongly correlated (Morrell, 2003, 2005). The following section analyzes how and why these political attitudes, efficacy, and trust in government, may be related to—and relevant for—financial literacy.

8.2 Financial Literacy May Inspire Higher Trust

There are a few reasons why higher financial literacy might connect to stronger feelings of political efficacy and trust in government. These attitudes all require a healthy level of confidence in major economic and political institutions. Researchers find that trust in financial institutions is vital for investing in the stock market (Guiso et al., 2008) and for retirement planning decisions like investing in a 401(k) plan or entering a private pension scheme (Agnew et al., 2012; Ricci & Caratelli, 2017). Also, scholars find that political trust can significantly affect people's financial lives. For instance, trust in government boosted financial well-being, especially during COVID-19 (Barrafrem et al., 2020).

Research on locus of control may also lend insight and connect these economic and political tendencies. Economists contend that having an internal locus of control, believing one has control over one's life, is an essential predictor of financial decision-making like saving, budgeting, and planning (Perry & Morris, 2005; Rotter, 1966). Similarly, political scientists have long connected internal locus of control to higher political efficacy and engagement (Deutchman, 1985; Sigel, 1975). Financial literacy, political efficacy, and trust in government share confidence in one's ability and an appreciation of external forces' importance. This nexus is similar to what researchers call a blended locus of control or "bi-local expectancy," associated with higher overall health and well-being (April et al., 2012).

Literacy May Moderate How Inequality Affects Trust in Government.

Over the past few decades, researchers have disagreed about how economic inequality shapes people's perceptions of government. Some scholars find that rising inequality encourages people to support the government in getting involved to "fix" the problem (Hayes, 2014; Meltzer & Richard, 1981). These studies find that inequality does not necessarily cause citizens to recoil from the government, and, instead, they may support intervention to correct the imbalance. Nevertheless, another body of research uncovers the opposite dynamic, finding that disparities can reinforce themselves rather than self-correct (Benabou, 2000; Kelly & Enns, 2010). Similarly, studies have found that inequality harms confidence in government alongside a loss of trust in fellow citizens (Rahn & Rudolph, 2005; Stephany, 2017; Uslaner & Brown, 2005). This project analyzes the relationship between financial literacy and political attitudes and explores whether literacy moderates the relationship between inequality and people's perceptions of government.

This project's survey data dive into this debate, exploring how financial literacy affects the relationship between people's perceptions of economic inequality and its influence on political trust. Even though empirical research on this dynamic is limited, financial literacy might encourage more of a self-correcting dynamic regarding economic inequality. Higher levels of financial literacy might ease economic anxiety from living in a community with extreme income disparities, reducing negativity toward the government. Upcoming sections evaluate whether this expectation is correct and explore broader connections between financial literacy, political efficacy, and confidence in government.

8.3 Methods

This article analyzes a survey of 1050 Amazon Mechanical Turk (MTurk) workers, examining respondents' demographics, beliefs about government, political attitudes, and views compatible with financial literacy. The survey data collection occurred during January and February of 2021, and participants were limited to adults 18 or over living in the United States. Although MTurk samples do not represent the general population as well as national probability samples, they are preferable to university students or convenience samples and tend to perform well on attention checks (Berinsky et al., 2012; Hauser & Schwarz, 2016; Paolacci & Chandler, 2014).

Respondents resided in 49 states and reported ages ranging from 20 to 79. The gender distribution skewed slightly toward men, with 46% female and 54% male. About 72% of respondents were White/Caucasian, and 28% were Black, Hispanic, Asian, American Indian, or some other race. Respondents represented a range of socioeconomic backgrounds, with 37% reporting household incomes under \$50,000 and 63% reporting incomes equal to or greater than \$50,000. Education varied as well, with about 18% reporting graduate degrees, 49% reporting Bachelor's degrees, and 33% reporting an Associate's degree or less.

Analyses utilize multiple linear regression to evaluate respondents' financial literacy and trust in government. Models include several independent variables measuring perceptions of inequality, political attitudes, household income, reported net worth, and relevant demographic variables. Basic descriptive statistics appear in Table 8.1.

The questionnaire asked respondents to share their attitudes toward and strategies for managing money. Respondents agreed/disagreed with statements like "I am much more of a saver than a spender" and "I try to maintain a standard of living below my income level." The items focus on basic concepts like saving and investing and capture respondents' knowledge and application of personal finance strategies that tend to build long-term wealth (Huston, 2010). The results section combines these items into a composite *Financial Literacy Index.* This index is an average of 6 different items, has an alpha of .73, and ranges from 1 (low literacy) to 4 (high literacy).

The questionnaire also analyzes political attitudes, borrowing similar question wordings as the well-established General Social Survey (GSS) run by NORC at the University of Chicago. Measures evaluate *Political Ideology* using a scale of 1 (very liberal) to 5 (very conservative) and *Party Identification* using a scale of 1 (strong Republican) to 5 (strong Democrat). *Trust in Government* is evaluated using the item "What percent of the time do you think you can trust the federal government in Washington to do what is best for the country?" Respondents report that they trust the government an average of 45% of the time. Another trust measure evaluates *Interpersonal Trust* and asks respondents whether they can or cannot trust others with the item "Generally speaking, would you say that most people can be trusted or that you can't be too careful in dealing with people?" *Political Internal Efficacy* is measured with a two-item index asking whether the respondent has "a pretty good understanding of the important political issues facing [their] country." The index has an alpha of .69.

Models include independent variables measuring perceptions of inequality, reported income and net worth, and demographic characteristics.

Table 8.1 Descriptive statistics of survey variables

Variable description	Alpha (for indices)	Mean	Minimum	Maximum
Financial literacy index	.73	2.54	0 (low literacy)	4 (high literacy)
Political efficacy index	.69	4.43	0 (low efficacy)	6 (high efficacy)
Trust in Government		45.66	0 (never trust)	100 (always trust)
Interpersonal trust		1.10	0 (cannot trust)	2 (can trust)
Perceived inequality index	.74	62.56	0 (perceive inequality as low)	100 (perceive inequality as high)
Perceived diversity		51.00	0 (perceive diversity as low)	100 (perceive diversity as high)
Household income		3.20	0 (under $10,000)	9 ($200,000 +)
Household net worth		4.11	0 (under $0)	11 ($ten million +)
Political ideology		1.70	0 (very liberal)	4 (very conservative)
Party identification		2.30	0 (strong republican)	4 (strong democrat)
Children under 17		.89	0 (no kids)	5 (more than 4)
Age		41	20	79
Education (highest year of school completed)		3.53	0 (less than high school)	5 (graduate degree)
Race		.28	0 (White)	1 (Black, Asian, Native American, Native Hawaiian, or Biracial)
Gender		.46	0 (male)	1 (female)

The *Perceived Inequality Index* is an average of two items, with an alpha of .74, evaluating each respondent's perceptions of inequality (1) in their city/town and (2) in the entire nation. The items appear: "Think about how income is distributed in the United States/your city or town. On a scale from 0 to 100, would you say that inequality in this country is very low, very high, or somewhere in between?" Because economic inequality and racial diversity are often related, another measure, *Perceived Diversity*, evaluates respondents' perceptions of racial diversity with the item "Think about racial diversity in your city or town. On a scale from 0 to 100, would you say that racial diversity where you live is very low, very high, or somewhere in between?"

Survey items ask respondents to share their *Household Income* with the question "How much total combined money did all members of your household earn last year?" and *Household Net Worth* with the question "Please choose the answer that best represents your household's approximate net worth. Net worth is all of your household's assets (e.g., money saved, home equity, or retirement investments) minus liabilities (e.g., home, student, or credit card debt)." The remaining independent variables include demographic measures of race, gender, age, education, and number of kids under 18. Again, descriptive information about these variables is listed below. Before launching the final survey, the project received Institutional Review Board (IRB) approval, and a pilot study of the complete questionnaire was conducted using university students.

8.4 Results and Discussion

It is helpful to begin by evaluating the intersection of financial literacy and political trust. Based on the earlier discussion, higher literacy levels could be associated with more substantial confidence in government. Figure 8.1 presents a bar graph showing "low financial literacy" respondents who score below average on the literacy index, compared to "high financial literacy" respondents who score above average. About 40% of the low scorers trust the government more than half the time, compared to about 60%

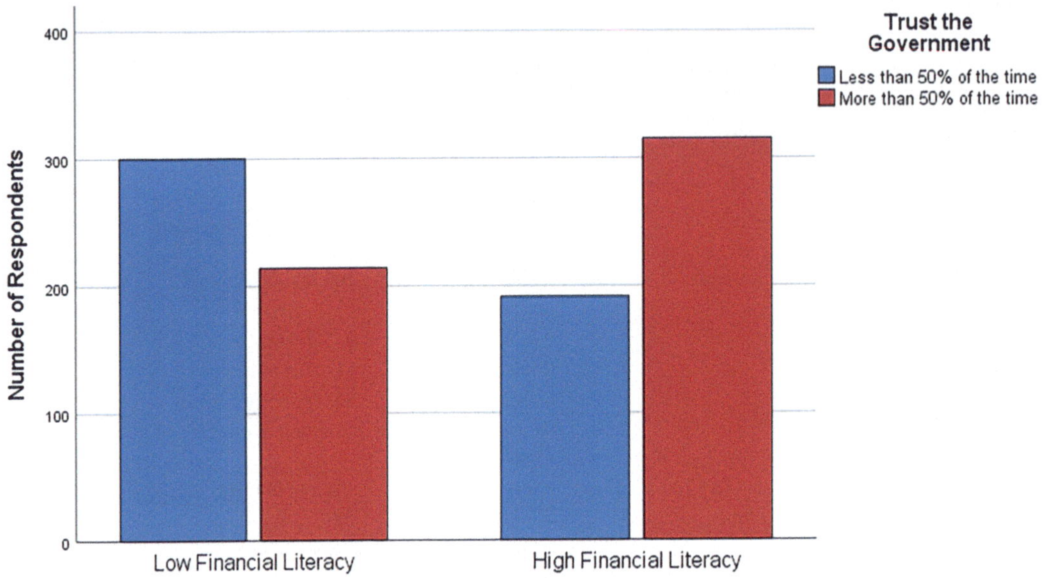

Fig. 8.1 Trust government by financial literacy

of high scorers. This helpful visual provides evidence that literacy and trust in government are related.

Table 8.2 below includes results from multiple linear regression analyses exploring the impact of several individual-level variables on respondents' levels of financial literacy and trust in government. Models include several independent variables measuring perceptions of inequality, political attitudes, household income, reported net worth, and relevant demographic variables like gender, age, and race. For each model, the table lists unstandardized regression coefficients and standard errors.

Table 8.2's first model evaluates financial literacy and reveals several intriguing findings. First, it suggests compelling connections between financial and political attitudes. People who trust in government are more likely to report higher levels of financial literacy. Similarly, people with high political efficacy are also high in literacy. In a democratic context, political trust and efficacy are generally beneficial and supportive of a healthy democracy, just like financial literacy is advantageous in that it helps average citizens to build wealth. Also, these attitudes are similar in that they exhibit a firm vote of confidence in the broader economic and political system.

The first model in Table 8.2 also reveals that ideologically conservative Americans tend to have higher levels of financial literacy. However, there is no relationship between literacy and party identification. The ideological connection may stem from right-leaning US citizens, leaders, and parties emphasizing the rhetoric of fiscal conservatism and limited government spending (Pew Research Center, 2015). The coefficient for net worth is large, positive, and statistically significant, which is not surprising given that the literacy measure captures individual-level attitudes compatible with building wealth. Interestingly, income is inversely related to financial literacy, and education is not influential. The direction of this relationship suggests that, controlling for wealth, higher socioeconomic status does not predict higher literacy. Coursework in financial literacy is present but still needs improvement in American public schools and colleges, so it makes sense that educational level is not predictive (Council for Economic Education, 2020).

Finally, a few demographic variables are noteworthy in the first model. People with kids are

Table 8.2 Analyzing financial literacy and trust in government

	Model 1: financial literacy	Model 2: trust in government	Model 3: trust in government (with interaction term)
Perceived inequality	–	.058 (.043)	−.351* (.142)
Perceived diversity	–	.187*** (.031)	.179*** (.031)
Ideology (Conservative)	.065** (.019)	−.728 (.674)	−.840 (.672)
Party ID (Democrat)	.010 (.018)	1.274* (.622)	1.332* (620)
Government trust	.005*** (.001)	–	–
Political efficacy	.106*** (.017)	–	–
Interpersonal trust	.011 (.030)	7.743*** (1.000)	7.880*** (.997)
Financial literacy	–	6.138*** (1.110)	−3.651 (3.414)
Interaction: Inequality x financial literacy	–	–	.155** (.051)
Household income	−.027*(.015)	−1.017 (.533)	−1.089* (.531)
Household net worth	.072*** (.010)	.342 (.377)	.352 (.375)
Age	−.003 (.002)	−.230** (.066)	−.213** (.066)
Education	.030 (.020)	4.030*** (.683)	4.171*** (.681)
Race (Non-white)	.241*** (.050)	2.779 (1.773)	2.565 (1.766)
Gender (Female)	−.183*** (.044)	1.196 (1.558)	1.307 (1.551)
Kids under 18	.047* (.021)	4.465*** (.726)	4.340*** (.724)
Constant	1.432*** (.141)	−5.362 (5.514)	20.018* (10.014)
R squared	0.245	0.297	0.305
N	939	942	942

Table cells list unstandardized regression coefficients with standard errors in parentheses
$^{***}p < .001$, $^{**}p < .01$, $^{*}p < .05$

more likely to report higher levels of financial literacy, which may be because children pressure parents to plan more effectively for the future and become more financially responsible. When controlling for all aspects of wealth and socioeconomic status, people of color are considerably more likely than White Americans to have high literacy levels. This finding may come from necessity, in that marginalized groups have lacked financial advantages (including inherited wealth) over time in the United States, that it becomes more pivotal for them to behave in financially responsible ways and to develop these skills. Compared to men, women score lower in financial literacy. This is similar to previous studies and is probably a remnant of traditional gender roles that have long encouraged men to work outside the home and play a dominant role in earning and managing family resources (Paxton & Hughes, 2017).

Table 8.2's second model analyzes respondents' trust in government. The third model examines the same variables as Model 2 plus a term for the interaction between financial literacy and perceptions of inequality. In the second model, inequality does not significantly influence trust in government. However, it is interesting that perceptions of diversity are positively and strongly related to government confidence. This finding might reflect a self-selection effect in that people comfortable with (and who opt for) racial diversity in their surrounding community are more likely to trust the government. Party identification is also positive and influential, suggesting that Democrats are more likely than Republicans to trust the government. This makes sense, given that government skepticism is common among Republicans in the United States (Pew Research Center, 2015). Not surprisingly, people who are trusting of people generally are also trusting of the government. Moreover, similar to the previous model/discussion, people with higher levels of financial literacy also have more confidence in government. Finally, respondents who are younger, highly educated, and have children are all more trusting.

Although scholars are conflicted about how economic inequality shapes people's views of government, several studies find that inequality exerts a negative influence on confidence in government (Rahn & Rudolph, 2005; Stephany, 2017; Twenge et al., 2014). However, scholars have yet to analyze whether financial attitudes (like literacy) influence the relationship. The hypothesis was that respondents who are lower in literacy might feel more threatened by inequality's growth and, thus, exhibit lower trust in government. As predicted, the interaction term is significant in the model. With the term included, the "perceived inequality" coefficient reflects the influence of perceptions of inequality on trust in government when financial literacy equals zero. Because this coefficient is negative and statistically significant, we know that when literacy is incredibly low, people who perceive higher levels of inequality are less likely to trust the government. Also, when literacy levels become higher, the relationship reverses. Indeed, financial literacy moderates the relationship between inequality and trust.

In sum, models from Table 8.2 illustrate meaningful connections between financial and political attitudes, with financial literacy strongly and positively connected to both political trust and efficacy. Financial literacy also moderates the relationship between perceptions of economic inequality and confidence in government. People with higher literacy levels seem less likely to turn against the government when facing higher inequality. This finding suggests that literacy could be an important "buffer" that helps to create a self-correcting dynamic between inequality and attitudes about government. Given that government intervention may be crucial for reducing the wealth gap, and trust moderates the relationship between inequality and support for redistribution, it is beneficial to keep trust from dipping too low (Macdonald, 2020). If financial literacy is associated with stronger trust and more resilience in the context of economic challenges like inequality, it has significant political implications that need more exploration and attention.

8.5 Conclusions and Implications

Economic inequality has been a growing problem worldwide, leading to United Nations Sustainable Development Goal 10, aiming to "reduce inequality within and between countries." In addition, sub-goal 8.10 seeks to further democratize economic empowerment by increasing "…access to banking, insurance, and financial services for all." This chapter analyzes the political implications of attitudes about financial inclusion and literacy, especially as they connect to this formidable global challenge of economic inequality. This chapter investigates intersections between financial and political views by combining theory and findings from political science, economics, and related fields.

Results suggest that financial literacy is strongly and positively connected to attitudes widely considered healthy in democratic contexts, including confidence in government and political efficacy. Financial literacy also moderates economic inequality's influence on people's confidence in government. Those with lower literacy levels are more likely to turn against the government when challenged by inequality, which could encourage a reinforcing dynamic between inequality and opposition to government intervention. Given that government involvement may be crucial for reducing the wealth gap, political trust (especially as it is reinforced by financial literacy) may encourage more robust support for political leaders trying to enact redistributive policies. The clear finding is that financial literacy encourages trust and resilience in the context of widening economic inequality. In that case, it adds to existing arguments favoring democratizing financial education and widening access to this skillset. Moreover, it emphasizes the need for more research exploring connections between political and financial attitudes.

References

Agnew, J., Szykman, L., Utkus, S., & Young, J. (2012). Trust, plan knowledge, and 401(k) savings behavior. *Journal of Pension Economics and Finance, 11*(1), 1–20.

April, K., Dharani, B., & Peters, K. (2012). Impact of locus of control expectancy on level of well being. *Review of European Studies, 4*(2). https://doi.org/10.5539/res.v4n2p124

Barrafrem, K., Västfjäll, D., & Tinghög, G. (2020). Financial well-being, COVID-19, and the financial better-than-average effect. *Journal of Behavioral and Experimental Finance, 28*, 100410.

Bartels, L. (2008). *Unequal democracy: The political economy of a new gilded age*. Princeton University Press.

Benabou, R. (2000). Unequal societies: Income distribution and the social contract. *American Economic Review, 90*(1), 96–129.

Berinsky, A. J., Huber, G. A., & Lenz, G. S. (2012). Evaluating online labor Markets for Experimental Research: Amazon.com's mechanical Turk. *Political Analysis, 20*, 351–368.

Condon, M., & Wichowsky, A. (2020). *The economic other: Inequality in the American political imagination*. University of Chicago Press.

Council for Economic Education. (2020). *Survey of the states: Economic and personal finance education in our nation's schools*. Retrieved from https://www.councilforeconed.org/wp-content/uploads/2020/02/2020-Survey-of-the-States.pdf

Dabla-Norris, E., Kochhar, K., Suphaphiphat, N., Ricka, F., & Tsounta, E. (2015). *Causes and consequences of income inequality: A global perspective*. Retrieved from https://www.imf.org/external/pubs/ft/sdn/2015/sdn1513.pdf

Deutchman, I. E. (1985). Internal-external locus of control, power, and political participation. *Psychological Reports, 57*(3), 835–843.

Guiso, L., Sapienza, P., & Zingales, L. (2008). Trusting the stock market. *The Journal of Finance, 63*, 2557–2600.

Hauser, D. J., & Schwarz, N. (2016). Attentive Turkers: MTurk participants perform better on online attention checks than do subject pool participants. *Behavior Research Methods, 48*, 400–407.

Hayes, T. (2014). Do citizens link attitudes with preferences? Economic inequality and government spending in the "New Gilded Age". *Social Science Quarterly, 95*(2), 468–485.

Huston, S. (2010). Measuring financial literacy. *The Journal of Consumer Affairs, 44*(2), 296–316.

Kelly, N., & Enns, P. (2010). Inequality and the dynamics of public opinion: The self-reinforcing link between economic inequality and mass preferences. *American Journal of Political Science, 54*(4), 855–870.

Kinder, D., & Kiewiet, D. (1981). Sociotropic politics: The American case. *British Journal of Political Science, 11*(2), 129–161.

Klapper, L., Lusardi, A. & Oudheusden, P. (2016). Financial literacy around the world. . Retrieved from https://gflec.org/wp-content/uploads/2015/11/3313-Finlit_Report_FINAL-5.11.16.pdf?x28148

Kraus, M., Park, J., & Tan, J. (2017). Signs of social class: The experience of economic inequality in everyday life. *Perspectives on Psychological Science, 12*(3), 422–435.

Macdonald, D. (2020). Trust in government and the American public's responsiveness to rising inequality. *Political Research Quarterly, 73*(4), 790–804.

McCarty, N. M., Poole, K. T., & Rosenthal, H. (2006). *Polarized America: The dance of ideology and unequal riches*. MIT Press.

Meltzer, A., & Richard, S. (1981). A rational theory of the size of government. *Journal of Political Economy, 89*(5), 914–927.

Morrell, M. (2003). Survey and experimental evidence for a reliable and valid measure of internal political efficacy. *Public Opinion Quarterly, 67*, 589–602.

Morrell, M. (2005). Deliberation, democratic decision making, and internal political efficacy. *Political Behavior, 27*(1), 49–69.

Muller, E., & Seligson, M. (1994). Civic culture and democracy: The question of causal relationships. *American Political Science Review, 88*, 635–652.

Mutz, D., & Mondak, J. (1997). Dimensions of sociotropic behavior: Group-based judgments of fairness and well-being. *American Journal of Political Science, 41*(1), 284–308.

OECD. (2014). Relationship between financial literacy and student background. In *PISA 2012 results: Students and money (volume VI): Financial literacy skills for the 21st century*. OECD Publishing.

Paolacci, G., & Chandler, J. (2014). Inside the turk: Understanding mechanical turk as a participant pool. *Current Directions in Psychological Science, 23*, 184–188.

Paxton, P., & Hughes, M. (2017). *Women, politics, and power*. Sage Publications.

Perry, V. G., & Morris, M. D. (2005). Who is in control? The role of self-perception, knowledge, and income in explaining consumer financial behavior. *Journal of Consumer Affairs, 39*(2), 299–313.

Pew Research Center. (2015). *Beyond distrust: How Americans view their government*. Retrieved from http://www.people-press.org/2015/11/23/beyond-distrust-how-americans-view-their-government/

Pew Research Center. (2017). *Public trust in government: 1958–2017*. Retrieved from http://www.people-press.org/2017/12/14/public-trust-in-government-1958-2017/

Rahn, W., & Rudolph, T. (2005). A tale of political trust in American cities. *Public Opinion Quarterly, 69*(4), 530–560.

Ricci, O., & Caratelli, M. (2017). Financial literacy, trust, and retirement planning. *Journal of Pension Economics and Finance, 16*(1), 43–64.

Rotter, J. B. (1966). Generalized expectancies for internal versus external control of reinforcement.

Psychological Monographs: General and Applied, 80(1), 1–28.

Sigel, R. (1975). Psychological antecedents and political involvement: The utility of the concept of locus of control. *Social Science Quarterly, 56*(2), 315–323.

Stephany, F. (2017). Who are your joneses? Socio-specific income inequality and trust. *Social Indicators Research, 134*(3), 877–898.

Twenge, J., Campbell, W., & Carter, N. (2014). Declines in trust in others and confidence in institutions among American adults and late adolescents, 1972–2012. *Psychological Science, 25*(10), 1914–1923.

United Nations. (2020). *The sustainable development goals report*. Retrieved from https://unstats.un.org/sdgs/report/2020/The-Sustainable-Development-Goals-Report-2020.pdf

Uslaner, E. M., & Brown, M. (2005). Inequality, trust, and civic engagement. *American Politics Research, 33*(6), 868–894.

9 Financial Literacy, Access to Microfinance, and Sustainable Development

Morshadul Hasan and Ariful Hoque

Abstract

This study aims to explore the impact of financial knowledge on getting microfinance access. An experimental methodology is employed, particularly probabilistic models, such as Logit and Probit, using data from a primary survey. The results of this study confirm that knowledge of savings and loans, savings and loan interest rates, installment payments, and security deposits have positive and significant effects on access to microfinance. Access to microfinance promotes *Sustainable Development* through reducing poverty (SDG 1), achieving zero hunger (SDG 2), promoting gender equality (SDG 5), creating decent employment, promoting economic growth, and reducing income inequality (SDG 10).

This book chapter is one of the sections of *Morshadul Hasan's* Master's thesis (Nanjing Audit University, Jiangsu, China). The thesis title is "*Does financial literacy influence on inclusive finance? Evidence from Bangladesh.*"

M. Hasan (✉) · A. Hoque
Murdoch Business School, Murdoch University, Perth, WA, Australia
e-mail: m.hasan@murdoch.edu.au; a.hoque@murdoch.edu.au

Keywords

Financial literacy · Microfinance · Financial knowledge · Sustainable development · Access to finance · Financial inclusion

9.1 Introduction

Microfinance is one of the most effective ways of financial access after banking access. It is an effective and powerful tool for poverty reduction. In addition, it provides financial services for people experiencing poverty, including savings, transfers, insurance, and credit (Cull et al., 2009; Morduch, 1999). Microfinance is a versatile and essential element of inclusive finance and rural development, playing a crucial role in achieving financial inclusion and fostering economic growth (Hasan et al., 2020a, b). Its primary objective is to alleviate poverty, improve social and financial efficiency, reduce income inequality, and promote an all-around prosperous society to contribute United Nations' Sustainable Development Goals (Asian Development Bank, 2000; Chauhan, 2021; Khan et al., 2021; OECD, 2007; Subramaniam et al., 2021).

In this twenty-first century, 43% of the world's population still lives below the poverty line ($5.50 a day).[1] That indicates the necessity of the

[1] https://www.worldbank.org/en/topic/poverty/overview

G. W. Muschert et al. (eds.), *Financial Inclusion*, Sustainable Development Goals Series,
https://doi.org/10.1007/978-3-031-68803-4_9

development of microfinance and, most importantly, sound microfinance strategy as one of the most powerful tools to drive economic development. Therefore, the demand for model and technology-based microfinance is everywhere, most importantly in the least developed countries. By offering savings, insurance, and payment services to underserved communities, microfinance promotes financial inclusion and supports the credit markets of self-employed individuals. This effect is significant in rural areas, where microfinance provides a vital means of accessing financial services (Cull & Morduch, 2018). Financial literacy is also considered a strong force in promoting access to microfinance. Nawaz (2015) examined the relationship between financial literacy and women's empowerment, noting that women with a high level of financial competence can achieve greater socioeconomic empowerment. Many microfinance agencies and NGOs offer training programs to help women develop these skills, including how to keep accounts, utilize money effectively, advise their husbands, and inform other family members about economic activities. By improving their financial literacy, women are better equipped to use the money for productive purposes and take control of their family's overall financial situation. Hasan et al. (2021, 2022) strongly recommend that financial training be a mandatory component of all microfinance programs. Financial education is important to be literate in financial matters. However, it improves people's understanding of different financial products and concepts through different instructions, information, and advice to develop the skills to make more aware of financial risks and opportunities (Berry et al., 2018; Fan & Chatterjee, 2018; Frisancho, 2019; Opletalová, 2015; Sun et al., 2020; Urban et al., 2018).

The significance of financial knowledge is emerging for financial development. Financial literacy has always been necessary, especially the saliency and relevance of financial education regarding financial products, services, and activities (Fernandes et al., 2014; Sun et al., 2020). However, access is still an important issue in rural finance in developing countries, especially those with overpopulation (Le et al., 2019; O'Toole & Newman, 2017). Some studies highly emphasized the importance of financial literacy from different perspectives (Hasan et al., 2021, 2022). However, financial literacy literature on microfinance access needs further exploration.

Nonetheless, the research based on financial literacy and modern microfinance access has high demand in Bangladesh and other countries, where financial inclusion is still an emerging issue. More significant work still needs to be done on how financial literacy impacts technology-based microfinance. To address these noted research gaps, this study specifies the objective of determining the impact of financial literacy on access to microfinance. After identifying this question and considering these scenarios, the researchers decided to measure the impact of financial literacy on access to microfinance. This study has high significance and practical significance.

9.2 Theoretical Framework

This study mainly highlights the knowledge of different factors of microfinance activities. The knowledge or capabilities of microfinance services is regarded here as the financial literacy consumers. The first term, *financial literacy*, generally refers to the typical education or knowledge about financial products and services. The theoretical structure of this research included seven hypotheses, including control variables. The theoretical framework appears in the following sections with the below-stated hypotheses.

H1: Knowledge of savings and loans positively influences access to microfinance.

H2: Knowledge of savings and loan interest rates positively influences access to microfinance.

H3: Knowledge of installment positively influences access to microfinance.

H4: Knowledge of security money positively influences access to microfinance.

H5: Knowledge of personal investment positively influences access to microfinance.

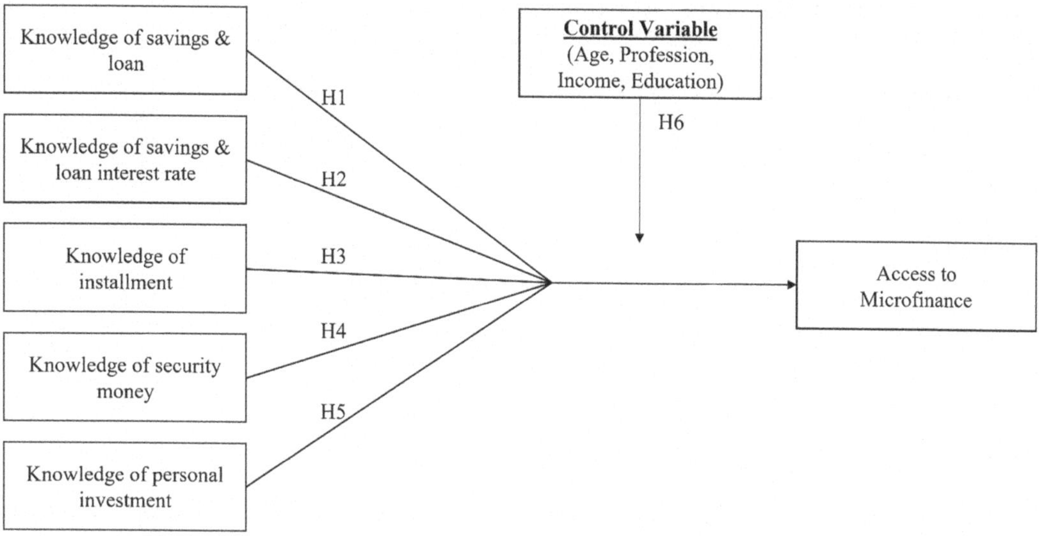

Fig. 9.1 The theoretical framework of this research

H6a: The age of the respondents positively influences access to microfinance.
H6b: The profession of the respondents negatively influences access to microfinance.
H6c: The income of the respondents negatively influences access to microfinance.
H6d: Education has a negative influence on access to microfinance (Fig. 9.1).

9.3 Methodology

Previous studies related to financial education and financial literacy (Bumcrot et al., 2013; Clark et al., 2017; Klapper et al., 2013; Lusardi, 2012, 2019) followed some fundamental questions relating to finance literature such as savings, interest rates, inflation, taxes, risks, discounts, stock market, and so on to ask the respondents. One example is, "Suppose you had 100 in a savings account, and the interest rate was 2% per year. After 5 years, how much do you think you would have in the account if you left the money to grow? Answer options: (i) more than 102, (ii) exactly 102, (iii) less than 102, (iv) don't know, (v) prefer not to say" (Bumcrot et al., 2013). The respondents were mainly educated or could answer those questions. However, those questions used in measuring financial literacy conflict with a survey where rural, uneducated people are the targeted respondents, as rural people with little income will rarely complete primary education. This study considers the level of knowledge regarding essential financial services such as savings, deposits, withdrawals, interest rates, etc. This study followed these questions to measure the respondent's knowledge of different financial services instead of core financial terms. Therefore, a complete survey-based empirical methodology estimates the expected results. The systematic process of the methods of this study is given sequentially in the following parts of this section.

9.3.1 Study Area, Data Collection, and Sample Distribution

This study follows two criteria for selecting the respondents for the survey. These respondents should be over 18 years old and have at least earning sources. That means an adult earning person can be a respondent to this survey. Based on these two criteria, the study area is Bangladesh's three main populous cities. These are Dhaka, the capital of Bangladesh; Gazipur

Table 9.1 Demographic profile of the respondents

Demographic profiles	%	Demographic profiles	%
Age		**Profession**	
Below 25 years	26.76	Business (self-employed)	40.49
26 to 30 years	29.58	Service	59.51
31 to 35 years,	21.13		
36–40 years,	11.27		
And 40 years plus	11.27		
Education level		**Income level**	
No education	0.70	Below 10,000	21.83
Primary level	26.76	10,000–20,000	53.52
Secondary education	41.55	20,000–30,000	14.79
Higher secondary	19.01	30,000–50,000	7.04
Graduation	11.97	50,000 plus	2.82

and Narayangonj are another two industrially developed cities. The random sampling process is followed here in the entire data collection process. The questionnaire was presented to targeted respondents randomly after a direct presence in some local residency areas, and the data collection period started from January 2019 to March 2020. Finally, this study uses 852 respondents' data to experiment with the impact of financial literacy. The demographic profile of the respondents is in the following Table 9.1. Data reliability is confirmed, as the value of Cronbach's Alpha of the overall model is 0.868, representing good reliability.

9.3.2 Econometrics Model

This study follows the methods from Hassan Al-Tamimi and Anood Bin Kalli (2009), Fernandes et al. (2014), Kiliyanni and Sivaraman (2018), Agyei (2018), Ouma et al. (2017), and Feng et al. (2019), who used a probabilities regression model to estimate the impact of financial literacy in different circumstances. This study follows probability distributions, and two commonly used models for binary dependent variables are Logit and Probit.

The algorithm of the Probit model appears below.

$$Pr(Y=1 \mid X) = \Phi(\beta_0 + \beta_1 X_1 + \beta_2 X_2 + \ldots\ldots\ldots\ldots\ldots + \beta_n X_n) \quad (9.1)$$

Another model that gives numerical results is very much like those from the Probit model. That model is computationally somewhat easier; it is the logistic or Logit model. The formal equation stands in the following algorithm.

$$Log(p) = Log(\frac{\text{p}}{\text{(1-p)}}) = \beta_0 + \beta_1 X_1 + \beta_2 X_2 + \ldots\ldots\ldots\ldots\ldots + \beta_n X_n \quad (9.2)$$

The term $Log(\frac{\text{p}}{\text{(1-p)}})$ is sometimes called the Logit function and has a natural interpretation as the logarithm of odds. The logistic model is widely used for binomial data in many statistical programs. The estimated model appears in the following sections.

$$\begin{aligned} Logit(MFAcc) = \beta_0 + \beta_1 * AGE + \beta_2 * PRO \\ + \beta_3 * INC + \beta_4 * EDU + \beta_5 * SL \\ + \beta_6 * SLIR + \beta_7 * INS + \beta_8 * SM \\ + \beta_9 * INV \end{aligned} \quad (9.3)$$

According to the above equation, the dependent variable is a microfinance or nonbanking account (MFAcc) held by respondents. The key independent variables are knowledge of savings and loans (SL), knowledge of savings and loan interest rates (SLIR), knowledge of installments (INS), knowledge of security money (SM), and knowledge of personal investment (INV). The control variables are the respondents' age (AGE), profes-

Table 9.2 Descriptive statistics

Variable	Obs	Mean	Std. Dev.	Min	Max
Microfinance accounts	852	0.451	0.498	0	1
Age	852	2.507	1.299	1	5
Profession	852	0.595	0.491	0	1
Income	852	2.155	0.937	1	5
Education	852	3.148	0.972	1	5
Savings and loan	852	2.359	1.190	1	5
Savings and loan interest rate	852	2.021	1.129	1	5
Instalment	852	2.493	1.289	1	5
Security money	852	2.169	1.187	1	5
Personal investment	852	2.254	1.292	1	5

sion (PRO), income (INC), and education (EDU). The dependent variable, the *microfinance account*, is a dummy variable with binary code. If the respondent has a nonbanking financial account, the response is yes; if not, the answer is no. The coding is binary (yes = 1 and no = 0). The other independent variables of this equation (Eq. 9.3), knowledge of savings and loans, knowledge of savings and loan interest rates, knowledge of installments, knowledge of security money, and knowledge of personal investment, are categorical. A rating scale with five different levels; 1 = Poor, 2 = Fair, 3 = Good, 4 = Very Good, 5 = Excellent.

9.4 Findings and Analysis

9.4.1 Descriptive Statistics

The findings and analysis of this study appear in four sections. First, the descriptive study is in Table 9.2, the empirical results in Table 9.3, the robustness experiment in Table 9.4, and the hypotheses overview in Table 9.5.

Bangladesh is the mother country for microfinance, as it has started its journey from here. According to the survey, only 45% of respondents have access to microfinance accounts. Therefore, most of the respondents lack access to microfinance. Usually, people managing a small business are more interested in microfinance than banking. Also, this study found that most microfinance account holders use services compared to businesses. Education influences microfinance access. Almost 27% of the microfinance account holders have at least primary education. However, most have secondary schooling experiences (42%). Since the respondents were selected based on income level criteria, it is worth noting that 21.83%, 53.52%, 14.79%, 7.04%, and 2.82% of microfinance accounts owners have an income level below 10,000 ($120), 10,000–20,000 ($120–$240), 20,000–30,000 ($240–$360), 30,000–50,000 ($360–$600), and 50,000 plus (more than $600), respectively.[2] Whoever can access microfinance has some knowledge about withdrawals and deposits from microfinance accounts. The mean value of knowledge regarding savings and loans is almost more than fair (2.359); however, knowledge regarding savings and loan interest rates is just fair (2.02). The respondents have good knowledge regarding installments. The mean value of knowledge regarding security money and personal investment has an excellent average mean value (mean value of 2.17 and 2.25, respectively).

9.4.2 Empirical Results

Based on the data, the regression results are analyzed based on two separate equations, Logit and Probit. Each model has two structures, such as section A and section B. The first part A of each model, consists only of the covariates relating to the dependent variable. In section B, all covariates have been included. After getting the result,

[2] Standard BDT to USD rate in December 2019.

Table 9.3 Regression result of financial literacy and microfinance access

	Logit		Probit	
Variable	A	B	A	B
Age		0.115		0.064
		(0.140)		(0.077)
Profession		−0.264		−0.167
		(0.332)		(0.185)
Income		−1.474***		−0.798***
		(0.233)		(0.127)
Education		−0.908***		−0.520***
		(0.221)		(0.119)
Savings & Loan	0.658***	0.477**	0.351***	0.260**
	(0.193)	(0.212)	(0.112)	(0.117)
Savings & Loan Interest Rate	0.089	0.820***	0.062	0.467***
	(0.186)	(0.228)	(0.100)	(0.119)
Instalment	1.243***	1.309***	0.748***	0.770***
	(0.171)	(0.201)	(0.101)	(0.117)
Security money	1.001***	1.329***	0.581***	0.724***
	(0.181)	(0.207)	(0.102)	(0.116)
Personal investment	−0.068	0.149	−0.037	0.079
	(0.155)	(0.181)	(0.080)	(0.093)

Logit model—Section A: Log likelihood = −233.12; LR $Chi^2(5)$ = 706.58; Prob > Chi^2 = 0.000; Pseudo R^2 = 0.603. *Logit model—Section B:* Log likelihood = −172.68; LR Chi^2 (9) = 827.47; Prob > Chi^2 = 0.000; Pseudo R^2 = 0.706. *Probit model—Section A:* Log likelihood = −228.471; LR Chi^2 (5) = 715.89; Prob > Chi^2 = 0.000; Pseudo R^2 = 0.610. *Probit model—Section B:* Log likelihood = −169.99; LR Chi^2 (9) = 832.48, Prob > Chi^2 = 0.000, Pseudo R^2 = 0.710

Table 9.4 Regression output of Clog-log model

	Clog-log	
Variables	Section A	Section B
Age		0.042
		(0.078)
Profession		−0.209
		(0.195)
Income		−0.823***
		(0.130)
Education		−0.606***
		(0.121)
Savings & Loan	0.426***	0.347***
	(0.138)	(0.121)
Savings & Loan interest rate	0.057	0.498***
	(0.095)	(0.114)
Instalment	1.030***	0.995***
	(0.121)	(0.135)
Security money	0.575***	0.680***
	(0.108)	(0.116)
Personal investment	−0.105	0.056
	(0.076)	(0.083)

Clog-log Section A: Log likelihood = −236.52; LR Chi^2 (5) = 699.79; Prob > Chi^2 = 0.000; *Clog-log Section B:* Log likelihood = −182.61903; LR Chi^2 (9) = 807.59, Prob > Chi^2 = 0.0000, The Log likelihood is −15.86; LR Chi^2 (9) = 359.23 and Prob > Chi^2 = 0.000

Table 9.5 Hypothesis investigation result

Hypothesis		Significance	Measure	Results
H1	Knowledge of savings and loan positively influences access to microfinance	Positive	Moderate	Accepted
H2	Knowledge of savings and loan interest rate positively influences access to microfinance	Positive	High	Accepted
H3	Knowledge of installment positively influences access to microfinance	Positive	High	Accepted
H4	Knowledge of security money positively influences access to microfinance	Positive	High	Accepted
H5	Knowledge of personal investment positively influences access to microfinance	Insignificant	–	Rejected
H6a	The age of the respondents positively influences access to microfinance	Insignificant	–	Rejected
H6b	The profession of the respondents negatively influences access to microfinance	Insignificant	–	Rejected
H6c	The income of the respondents negatively influences access to microfinance	Negative	High	Accepted
H6d	Education has a negative influence on access to microfinance	Negative	High	Accepted

the best one is accepted here to analyze the results for further analysis. However, according to Table 9.3, the log-likelihood value, pseudo R^2, and LR chi^2 value in section B of both models are highly acceptable. Therefore, this study recommends the output of section B for both models, Logit and Probit.

The age and profession are insignificant in getting microfinance access in both models. The third variable, the income level of the respondent, is negatively significant; the coefficient is −1.474 (Section B, Logit model) and −0.798 (Section B, Probit model). This output indicates that the higher the income level, the less the possibility of microfinance involvement. Usually, lower-income people prefer getting support from microfinance.

In general cases, as much as respondents have information or knowledge regarding deposits, withdrawals, savings, and loans, they highly prefer microfinance access. In this study, both models' knowledge regarding savings and loans is also moderately significant in section B. Coefficient is 0.477 in the Logit model and 0.260 in the Probit model. This output refers to the respondents with higher knowledge regarding savings and loans who prefer to get microfinance access. However, in section A for both models, knowledge regarding savings & loans is highly significant. Knowledge regarding savings & loan interest rates is highly significant in section B for both models. The coefficient is 0.820 in the Logit model and 0.467 in the Probit model. This output refers that respondents with higher knowledge regarding savings & loan interest rates prefer to get microfinance access. However, in section A for both models, knowledge regarding savings and loan interest rates is insignificant. Knowledge regarding installments is also highly significant in all sections for both models. The coefficient is 1.309 in the Logit model (Section B) and 0.770 in the Probit model (Section B). This output refers that respondents with higher knowledge regarding microfinance installment prefer to get microfinance access.

Knowledge regarding security money is one of the most critical issues in microfinance access. People usually care about collateral, which is one of the most influencing issues if taking a loan from a microfinance institution. The knowledge regarding security money positively influences getting microfinance access. The coefficient is 1.329 in the Logit model (Section B) and 0.724 in the Probit model (Section B). This output refers that respondents with higher knowledge regarding security money for a loan prefer to get microfinance access. Another one of the most critical issues is knowledge of personal investment. In this study, knowledge regarding personal investment is insignificant in getting microfinance access. However, it was assumed that people with better investment knowledge would prefer microfinance access.

9.4.3 Robustness Checking

Using both the Logit and Probit models is typical for analysis involving binary dependent variables. This study also experimented with another model, complementary log–log regression, to check the robustness. This model is an alternative to the stated two models. In the complementary log–log regressions in sections A and B, this study finds that the result of those Logit and Probit modes are acceptable for any decision-making and further experiments. Besides, the Clog-log model shows a similar result to the Probit model. Here, the Logit model shows a higher impact; however, Probit also shows the same significance but is very close to the clog-log model. Therefore, the result of the Probit model is recommended for any policy implications. The result of Clog-log regression appears in the following Table 9.4.

9.4.4 Hypothesis Experimentation Overview

This study experiments with nine hypotheses, including four subsections of Hypothesis 6 (H6a, H6b, H6c, H6d). The details of the outcome of the hypotheses are mentioned in the following Table 9.5.

9.5 Access to Microfinance and Sustainable Development

Knowledge regarding financial literacy promotes access to microfinance and access to finance, which also significantly promote sustainable development (Hasan et al., 2024).

- Access to microfinance provides financial services to low-income level individuals to support their businesses, which finally alleviates poverty *(SDG 1: No Poverty).*[3]
- Access to microfinance also helps rural farmers with supporting access to loans, agri-insurance, savings, and other ways that upsurge their productivity and level of income, as well as alleviate hunger *(SDG 2: Zero Hunger).*[4]
- Access to microfinance support woman by encouraging gender equality with many financial services, including rural credit, which increase woman's financial independence and efficient decision-making power *(SDG 5: Gender Equality).*[5]
- Access to microfinance promotes financial well-being with credit access to lower-income individuals, marginalized communities, and women to reduce income inequalities as well as social inequalities *(SDG 10: Reduced inequalities).*[6]

9.6 Discussion and Conclusion

Nowadays, financial literacy is one of the vital factors in promoting financial inclusion. Microfinance is the core and most important part of rural finance. Based on the concept, this study specified the purposes of this study and specified objectives. The study employed Logit and Probit models to clarify the likelihood of microfinance access. Usually, the Probit model and a complementary log–log regression model are the two standard models for the alternative to logistic regression. The microfinance model in Bangladesh has launched a pilot project for poverty alleviation. Credit poverty alleviation funds have been in place, and poor households have achieved remarkable results in poverty allevia-

[3] Please see (Imai et al., 2010; Khan et al., 2021; Mushtaq & Bruneau, 2019; Nawaz, 2010; Subramaniam et al., 2021; Tavanti, 2013; Weiss & Montgomery, 2005) for further reading.

[4] Please see (Kim et al., 2019) for further reading.

[5] Please see (Garikipati et al., 2017; Warnecke, 2015; Zulfiqar, 2017) for further reading.

[6] Please see (Hermes, 2014; Lacalle-Calderon et al., 2019; Mushtaq & Bruneau, 2019) for further reading.

tion. Since then, international multilateral and bilateral development organizations and nongovernmental organizations have adopted many microfinance poverty alleviation models worldwide and have expanded from rural poverty alleviation to urban assistance for laid-off and unemployed people in reemployment and entrepreneurship. Based on the hypotheses of this study, the result indicated that H1, H2, H3, H4, H6c, and H6d are accepted. As per the *p*-value of the regression output, this study rejects H5, which means that knowledge about personal investment does not impact microfinance access. The significance of this study's result is consistent with (Lyons & Kass-Hanna, 2019) and (Morgan & Trinh, 2019). These two studies experimented that people with higher financial knowledge are more likely to participate in active saving behaviors and are less likely to borrow money, especially from informal financial channels, and finally support the development of financial inclusion. Also, the result of this study gained support from two other studies, Hasan et al. (2021) and Hasan et al. (2022), which show that a higher degree of knowledge of different financial services promotes access to financial services.

Since this study has a high practical implication for rural finance and the overall economy, there is a big platform for new research in this area. There is also a platform for measuring the financial knowledge of rural people and comparatively high-income educated people. In addition, there is a need to create an index for financial literacy, especially in underdeveloped and developing countries. In the future, new research must persist in combining data from other underdeveloped and developing countries, where financial inclusion is still an emerging issue.

References

Agyei, S. K. (2018). Culture, financial literacy, and SME performance in Ghana. *Cogent Economics and Finance, 6*(1). https://doi.org/10.1080/23322039.2018.1463813

Asian Development Bank. (2000). *Finance for the poor: Microfinance development strategy 2000 finance for the poor: Microfinance development strategy*. Asian Development Bank.

Berry, J., Karlan, D., & Pradhan, M. (2018). The impact of financial education for youth in Ghana. *World Development, 102*, 71–89. https://doi.org/10.1016/j.worlddev.2017.09.011

Bumcrot, C., Lin, J., & Lusardi, A. (2013). The geography of financial literacy. *Numeracy, 6*(2). https://doi.org/10.5038/1936-4660.6.2.2

Chauhan, S. (2021). Social and financial efficiency: A study of Indian microfinance institutions. *IIM Kozhikode Society and Management Review, 10*(1), 31–43. https://doi.org/10.1177/2277975220953311

Clark, R., Lusardi, A., & Mitchell, O. S. (2017). Financial knowledge and 401(k) investment performance: A case study. *Journal of Pension Economics and Finance, 16*(3), 324–347. https://doi.org/10.1017/S1474747215000384

Cull, R., & Morduch, J. (2018). Microfinance and economic development. In *Handbook of finance and development* (pp. 550–572). https://doi.org/10.4337/9781785360510.00030

Cull, R., Demirgüç-Kunt, A., & Morduch, J. (2009). Microfinance meets the market. *Journal of Economic Perspectives, 23*(1), 167–192. https://doi.org/10.1257/jep.23.1.167

Fan, L., & Chatterjee, S. (2018). Application of situational stimuli for examining the effectiveness of financial education: A behavioral finance perspective. *Journal of Behavioral and Experimental Finance, 17*, 68–75. https://doi.org/10.1016/j.jbef.2017.12.009

Feng, X., Lu, B., Song, X., & Ma, S. (2019). Financial literacy and household finances: A Bayesian two-part latent variable modeling approach. *Journal of Empirical Finance, 51*(February), 119–137. https://doi.org/10.1016/j.jempfin.2019.02.002

Fernandes, D., Lynch, J. G., & Netemeyer, R. G. (2014). The effect of financial literacy on downstream financial behaviors. *Management Science, 60*, 1861. https://doi.org/10.1287/mnsc.2013.1849

Frisancho, V. (2019). The impact of financial education for youth. *Economics of Education Review*. https://doi.org/10.1016/j.econedurev.2019.101918

Garikipati, S., Johnson, S., Guérin, I., & Szafarz, A. (2017). Microfinance and gender: Issues, challenges and the road ahead. *Journal of Development Studies, 53*(5), 641–648. https://doi.org/10.1080/00220388.2016.1205736

Hasan, M. M., Yajuan, L., & Khan, S. (2020a). Promoting China's inclusive finance through digital financial services. *Global Business Review*, 1–23. https://doi.org/10.1177/097215091989534

Hasan, M. M., Yajuan, L., & Mahmud, A. (2020b). Regional development of China's inclusive finance through financial technology. *SAGE Open, 10*(1), 215824401990125. https://doi.org/10.1177/2158244019901252

Hasan, M., Le, T., & Hoque, A. (2021). How does financial literacy impact on inclusive finance?

Financial. Innovations, 7(1). https://doi.org/10.1186/s40854-021-00259-9

Hasan, M., Noor, T., Gao, J., Usman, M., & Abedin, M. Z. (2022). Rural consumers' Financial literacy and access to FinTech services. *Journal of the Knowledge Economy*. https://doi.org/10.1007/s13132-022-00936-9

Hasan, M., Hoque, A., Abedin, M. Z., & Gasbarro, D. (2024). FinTech and sustainable development: A systematic thematic analysis using human-and machine-generated processing. *International Review of Financial Analysis, 95*, 103473.

Hassan Al-Tamimi, H. A., & Anood Bin Kalli, A. (2009). Financial literacy and investment decisions of UAE investors. *Journal of Risk Finance, 10*(5), 500–516. https://doi.org/10.1108/15265940911001402

Hermes, N. (2014). Does microfinance affect income inequality? *Applied Economics, 46*(9), 1021–1034. https://doi.org/10.1080/00036846.2013.864039

Imai, K. S., Arun, T., & Annim, S. K. (2010). Microfinance and household poverty reduction: New evidence from India. *World Development, 38*(12), 1760–1774. https://doi.org/10.1016/j.worlddev.2010.04.006

Khan, A. A., Khan, S. U., Fahad, S., Ali, M. A. S., Khan, A., & Luo, J. (2021). Microfinance and poverty reduction: New evidence from Pakistan. *International Journal of Finance and Economics, 26*(3), 4723–4733. https://doi.org/10.1002/ijfe.2038

Kiliyanni, A. L., & Sivaraman, S. (2018). A predictive model for financial literacy among the educated youth in Kerala, India. *Journal of Social Service Research, 44*(4), 537–547. https://doi.org/10.1080/01488376.2018.1477699

Kim, Y., Sohn, H. S., & Park, B. (2019). Make the village better: An evaluation of the Saemaul zero hunger communities project in Tanzania and Bangladesh. *World Development, 124*, 104652. https://doi.org/10.1016/j.worlddev.2019.104652

Klapper, L., Lusardi, A., & Panos, G. A. (2013). Financial literacy and its consequences: Evidence from Russia during the financial crisis. *Journal of Banking and Finance, 37*(10), 3904–3923. https://doi.org/10.1016/j.jbankfin.2013.07.014

Lacalle-Calderon, M., Larrú, J. M., Garrido, S. R., & Perez-Trujillo, M. (2019). Microfinance and income inequality: New macrolevel evidence. *Review of Development Economics, 23*(2), 860–876. https://doi.org/10.1111/rode.12573

Le, T. H., Chuc, A. T., & Taghizadeh-Hesary, F. (2019). Financial inclusion and its impact on financial efficiency and sustainability: Empirical evidence from Asia. *Borsa Istanbul Review, 19*(4), 310–322. https://doi.org/10.1016/j.bir.2019.07.002

Lusardi, A. (2012). Numeracy, financial literacy, and financial decision-making. *Numeracy, 5*(1), 1–12. https://doi.org/10.5038/1936-4660.5.1.2

Lusardi, A. (2019). Financial literacy and the need for financial education: Evidence and implications. *Swiss Journal of Economics and Statistics, 155*(1), 1–8. https://doi.org/10.1186/s41937-019-0027-5

Lyons, A. C., & Kass-Hanna, J. (2019). Financial inclusion, financial literacy and economically vulnerable populations in the Middle East and North Africa. *Emerging Markets Finance and Trade, 1–40*. https://doi.org/10.1080/1540496X.2019.1598370

Morduch, J. (1999). The microfinance promise. *Journal of Economic Literature, 37*(4), 1569–1614. https://doi.org/10.1257/jel.37.4.1569

Morgan, P., & Trinh, L. (2019). Determinants and impacts of financial literacy in Cambodia and Viet Nam. *Journal of Risk and Financial Management, 12*(1), 19. https://doi.org/10.3390/jrfm12010019

Mushtaq, R., & Bruneau, C. (2019). Microfinance, financial inclusion, and ICT: Implications for poverty and inequality. *Technology in Society, 59*, 101154. https://doi.org/10.1016/j.techsoc.2019.101154

Nawaz, S. (2010). Microfinance and poverty reduction: Evidence from a village study in Bangladesh. *Journal of Asian and African Studies, 45*(6), 670–683. https://doi.org/10.1177/0021909610383812

Nawaz, F. (2015). Microfinance, financial literacy, and household power configuration in rural Bangladesh: An empirical study on some credit borrowers. *Voluntas: International Journal of Voluntary and Nonprofit Organizations, 26*(4), 1100–1121. https://doi.org/10.1007/s11266-015-9585-z

O'Toole, C., & Newman, C. (2017). Investment financing and financial development: Evidence from Viet Nam. *Review of Finance, 21*(4), 1639–1674. https://doi.org/10.1093/rof/rfw017

OECD. (2007). The financial Sector's contribution to pro-poor growth. In *Promoting pro-poor growth: Policy guidance for donors* (p. 319). OECD Publishing. https://doi.org/10.1787/9789264024786-11-en

Opletalová, A. (2015). Financial education and financial literacy in the Czech education system. *Procedia – Social and Behavioral Sciences, 171*, 1176–1184. https://doi.org/10.1016/j.sbspro.2015.01.229

Ouma, S. A., Odongo, T. M., & Were, M. (2017). Mobile financial services and financial inclusion: Is it a boon for savings mobilization? *Review of Development Finance, 7*(1), 29–35. https://doi.org/10.1016/j.rdf.2017.01.001

Subramaniam, Y., Masron, T. A., Wahab, M. A., & Mia, M. A. (2021). The impact of microfinance on poverty and income inequality in developing countries. *Asian-Pacific Economic Literature, 35*(1), 36–48. https://doi.org/10.1111/apel.12326

Sun, H., Yuen, D. C. Y., Zhang, J., & Zhang, X. (2020). Is knowledge powerful? Evidence from financial education and earnings quality. *Research in International Business and Finance, 52*, 101179. https://doi.org/10.1016/j.ribaf.2019.101179

Tavanti, M. (2013). Before microfinance: The social value of microsavings in Vincentian poverty reduction. *Journal of Business Ethics, 112*(4), 697–706. https://doi.org/10.1007/s10551-012-1566-1

Urban, C., Schmeiser, M., Michael Collins, J., & Brown, A. (2018). The effects of high school personal financial education policies on financial behavior.

Economics of Education Review, 78, 101786. https://doi.org/10.1016/j.econedurev.2018.03.006
Warnecke, T. (2015). "Greening" gender equity: Microfinance and the sustainable development agenda. *Journal of Economic Issues, 49*(2), 553–562. https://doi.org/10.1080/00213624.2015.1042803
Weiss, J., & Montgomery, H. (2005). Great expectations: Microfinance and poverty reduction in Asia and Latin America. *Oxford Development Studies, 33*(3–4), 391–416. https://doi.org/10.1080/13600810500199210
Zulfiqar, G. (2017). Does microfinance enhance gender equity in access to finance? Evidence from Pakistan. *Feminist Economics, 23*(1), 160–185. https://doi.org/10.1080/13545701.2016.1193213

Part III

Global and Regional Topics

Financial Inclusion of Rural Women in the Global South

10

Carolina Herrera-Cano, Arley Pino-Villegas, Carlos Felipe Munera-Alzate, and Maria Alejandra Gonzalez-Perez

Abstract

This chapter aims to understand the importance of the financial inclusion of rural women in the Global South as a sustainable development concern within SDG 8.10. We analyzed the importance of a gender perspective within the 2030 Development Agenda. Further, we explored the current situation of rural women's access to basic financial services based on recent studies on the matter. After revising the causes of the financial exclusion of women in rural communities, we found that entrenched gender norms and socioeconomic segregation prevent women in these contexts from participating in the financial system. Lack of financial literacy, time poverty (due to the care work burden), gender violence, the mobile gender gap, and few customized services for rural women are perpetuating the exclusion of women, despite private and public efforts. By promoting strategies toward financial inclusion, society can not only support gender equality but also economic growth and sustainable development. Thus, this chapter collects best practices historically and currently implemented in the Global South and proposes recommendations for businesses, governments, and civil society. Today, the world relies more on the financial system to handle crises, and ignoring rural women makes it harder *to leave no one behind.*

Keywords

Financial inclusion · Gender equality · Rural women · Global South · Sustainable development · Economic growth · SDG 8.10 · SDG 5

10.1 Introduction

Financial inclusion is a significant concern for the 2030 Development Agenda. According to the Global Findex Report (Demirgüç-Kunt et al., 2018), access to financial services helps reduce poverty and hunger, supports households in income management and planning (Ruiz, 2013), and increases investments in education, businesses, and healthcare (Arnold & Gammage, 2019; Munyanyi, 2014). Providing efficient and safe infrastructure and institutions has become imperative in global development initiatives. Indeed, target 10 in Sustainable Development

C. Herrera-Cano
EAFIT University, Medellín, Colombia

WiSE Centre for Economic Justice, Glasgow Caledonian University, Glasgow, United Kingdom
e-mail: carolina.herreracano@gcu.ac.uk

A. Pino-Villegas · C. F. Munera-Alzate · M. A. Gonzalez-Perez (✉)
EAFIT University, Medellín, Colombia
e-mail: apinov@eafit.edu.co; cmunera7@eafit.edu.co; mgonza40@eafit.edu.co

G. W. Muschert et al. (eds.), *Financial Inclusion*, Sustainable Development Goals Series,
https://doi.org/10.1007/978-3-031-68803-4_10

Goal 8 stresses the need to join forces "to encourage and expand access to banking, insurance and financial services for all" (United Nations, 2015, p. 24). With the above in mind, in this chapter, we seek to identify historical and current opportunities and challenges for rural women in the global south and then, based on that, show how financial inclusion can help alleviate their vulnerabilities.

For this purpose, in the first part of the chapter, we will explain why financial inclusion is a sustainable development concern and why it is critical for the achievement of the 2030 Agenda and why a gender perspective of Sustainable Development Goal (SDG) 8.10 (as well as the rest of the development agenda) is crucial. Second, we will provide facts and figures showing rural women's financial exclusion and its consequences. Then, we will analyze the causes of such financial exclusion and, subsequently, the benefits of financial inclusion. Finally, we will describe current and historical programs that have successfully promoted the financial inclusion of rural women in the form of recommendations for governments, private organizations, and civil society.

10.2 Financial Inclusion and Gender Equality: A Development Concern

The 2030 Development Agenda reflects grand challenges and is systemic, encompassing social, environmental, and economic dimensions. It implies that we must first understand synergistic relationships and potential trade-offs to define priorities and implement strategies. Furthermore, it is critical to recognize that the 17 SDGs may represent wicked problems (van Tulder, 2018). According to van Tulder (2018, p. 36), the SDGs are "systemic in nature, complexly interconnected, and materialised at the interface between public–private and profit–nonprofit interests." More specifically, in terms of gender equality (SDG5), "the wicked" (systemic, ambiguous, complex, symptomatic of another problem, and conflictual) is notoriously intertwined as it comprises deeply rooted gender norms. Therefore, financial inclusion (specific to SDG 8.10) of rural women (SDG 5) implies the understanding and mobilizing of people and resources across the Sustainable Development Agenda. Ten of the 17 SDGs include gender-specific targets, as shown in Table 10.1 below, and 22% of SDG targets are gender-specific (United Nations, 2019). Thus, there are multiple ways people, businesses, and governments can contribute to gender equality through targets in other development goals.

It is critical to acknowledge how financial inclusion does not solely resolve the issues of gender equality or women's empowerment. SDG 5 aims to achieve equality between men and women regarding rights and opportunities, valuation and treatment, and empowerment (i.e., fostering women's voice and agency). However, through a gender-sensitive analysis of SDG8.10, we can contribute to expanding women's rights in the Global South.

10.3 Rural Women's Access to Financial Services

The World Bank estimates that 1.7 billion adults lack access to financial services globally; these calculations also reveal that in high-income countries, "account ownership is nearly universal" (Demirgüç-Kunt et al., 2018, p. 4), and in every context, the urban population benefits more from financial services than the rural. Hence, obstacles to bank account ownership and digital payment are a significant concern for the Global South. The inability to take advantage of basic financial services is a reality, preventing millions from escaping poverty and investing in their future. Within such a group, women and girls face particular vulnerabilities.

Women and girls face particular vulnerabilities. Although rural women's financial inclusion drives poverty reduction and economic empowerment, rural women's access to financial products and services has been unfavorable (Manta, 2019). Globally, rural women suffer gendered financial exclusion restrictions

Table 10.1 Gender-specific targets in the Sustainable Development Goals (SDGs)

SDG	Gender-specific targets
Goal 1 End poverty in all its forms everywhere	1.2 By 2030, reduce at least by half the proportion of men, women, and children of all ages living in poverty in all its dimensions according to national definitions 1.4 By 2030, ensure that all men and women, in particular the poor and the vulnerable, have equal rights to economic resources, as well as access to basic services, ownership and control over land and other forms of property, inheritance, natural resources, appropriate new technology, and financial services, including microfinance 1. b Create sound policy frameworks at the national, regional, and international levels, based on pro-poor and gender-sensitive development strategies, to support accelerated investment in poverty eradication actions
Goal 2 End hunger, achieve food security and improved nutrition, and promote sustainable agriculture	2.2 By 2030, end all forms of malnutrition, including achieving, by 2025, the internationally agreed targets on stunting and wasting in children under 5 years of age, and address the nutritional needs of adolescent girls, pregnant and lactating women, and older persons 2.3 By 2030, double the agricultural productivity and incomes of small-scale food producers, in particular women, indigenous peoples, family farmers, pastoralists, and fishers, including through secure and equal access to land, other productive resources, and inputs, knowledge, financial services, markets and opportunities for value addition and nonfarm employment
Goal 3 Ensure healthy lives and promote well-being for all at all ages	3.1. By 2030, reduce the global maternal mortality ratio to less than 70 per 100,000 live births
Goal 4 Ensure inclusive and equitable quality education and promote lifelong learning opportunities for all	4.1. By 2030, ensure that all girls and boys complete free, equitable, and quality primary and secondary education leading to relevant and effective learning outcomes 4.2. By 2030, ensure that all girls and boys have access to quality early childhood development, care, and pre-primary education so that they are ready for primary education 4.3. By 2030, ensure equal access for all women and men to affordable and quality technical, vocational, and tertiary education, including university 4.6 By 2030, ensure that all youth and a substantial proportion of adults, both men and women, achieve literacy and numeracy

(continued)

Table 10.1 (continued)

SDG	Gender-specific targets
Goal 5 Achieve gender equality and empower all women and girls	5.1. End all forms of discrimination against all women and girls everywhere 5.2. Eliminate all forms of violence against all women and girls in the public and private spheres, including trafficking and sexual and other types of exploitation 5.3. Eliminate all harmful practices, such as child, early and forced marriage, and female genital mutilation 5.4. Recognize and value unpaid care and domestic work through the provision of public services, infrastructure, and social protection policies and the promotion of shared responsibility within the household and the family as nationally appropriate 5.5. Ensure women's full and effective participation and equal opportunities for leadership at all levels of decision-making in political, economic, and public life 5.6. Ensure universal access to sexual and reproductive health and reproductive rights as agreed in accordance with the Programme of Action of the International Conference on Population and Development and the Beijing Platform for Action and the outcome documents of their review conferences 5.a. Undertake reforms to give women equal rights to economic resources, as well as access to ownership and control over land and other forms of property, financial services, inheritance, and natural resources, in accordance with national laws 5.b. Enhance the use of enabling technology, in particular information and communications technology, to promote the empowerment of women 5.c. Adopt and strengthen sound policies and enforceable legislation for the promotion of gender equality and the empowerment of all women and girls at all levels
Goal 6 Ensure availability and sustainable management of water and sanitation for all	6.2. By 2030, achieve access to adequate and equitable sanitation and hygiene for all and end open defecation, paying special attention to the needs of women and girls and those in vulnerable situations
Goal 8 Promote sustained, inclusive, and sustainable economic growth, full and productive employment, and decent work for all	8.5. By 2030, achieve full and productive employment and decent work for all women and men, including for young people and persons with disabilities, and equal pay for work of equal value 8.8. Protect labour rights and promote safe and secure working environments for all workers, including migrant workers, in particular women migrants, and those in precarious employment
Goal 11 Make cities and human settlements inclusive, safe, resilient, and sustainable	11.2. By 2030, provide access to safe, affordable, accessible, and sustainable transport systems for all, improving road safety, notably by expanding public transport, with particular attention to the needs of those in vulnerable situations, women, children, persons with disabilities and older persons 11.7. By 2030, provide universal access to safe, inclusive, and accessible green and public spaces, in particular for women and children, older persons, and persons with disabilities
Goal 13 Take urgent action to combat climate change and its impacts	13.b. Promote mechanisms for raising capacity for effective climate change-related planning and management in least-developed countries and small-island developing States, including focusing on women, youth, and local and marginalized communities
Goal 17 Strengthen the means of implementation and revitalize the global partnership for sustainable development	17.18. By 2020, enhance capacity-building support to developing countries, including for least developed countries and small island developing States, to increase significantly the availability of high-quality, timely, and reliable data disaggregated by income, gender, age, race, ethnicity, migratory status, disability, geographic location and other characteristics relevant in national contexts

(idem). Women and low-income communities face more barriers to financial services than men (Arnold & Gammage, 2019). 65% of the world's unbanked population are women, and only 65% of women (compared to 72% of men) own a bank account (Demirgüç-Kunt et al., 2018). By not participating in the banking system, people experience higher risks (cash can be unsafe and challenging to manage), more obstacles to overcome financial emergencies (job loss and crop failure) (idem), and increased chances of falling into poverty traps (Munyanyi, 2014). In the case of women, financial exclusion translates into less bargaining power and, thus, reduced ability to influence spending decisions at the household level (idem). Women who have access to financial institutions still suffer a significant disadvantage compared to their male counterparts. Even when facing better conditions than their counterparts with no access to financial institution support, women are still less likely to participate in or start entrepreneurial projects than men (Fareed et al., 2017).

Women also have limited access to mobile phones, which restricts their access even more to essential financial services. According to the 2021 Mobile Gender Gap Report, in low- and middle-income economies, women are 7% less likely to own a mobile phone and 15% less likely to own a smartphone. This "translates into 143 million fewer women mobile owners than men" (Carboni et al., 2021, p. 5). Both lower bank account ownership and mobile usage are particularly worrisome in a world that increasingly trusts the banking system and mobile transactions (Davidovic et al., 2020). In fact, during the COVID-19 pandemic, low- and middle-income countries rapidly adopted online banking for money transfers and remote financial services (Bazarbash et al., 2020). Women's bank account ownership has remained unchanged since 2011, despite recent governmental and private efforts in favor of access to financial services (Demirgüç-Kunt et al., 2018).

10.4 Rural Women's Financial Inclusion and Gender Norms

To achieve full financial inclusion, institutions need to understand the specific needs women and girls face in rural regions that differ from those of men (also in vulnerable conditions). Gender norms explain additional challenges that limit women's financial inclusion and, for instance, make them more prone to use the informal economy (Munyanyi, 2014) and cash (Arnold & Gammage, 2019). Women not reached by financial institutions are less likely to create entrepreneurial projects and carry them out successfully, primarily due to their financial illiteracy and lack of monetary and financial support (Fareed et al., 2017).

Social norms of behavior limit women's financial autonomy, as they not only dictate the economic activities in which both genders can engage but also assign women the burden of childcare and non-remunerated domestic activities (Yalman et al., 2019). Lack of physical access to banking services for women can happen due to time poverty and mobility constraints (Koning et al., 2021). This situation prevents women from engaging in the "productive" economy and creates time poverty. Moreover, in specific contexts, such as Northern India, women are either not allowed to interact with financial institutions if unescorted by a man (Field et al., 2021)—lack of female identity documentation is also a common issue—or discouraged from doing so because of safety issues (Koning et al., 2021).

Women tend to have lower financial literacy (Bhatia & Singh, 2019) because, from childhood, they receive less financial knowledge at home and school (Chambers et al., 2019). They also tend to be more risk-averse than men (Fletschner & Kenney, 2014). These situations lower women's access to formal financial products, confidence level when dealing with transactions, coping mechanisms to deal with financial shocks (OECD, 2013), and willingness to engage in financial services.

There is also evidence that, in addition to gender, race and socioeconomic status influence the capacity to obtain credit and access to other financial services (Kara et al., 2021). Minorities, disabled individuals, and immigrants are more likely to be denied formal credit, and traditional financial institutions are less willing to lend to those with lesser incomes and wealth (Kara et al., 2021). Women face more rejection, less formal credit, and more costs in the developing world by formal financial institutions, which do not see them as legitimate clients as they lack access to conventional forms of collateral since men are more likely to own or manage the family's property and income (Yalman et al., 2019). Such institutional discrimination also appears in the lack of gender-tailored financial products and services: rural financial programs designed and implemented with a male head of household as the target client, ignoring women as active, productive, and engaged economic agents (Fletschner & Kenney, 2014).

Additionally, when women often access financial services in different contexts, they come from informal organizations that often perpetuate poverty and even violence. For instance, as microfinance expands and hardens lending networks, rural women in Bangladesh often find themselves severely in debt to multiple unscrupulous creditors (Karim, 2011). All these conditions also reduce women's levels of trust in the financial system. Indeed, figures show that even when the government and institutions manage to increase bank accounts owned by women, in 2019, "half the accounts opened for women… remain[ed] dormant" (Arnold & Gammage, 2019, p. 969). Interventions aiming at financial inclusion can backlash and aggravate women's conditions: for instance, giving women access to a wallet can increase gender violence, as it disrupts household gender-power relations (Koning et al., 2021). These circumstances show the need to foster conditions in favor of the introduction and sustained participation of women in financial services, but they also show the need to understand the context in which they occur to provide appropriate solutions.

10.5 Financial Inclusion: Beyond Gender Equality

Although financial inclusion contributes to gender equality, women are not the only stakeholders in this concern. Financial inclusion enables both gender equity and economic growth and poverty reduction objectives, which improve the livelihoods of multiple individuals, households, and entire countries (Napier et al., 2013). Financial institutions can benefit from broadening their markets to rural areas and new customers (Women's World Banking, 2014). Financial inclusion has proven to be one of the most potent tools to dynamize the economic growth and development of developing nations, as the inclusion of the poor into financial institutions grants them access to credit and other financial services. Moreover, the additional economic resources deposited into the system allow the banking organizations to lend more money at lower interest rates, creating a favorable economic cycle (Sethi & Acharya, 2018).

There is a positive relationship between a country's GDP per capita growth and the financial inclusion of women (Cabeza-García et al., 2019). Financial inclusion of women also helps to close the gender pay gap as it improves women's ability to create an economy (idem). That is, it provides them more accessible access to financial support for their entrepreneurial capacity and financial independence. The resulting gender equality and increased livelihoods from financial inclusion (Bhatia & Singh, 2019) are explained explicitly by the fact that it improves women's agency, control over resources and earnings (Gammage et al., 2017), autonomy and decision-making on their use of time, and intra-household communication (Arnold & Gammage, 2019). Furthermore, financial inclusion also translates into reducing poverty and income inequalities (Omar & Inaba, 2020) and a substantive improvement in children's health, nutrition, and education since women invest more in them (Fletschner & Kenney, 2014).

In general, the financial inclusion of women significantly improves their economic situation, helping their families and their participation in political and social activities (Bhatia & Singh, 2019). Since the process of financial inclusion occurs when empowering women in those three dimensions (idem), in the following, we will delve into the potential steps necessary to contribute to the financial inclusion of rural women, recognizing the historical and current efforts that have allowed progress in a task in which there is still much to be done.

10.6 Financial Inclusion Initiatives and Recommendations

Affordably priced financial services (such as savings, investments, borrowing, and insurance) from stable and sustainable institutions are considered financial inclusion (Subbarao, 2009). However, it is crucial to understand that financial inclusion must go beyond access to loans or saving accounts, as more is needed to meet the needs of rural women (Women's World Banking, 2014). Improving women's financial inclusion also requires overcoming the socio-cultural, economic/legal, and educational factors that lead to gender inequalities (Yalman et al., 2019). PMJDY bank accounts, for example, benefited Indian women by providing more competitive interest rates, better access to credit, and financial literacy resources (Bhatia & Singh, 2019).

Also, in India, Telenor launched a market strategy that encouraged wives and husbands to acquire paired SIMS (one each with credit/airtime benefits). This measure increased female SIM ownership to a similar number of men's (Koning et al., 2021). CARE implemented a program in Kenya and Tanzania that provided rural women with mobile strategies and training for saving products; this project increased women's decision-making and trust in online activities (Arnold & Gammage, 2019). In Egypt, United Nations agencies and the government worked collectively to encourage women to register for identification documents, thus dealing with the leading cause of bank account gender imbalances in the country (Koning et al., 2021).

The Global South has successfully understood the context and needs of rural women, and many of the actions already implemented can serve as inspiration to further increase financial inclusion. These efforts show how governments, private organizations, and civil society can cooperate to address the needs of women in specific regions and benefit from their access to financial services. Below are some recommendations to help achieve this goal:

- Create gender-sensitive financial products and services, focusing on clients' needs according to local contexts and gender dynamics (Yalman et al., 2019). For example, there must be financial services in rural areas because women's jobs at home make it hard for them to move around.
- Modify the requirements for borrowing from formal financial institutions, expanding the possibility of accepting as collateral assets that women are more likely to own or lending based on alternatives other collateral (Fletschner & Kenney, 2014) as microfinance institutions do.
- Establish the current requirements for savings at a level that rural women can afford by reducing or eliminating charges, fees, and the minimum initial deposit (idem).
- Train rural women in financial literacy and adapt how financial products and services are communicated so that they are easily understandable and women can make informed decisions (Yalman et al., 2019).
- Promote a women-friendly culture in all financial institutions, making it clear to staff that women are also worthy clients and considering the voice of female staff in the organization. (Fletschner & Kenney, 2014).
- Use innovative strategies to deliver financial products and services in rural areas, such as mobile technologies, which have enabled transactions or savings for actors traditionally excluded from the financial system (Yalman et al., 2019).

- Promote an enabling environment for women to have access to employment, entrepreneurship, and ownership of productive assets (Napier et al., 2013). Also, more must be done to ensure that the time-consuming and unpaid tasks around the house are shared fairly.
- Create enabling government policies and legal frameworks as they impact women's rights and gender equality (idem).

Overall, the needs of rural women in the Global South would be better served if financial institutions understood that access to savings and credit is different from access to finance. Moreover, financial inclusion requires an enabling environment in all spheres of life.

10.7 Conclusion

Similarly, to the inclusion of gender-specific targets in other global goals, SDG 8.10 should be promoted by bearing in mind a gender perspective that *leaves no one behind*. The financial inclusion of rural women, especially in the Global South, is an urgent need for the 2030 Development Agenda. Currently, a large part of the world's population still does not have access to basic financial services, and they are primarily women (this is especially worrisome during times of crisis when governments are increasingly relying on these means for the distribution of aid). This means a growing problem for their development as it limits their possibilities based on entrenched gender norms that restrict women's economic activities and, on many occasions, their fundamental rights. Including rural women in the economic system is visibly beneficial for their livelihood: increased economic autonomy, bargaining power, and agency. However, when taking into account the context and specific needs of the women involved, these initiatives constitute a fundamental tool to create new markets for private businesses and, ultimately, economic growth. Therefore, the programs that involve women in financial services need translation into objective instruments and indicators of sustainable development. As indicated in SDG 8.10, strengthening the capacity of domestic financial institutions encourages and expands access to financial services for all and, from a gender equality perspective, may achieve a more inclusive 2030 Agenda.

References

Arnold, J., & Gammage, S. (2019). Gender and financial inclusion: The critical role for holistic programming. *Development in Practice, 29*(8), 965–973. https://doi.org/10.1080/09614524.2019.1651251

Bazarbash, M., Moeller, J., Griffin, N., Carcel-Villanova, H., Chhabra, E., Fan, Y., & Shirono, K. (2020). *Mobile money in the COVID-19 pandemic*. International Monetary Fund. https://www.imf.org/~/media/Files/Publications/covid19-special-notes/en-special-series-on-covid-19-mobile-money-in-the-covid-19-pandemic.ashx?la=en

Bhatia, S., & Singh, S. (2019). Empowering women through financial inclusion: A study of an urban slum. *Vikalpa: The Journal for Decision Makers, 44*(4), 182–197. https://doi.org/10.1177/0256090919897809

Cabeza-García, L., Del Brio, E. B., & Oscanoa-Victorio, M. L. (2019). Female financial inclusion and its impacts on inclusive economic development. *Women's Studies International Forum, 77*, 102300. https://doi.org/10.1016/j.wsif.2019.102300

Carboni, I., Jeffrie, N., Lindsey, D., Shanahan, M., & Sibthorpe, C. (2021). *Connected women: The Mobile gender gap report 2021*. GSMA. https://www.gsma.com/r/gender-gap/

Chambers, R. G., Asarta, C. J., & Farley-Ripple, E. N. (2019). Gender, parental characteristics, and financial knowledge of high school students: Evidence from multi-country data. *Journal of Financial Counseling and Planning, 30*(1), 97–109. https://doi.org/10.1891/1052-3073.30.1.97

Davidovic, S., Prady, D., & Tourpe, H. (2020, June 22). *You've got money: Mobile payments help people during the pandemic*. IMFBlog, International Monetary Fund. https://blogs.imf.org/2020/06/22/youve-got-money-mobile-payments-help-people-during-the-pandemic/https://blogs.imf.org/2020/06/22/youve-got-money-mobile-payments-help-people-during-the-pandemic/

Demirguc-Kunt, A., Klapper, L., Singer, D., Ansar, S., & Hess, J. (2018). *The global Findex database 2017: Measuring financial inclusion and the Fintech revolution*. World Bank. https://doi.org/10.1596/978-1-4648-1259-0

Fareed, F., Gabriel, M., Lenain, P., & Reynaud, J. (2017). *Financial inclusion and women entrepreneurship: Evidence from Mexico* (OECD economics department working papers, 1411). https://doi.org/10.1787/2fbd0f35-en

Field, E., Pande, R., Rigol, N., Schaner, S., & Moore, C. T. (2021). On her own account: How strengthening women's financial control impacts labor supply and gender norms. *American Economic Review, 111*(7), 2342–2375. https://doi.org/10.1257/aer.20200705

Fletschner, D., & Kenney, L. (2014). Rural women's access to financial services: Credit, savings, and insurance. In A. Quisumbing, R. Meinzen-Dick, T. Raney, A. Croppenstedt, J. Behrman, & A. Peterman (Eds.), *Gender in agriculture: Closing the knowledge gap* (pp. 187–208). Springer. https://doi.org/10.1007/978-94-017-8616-4_8

Gammage, S., Kes, A., Winograd, L., Sultana, N., Hiller, S., & Bourgault, S. (2017). *Gender and digital financial inclusion: What do we know and what do we need to know?* International Center for Research on Women (ICRW). https://www.icrw.org/wp-content/uploads/2017/11/Gender-and-digital-financial-inclusion.pdf

Kara, A., Zhou, H., & Zhou, Y. (2021). Achieving the United Nations' sustainable development goals through financial inclusion: A systematic literature review of access to finance across the globe. *International Review of Financial Analysis, 77*, 101833. https://doi.org/10.1016/j.irfa.2021.101833

Karim, L. (2011). *Microfinance and its discontents: Women in debt in Bangladesh*. University of Minnesota Press. https://doi.org/10.5749/minnesota/9780816670949.001.0001

Koning, A., Ledgerwood, J., & Singh, N. (2021). *Addressing gender norms to increase financial inclusion: Designing for impact*. CGAP Publications. https://www.cgap.org/research/publication/addressing-gender-norms-increase-financial-inclusion-designing-impact

Manta, A. (2019). Financial inclusion and gender barriers for rural women. *International Journal of Management, 10*(5), 61–72. https://doi.org/10.34218/ijm.10.5.2019.006

Munyanyi, W. (2014). Women financial inclusion in Zimbabwe: A descriptive comparison of rural and urban populaces. *International Journal of Education and Research, 2*(6), 255–266. http://www.ijern.com/journal/June-2014/21.pdf

Napier, M., Melamed, C., Taylor, G., & Jaeggi, T. (2013). *Promoting women's financial inclusion: A toolkit*. British Department for International Development & Deutsche Gesellschaft für Internationale Zusammenarbeit. https://assets.publishing.service.gov.uk/government/uploads/system/uploads/attachment_data/file/213907/promoting-womens-financial-inclusion-toolkit.pdf

OECD. (2013). Addressing women's financial education needs through policies and dedicated programmes. In *Women and financial education: Evidence, policy responses and guidance* (pp. 59–83). OECD Publishing. https://doi.org/10.1787/9789264202733-7-en

Omar, M. A., & Inaba, K. (2020). Does financial inclusion reduce poverty and income inequality in developing countries? A panel data analysis. *Journal of Economic Structures, 9*(37). https://doi.org/10.1186/s40008-020-00214-4

Ruiz, C. (2013). *From pawn shops to banks: The impact of formal credit on informal households* (World Bank Working Papers No. 6634). https://openknowledge.worldbank.org/handle/10986/16851

Sethi, D., & Acharya, D. (2018). Financial inclusion and economic growth linkage: Some cross country evidence. *Journal of Financial Economic Policy, 10*(3), 369–385. https://doi.org/10.1108/jfep-11-2016-0073

Subbarao, D. (2009, December 9). *Financial inclusion: Challenges and opportunities* (Remarks by Governor of the Reserve Bank of India at the Bankers Club in Kolkata). Bank for International Settlements. https://www.bis.org/review/r091215b.pdf

United Nations. (2015). *Transforming our world: The 2030 agenda for sustainable development*. United Nations A/RES/70/1. https://sustainabledevelopment.un.org/post2015/transformingourworld/publication

United Nations. (2019). *Progress on the sustainable development goals: The gender snapshot 2019*. UN Women & Department of Economic and Social Affairs. https://www.unwomen.org/en/digital-library/publications/2019/09/progress-on-the-sustainable-development-goals-the-gender-snapshot-2019

van Tulder, R. (2018). *Business & the sustainable development goals: A framework for effective corporate involvement*. Rotterdam School of Management, Erasmus University. https://sdghub.com/project/business-and-the-sdgs-a-framework-for-effective-corporate-involvement/

Women's World Banking. (2014). *Making women's work visible: Finance for rural women*. http://www.womensworldbanking.org/wp-content/uploads/2014/11/Making-Womens-Work-Visible-Rural-Finance-WomensWorldBanking.pdf

Yalman, I., Kaaria, S., & Mhlanga, N. (2019). *Women's access to rural finance: Challenges and opportunities*. Food and Agriculture Organization of the United Nations (FAO). https://www.fao.org/policy-support/tools-and-publications/resources-details/en/c/1314607/

11 Assessing the Impact of Financial Inclusion on Economic Growth in Island Economies

Narvada Gopy-Ramdhany, Reena Bhattu-Babajee, and Boopen Seetanah

Abstract

Financial inclusion (FI) has become an essential strategy for enhancing economic growth and achieving inclusive growth in a country (Karlan et al., Rev Income Wealth 60:36–78, 2014). Although the theoretical relationship between financial inclusion and economic growth is acknowledged, more empirical evidence is necessary regarding the financial inclusion-growth nexus, particularly for small island economies. This book chapter thus aims to supplement the literature by providing an insightful analysis of the hypothesized link for a sample of small island developing states (SIDs). The research also investigates the reverse causal link: economic growth's role in financial inclusion. A multidimensional financial inclusion index is constructed using the principal component analysis (PCA) and included in a classical augmented Solow growth function. Using a panel autoregressive distributed lag approach (PARDL) to account for the dynamic nature, empirical results confirm that financial inclusion is growth-conducive. The study could not establish any reverse causality using the panel Granger Causality test, validating a unidirectional relationship from financial inclusion to economic growth. Findings from this research supplement the existing literature on this SDG 8.10 goal and provide important implications for SIDs governments to consolidate and accelerate financial inclusion to enhance prosperity.

Keywords

Financial inclusion · Multidimensional index · Small island developing states · Panel data · Economic growth

JEL Classifications

C33 · C43 · G28 · O47

11.1 Introduction

The importance of an inclusive financial system has increasingly gained recognition. According to Claessens and Perotti (2007) and Barajas et al. (2020), financial inclusion (FI)—which includes access to formal financial services like credit, insurance, and safe savings opportunities—may be a factor in economic growth. The 2030 Sustainable Development Goals have eight objectives that promote financial inclusion.

N. Gopy-Ramdhany · R. Bhattu-Babajee ·
B. Seetanah (✉)
University of Mauritius, Moka, Mauritius
e-mail: b.seetanah@uom.ac.mu

G. W. Muschert et al. (eds.), *Financial Inclusion*, Sustainable Development Goals Series,
https://doi.org/10.1007/978-3-031-68803-4_11

Financial inclusion is perceived as leading to economic growth. When targeting the lower-income segments of the population, financial inclusion with the ease of access to funds would promote more expenses and developments at the individual level.

A more financially inclusive system would enable a better allocation and use of funds and discourage the expansion of informal lending sources. Further, with an increased capacity to place their funds, these can enable further developments, thus leading to a multiplier effect in economic growth. Most studies on the relationship between financial inclusion and economic growth have depicted a positive link between the two variables (Sarma, 2008; Sarma & Pais, 2011; Sethi & Acharya, 2018; Kim et al., 2018; Nizam et al., 2020). Most of the prior studies have been conducted in developed countries' contexts, with only a few studies in developing countries and, as such, no emphasis on island economies.

This research thus aims at assessing the impact of financial inclusion as an ingredient of economic growth for a sample of 11 island economies (over the period 2004–2019), which have their characteristics and vulnerabilities. It initially constructs a multidimensional index of financial inclusion using principal component analysis (PCA). Subsequently, it extends a classical augmented Solow growth function with this index to gauge the economic effect of financial inclusion. Given the dynamic nature of economic growth modeling, a panel autoregressive distributed lag approach (PARDL) is suitable for the estimations. A panel Granger causality test establishes the existence of reverse causality.

The rest of this book chapter is organized as follows: Section 11.2 briefly reviews the related literature. Section 11.3 dwells into the model specification and methodology and Sect. 11.4 analyzes the results, while Sect. 11.5 concludes.

11.2 Related Literature

11.2.1 Theoretical Review

Financial inclusion referred primarily to "the ease of access to essential financial products and services within the formal financial system at all levels of society. In particular, the under-banked and unbanked population, fairly and transparently" (UNCTAD, 2014). A multifaceted definition given by Demirguc-Kunt and Singer (2017) perceived financial inclusion as the accessibility and usage of valuable and affordable instruments in the financial system that fulfill the demands for savings, credit, transactions, payments, and insurance and whose delivery occurs responsibly and sustainably under adequately regulated conditions. Sahay et al. (2015) added that a comprehensive definition of financial inclusion should account for factors like accessibility and availability, and effective utilization of financial services. Jahan and Mohan (2019) added the quality of the products and services in the financial system as a component of financial inclusion.

The theoretical underpinnings that link financial inclusion and economic growth rely primarily on the finance-growth nexus, namely the "supply-leading" hypothesis and the "demand-following" hypothesis. Also termed the finance-led hypothesis, the "supply-leading" hypothesis explained the banking sector's significant contribution to deploying financial resources to productive units for growth (Bagehot, 1873). According to the study above, entrepreneurs influence financial intermediaries to provide an efficient market for loanable funds. Once disbursed, these loans are engaged in "new combinations," spurring technological innovation and economic development (Schumpeter, 1911). Financial inclusion is critical to growth as financial intermediaries bring the excluded into the formal sector by expanding and bringing innovative financial products, thus, stimulating the supply level of financial inclusion. As such, the efficiency of the intermediation process is reinforced through sound investment patterns, which contribute to economic stability and upturn (Cull et al., 2012). According to the "supply-leading"

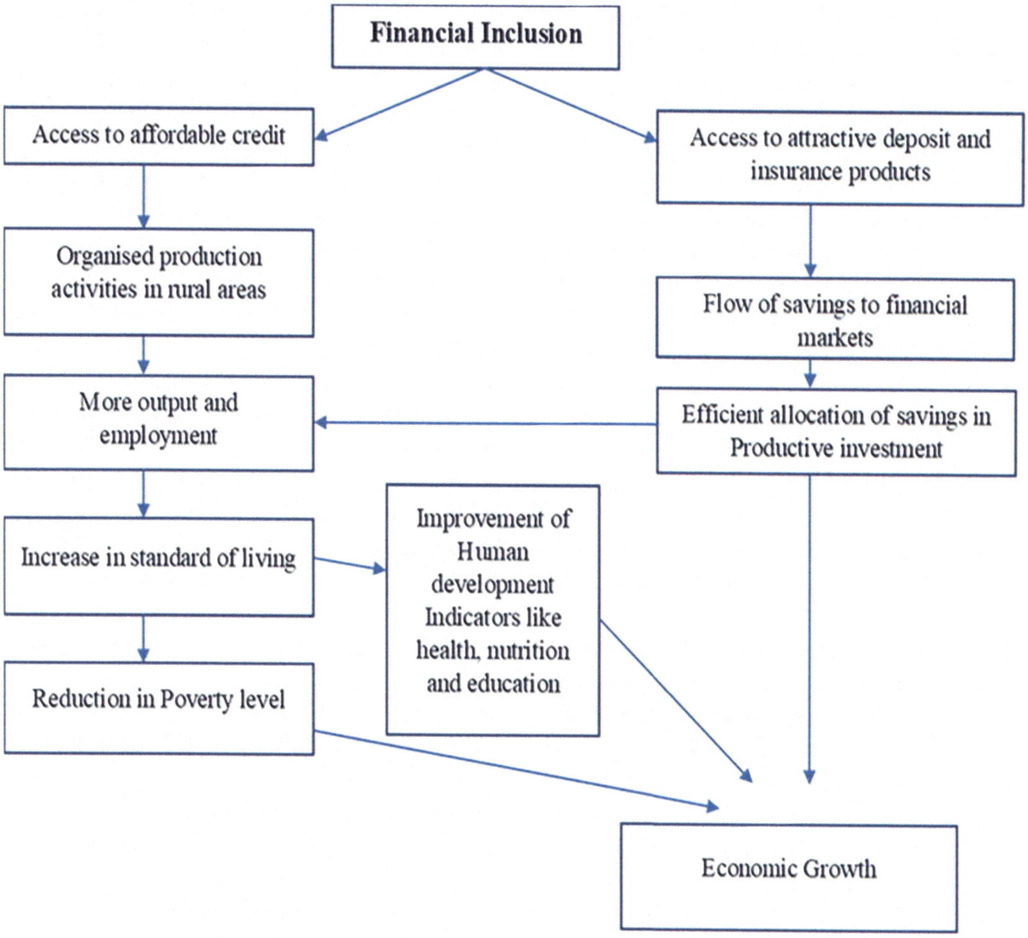

Fig. 11.1 Financial inclusion-economic growth theoretical framework. (Source: Sethi and Acharya (2018))

hypothesis framework, the theoretical approach to the financial inclusion-economic growth Nexus appears in Fig. 11.1.

Goldsmith (1969) added that capital flows are substantial for financial development, given that they foster entrepreneurship and technology transfer in expediting economic growth. Later, McKinnon (1973) stressed the inefficiency of banks that arise from government intervention, principally through interest rate ceilings and the exigency of collaterals, which induces financial repression. The academic, thus, stipulated that as a medium to curtail financial repression, the financial sector plays a fundamental role in fostering growth by promoting financial liberalization.

At its most basic level, the endogenous growth theory entails a rise in capital stock, creating favorable externalities that increase productivity. Romer (1986) advanced that the financial sector nurtures endogenous growth through capital accumulation, savings, and investment. Later, Romer (1990) accentuated the binding effect of improving innovations, research, and development, given that it promotes long-run economic growth. Technological progress occurs from manifold innovations in new markets, products, or processes such as ATM usage, cashless payments, and online banking. In response to these incentives, technological advancement thus emanates from economic activities and, ultimately, is a substantial driver of economic growth.

The "demand-following" hypothesis advocated by Robinson (1952) and Patrick (1966) indicates that growth spurs demand for financial goods. The financial system responds accordingly, inducing that finance is instead a positive outcome of economic growth. Lucas (1988) added that the effect of financial matters on EG is merely overestimated. A booming economy expedites the need for improved financial infrastructure, bringing a sustainable, stable, and inclusive financial structure (Demetriades & Hussein, 1996).

11.2.2 Empirical Review

Increased access to financial services positively correlates with economic development. Beck et al. (2007) explicitly examined the effect of an inclusive financial system on the growth of 99 countries from 2003 to 2004 through a questionnaire survey. Using a financial inclusion (FI) index that englobes variables like accessibility and usage of banking services, Sarma (2008) found a higher level of financial inclusion in economies with higher per capita GDP and lower levels of income inequality. Using a similar multidimensional FI index, Sarma and Pais (2011) depicted a positive link between these variables using linear regression in 49 countries in 2004. Lenka and Sharma (2017) constructed a FI index using principal component analysis (PCA). Subsequently, they established that FI positively affected economic growth in India from 1980 to 2014 by applying an autoregressive distributed lag (ARDL) approach. Through a sample of 31 developed and developing countries and using a panel cointegration approach, Sethi and Acharya (2018) concluded that financial inclusion positively affects economic growth. Using a sample of 188 countries in the Asia Pacific region, Yang et al. (2018) found a positive and significant correlation between FI and per capita income growth. However, the GMM results demonstrated that the degree of FI differed substantially across the nations.

Using a sample of emerging countries and a multidimensional financial inclusion index, Van et al. (2021) depicted a significant positive link between financial inclusion and economic growth. Kim et al. (2018) supported the solid and positive link between FI and growth in 55 OIC economies from 1990 to 2013. More recently, similar results have been confirmed by both Suidarma (2019) for 10 ASEAN countries for the period 2008 to 2015 and Chima et al. (2021) for 48 sub-Saharan African countries from 1995 to 2017.

Another strand of the empirical literature reported some contradictory results. Purwiyanta et al. (2020) found that financial access negatively impacted the growth in the Indonesian Provinces, mainly due to the existence of the paradox of thrift. Applying the VECM methodology to a sample of 15 nations of the Economic Community of West African States from 1970 to 2017, Nwogu (2021) found no long-run causality between financial inclusion and per capita GDP, though a short-run causal association emerged.

11.3 Methodology and Analysis

11.3.1 Data, Methodology, and Model Specification

11.3.1.1 Model Specification

The study thus attempts to assess the impact of financial inclusion on economic growth for a sample of small island developing states (SIDs). Following several past empirical works on growth modeling (see Seetanah, 2008; Seetanah et al., 2009 for the case of island economies, and more recently, Lenka and Sharma (2017) and Van et al. (2021), among others), an augmented Solow growth model is adopted and extended with a measure of financial inclusion. The inexistence of one particular measure for measuring financial inclusion necessitates the construction of a multidimensional index, which considers several indicators deemed necessary in measuring financial inclusion. The other growth ingredients in our model specification include trade, human capital, foreign direct investment, and private investments. The conceptual model is specified as follows:

$$\text{GDPPC} = \left(FI, FDI, \text{TRADE}, \text{HUMAN}, \text{GFCF}\right)$$

GDPPC, the gross GDP per capita, is a measure of economic growth; FI is a financial inclusion index as explained below; FDI represents the foreign direct investment net inflows as a percentage of GDP; TRADE measures the sum of imports and exports as a share of GDP and is a proxy for openness; HUMAN represents a measure of the level and quality of human capital, measured by the level of secondary school enrolment while GFCF is the gross fixed capital formation used as a proxy for private investment.

The above model can be written econometrically as follows:

$$\begin{aligned}\text{LGDPPC}_{it} &= \alpha + \pi_1 LFI_{it} + \pi_2 \text{LFDI}_{it} \\ &+ \pi_3 \text{LTRADE}_{it} + \pi_4 \text{LHUMAN}_{it} \\ &+ \pi_5 \text{LGFCF}_{it} + \varepsilon_{it}\end{aligned}$$

In the above equation, i and t denote the country and time dimensions. The model has transformed into its natural logarithmic specification for ease of estimates comparison and to generate the output elasticities. The Akaike Information Criteria (AIC) also validates the log-log specification.

Financial Inclusion Index Measurement

For the calculation of the financial inclusion (FI) index, an approach similar to Sarma (2008, 2012) and Park and Mercado (2018) is used. The estimated FI index is between 0 and 1, where a figure approaching 1 indicates a high level of financial inclusion, while a figure tending toward 0 suggests a low level of financial inclusion. Such an index allows easy comparison of a country's financial inclusion level over time and between different countries.

Access, availability, and utilization of certain financial services are the key criteria defining financial inclusion. Each dimension appears as two indicators. The first dimension, access, includes two indicators: the number of commercial bank loan accounts per 1000 adults (AC2) and commercial bank deposit accounts per 1000 adults (AC1). The indicators for the second dimension, accessibility, are the ratio of commercial bank branches to 100,000 adults (AV2) and the number of ATMs per 100,000 adults (AV1). While the third dimension usage measures indicators, outstanding commercial bank loans (U1) and outstanding commercial bank deposits (U2) are expressed as a percentage of GDP. For comparability, given the indicators appear in several units, the various indicators are factorized to range from 0 and 1.

An approach similar to Nguyen (2020), consisting of a two-stage principal component analysis (PCA), is suitable for computing the financial inclusion index. Applying the first stage of the PCA, each dimension is estimated using the respective weights of each indicator (namely α_1, α_2, …, φ_2) included in that particular dimension.

$$\text{Access dimension}_{i,t} = D_{1,i,t} = \alpha_1 AC_{1,i,t} + \alpha_2 AC_{2,i,t}$$

$$\begin{aligned}&\text{Availability dimension}_{i,t} \\ &= D_{2,i,t} = \beta_1 AV_{1,i,t} + \beta_2 AV_{2,i,t}\end{aligned}$$

$$\text{Usage dimension}_{i,t} = D_{3,i,t} = \varphi_1 U_{1,i,t} + \varphi_2 U_{2,i,t}$$

Although the three dimensions define financial inclusion, they might not have an equal contribution to the level of financial inclusion. Therefore, in the second stage of the PCA, the weight of each dimension is estimated (W_1, W_2, W_3), and the financial inclusion index is estimated for a particular country at different periods.

$$\begin{aligned}\text{Financial inclusion index}_{i,t} &= W_1 D_{1,i,t} + W_2 D_{2,i,t} \\ &+ W_3 D_{3,i,t}\end{aligned}$$

11.3.1.2 Data

Considering data availability, a sample of 11 SIDs is chosen for the study. Data from 2004 to 2019 has been used for calculating the financial inclusion index and all other variables used in this study. The primary data sources have been the International Monetary Fund (IMF) and the World Bank.

Table 11.1 Descriptive statistics

Variable: financial inclusion index	
Observation	176
Mean	0.489
Standard deviation	0.25
Minimum	0.035
Maximum	0.994

Table 11.2 Long-run ARDL estimates

ARDL long-run equilibrium model: dependent variable LGDPPC			
Variable	Coefficient	Z	P > |Z|
LFI	0.3*	1.7	0.09
LFDI	0.1**	2.36	0.02
LHUMAN	2.1***	7.83	0.00
LTRADE	0.32***	10.49	0.00
LGFCF	0.90***	3.39	0.00

***Denotes significance at 1% level, **Denotes significance at 5% level, * Denotes significance at 10% level

Table 11.3 Short-run error correction model estimates

ARDL error correction model: dependent variable LGDPPC			
Variable	Coefficient	Z	P > |Z|
CointEq(−1)	−0.16*	−1.72	0.09
LFI	−0.04	−0.91	0.36
LFDI	0.01	0.55	0.59
LHUMAN	−0.06	−0.32	0.75
LTRADE	0.04	0.26	0.79
LGFCF	−0.11	−0.52	0.60

***Denotes significance at 1% level, **Denotes significance at 5% level, and *Denotes significance at 10% level

Table 11.4 Pairwise Granger causality test results

	F statistic	*P* value
Financial inclusion granger causes economic growth	8.70	0.01
Economic growth granger causes financial inclusion	0.01	0.96

Data Analysis

Unit Root Test and Cointegration Results

Based upon the unit root test (Levin Lin Chu), it is found that the variables included in the model are either integrated of order 0 (I (0)) or order 1 (I (1)). Under such specifications and given that growth modeling is dynamic, the autoregressive distributed lag approach (ARDL) is the most suitable model for assessing both the long-run and short-run dynamic link between the main variables under study.

Pedroni and Kao cointegration tests are subsequently conducted, and both confirm that the data series are cointegrated. In the presence of both I(0) and I(1) variables, we subsequently test for a long-run link between the variables included in the model. The explanatory variables (financial inclusion, foreign direct investment, trade, human capital, and gross fixed capital formation) and GDP per capita can thus be inferred to have a long-run equilibrium connection. The long-term and short-term relationships between the dependent and explanatory variables are subsequently estimated.

Panel Autoregressive Distributed Lag Model (ARDL) or Pooled Mean Group (PMG)

Based on the unit root and cointegration tests, the panel ARDL framework is the most appropriate for analyzing the data. The model's specificity is the assessment of the dynamic links between the variables. For that purpose, the explanatory variables, their lagged value/s, and the lagged value/s of the dependent variable are used as the determinants of the dependent variable.

The dynamic panel model proposed by Pesaran et al. (1999), known as the pooled mean group (PMG), is used to estimate the long-run and short-run relationships between the variables given in the model below:

$$
\begin{aligned}
{}'' \text{LGDPPC}_{it} &= \varphi_i \text{LGDPPC}_{i,t-1} + \beta'_{1i}\text{LFI}_{it} \\
&+\beta'_{2i}\text{LFDI}_{it} + \beta'_{3i}\text{LTrade}_{it} + \beta'_{4i}\text{LHuman Capital}_{it} \\
&+\beta'_{5i}\text{LGFCF}_{it} + 0.5em\sum_{j=1}^{p-1}\lambda_{ij}^{*''}\text{ LGDPPC}_{i,t-j} \\
&+\sum_{j=0}^{q-1}\delta_{1ij}^{*'\,''}\text{ LFI}_{i,t-j} + \sum_{j=0}^{q-1}\delta_{2ij}^{*'\,''}\text{ LFDI}_{i,t-j} \\
&+\sum_{j=0}^{q-1}\delta_{3ij}^{*'\,''}\text{ LTrade}_{i,t-j} + \sum_{j=0}^{q-1}\delta_{4ij}^{*'\,''}\text{ LHuman Capital}_{i,t-j} \\
&+\sum_{j=0}^{q-1}\delta_{5ij}^{*'\,''}\text{ LGFCF} + \mu_i + \varepsilon_{it}
\end{aligned}
$$

The PMG uses the maximum likelihood approach for the estimation of parameter values. The PMG estimation method assumes that the long-run coefficients remain the same across the different cross sections.

11.4 Analysis

Table 11.1 summarizes the descriptive statistics of the variables in this study. The data available on financial inclusion indicates that the average level of financial inclusion for SIDs is approximately equal to 0.49,[1] indicating that in most of these countries, the access, availability, and usage of financial services and products are readily available and used by most individuals. In some island economies, such as Mauritius, Dominican Republic, Fiji, and Jamaica, among others, it is noted that for some years, the financial inclusion index has been above 0.9, indicating a high level of financial inclusion.[2] Considering the importance of financial inclusion in the SIDs, the study's primary purpose is to investigate whether financial inclusion has any consequence on the economy, mainly promoting economic growth.

11.4.1 ARDL Long-/Short-Run Analysis

The ARDL/PMG model establishes the long-term relationship between economic growth and financial inclusion and the short-term dynamics between the variables and their lagged values (Table 11.2).

The long-run regression results indicate that financial inclusion has a significant favorable influence on economic growth. A 1% increase in the financial inclusion index is estimated to result in a 0.3% increase in economic growth. As discussed, the results align with the theoretical links between financial inclusion and growth. As such, they are consistent with empirical literature in general. For instance, using a total sample of 152 countries and subsequently smaller samples of countries regrouped in income level, Van et al. (2021) also reported that a 1% increase in financial inclusion led to increased economic growth. While Sethi and Acharya (2018) also depicted a positive long-run relationship for a sample of developed and developing countries. Using a similar approach, country case studies from Lenka and Sharma (2017) confirmed a positive link between financial inclusion and economic growth in the long- and short run when studying the Indian context. Makina and Walle (2019) studied 42 African countries with relatively low levels of financial inclusion. Using a GMM approach, they detected a positive link between financial inclusion and economic growth.

The other explanatory variables induce growth, with human capital and private investment showing relatively higher predictive power for economic growth in this sample of small island economies.

The short-run estimates from the error correction model appear in Table 11.3. The error correction term's measurement of the model's rate of adjustment to long-term equilibrium is equal to −0.16, which means that 16% of the short-term disequilibrium must be corrected annually to reach long-term equilibrium. The negative and significant coefficient can prove a long-term causal relationship between the key variables under study. For small island developing states, no significant link appears in the short run between financial inclusion and economic growth. This finding suggests that the influence of financial inclusion on economic growth might not be immediate, but the two variables bear a more long-term relationship. It implies that financial inclusion will take some time to have its desired effects. Similar results appear for the other control variables included in the study, where the variables affect economic growth more in the long term. Our findings differ from those of

[1]The financial inclusion index varies between 0 and 1, with a value tending toward 1, implying a high level of financial inclusion, and a value approaching 0 suggests low financial inclusion.

[2]The financial inclusion index for the SIDs appears in Table 11.5 in the Appendix.

Sethi and Acharya (2018), who employed a cointegration strategy to discover a significant, positive relationship between financial inclusion and economic development, even in the near term. Financial inclusion and economic growth were also positively correlated in South Asian nations over the short term by Thathsarani et al. (2021).

11.4.2 Granger Causality

From Brooks (2019), "Granger causality measures the correlation between the actual value of a variable and the past (lagged) values of the other variables." The causality test establishes the direction of causation when studying the nexus of financial inclusion-economic growth. The pairwise Granger causality test subsequently determines if economic growth influences financial inclusion and if there is a reverse causality effect. The results of the Granger causality test appear in Table 11.4.

The findings demonstrate a one-way relationship between financial inclusion and economic growth for SIDs. This result is consistent with earlier investigations that discovered a one-way relationship between financial inclusion and economic growth (see Lenka & Sharma, 2017; Sethi & Sethy, 2019). It is noteworthy that Sethi and Acharya (2018) are among the few studies confirming a causal link between financial inclusion and economic growth in both directions.

11.5 Conclusions

This research investigated the impact of financial inclusion as an ingredient of economic growth for a sample of island economies with their characteristics and vulnerabilities. A multidimensional financial inclusion index is constructed using the principal component analysis (PCA) and included in a classical augmented Solow growth function. Using a panel autoregressive distributed lag approach (PARDL) to account for the dynamic nature, empirical results confirm that financial inclusion is growth-conducive. However, the study could not establish any reverse causality using the panel Granger causality test, validating the fact that there is a unidirectional link. This study has its limitations. For instance, one limitation is the relatively small sample of island economies and the study's time frame, which are highly dependent on data availability. Moreover, the economic growth model could also have included additional growth factors, and finally, the indirect effects of financial inclusions on growth, if any, could have been studied.

Findings from this research supplement the existing literature on this SDG 8.10 goal and provide important implications for SIDs governments to consolidate and accelerate financial inclusion to enhance prosperity. The prioritization and acceleration of financial literacy and education for the entire community, encouragement of banks (particularly public banks) to increase their services in remote areas through more bank branches, provision of higher levels of microcredit more formally for SMEs or lower-income earners and government intervention in credit allocation, among others, are believed to be essential steps towards the enhancement of financial inclusion.

Appendix

Table 11.5 Financial inclusion index for SIDs

Year	2005	2010	2015	2016	2017	2018	2019
Mauritius	0.23	0.76	0.86	0.69	0.64	0.41	0.44
The Bahamas	0.59	0.51	0.37	0.5	0.36	0.36	0.27
Belize	0.29	0.59	0.28	0.45	0.278	0.38	0.4
Comoros	0.15	0.17	0.69	0.83	0.77	0.8	0.88
Dominican Republic	0.17	0.36	0.83	0.88	0.88	0.87	0.96
Fiji	0.16	0.43	0.84	0.93	0.89	0.86	0.83
Jamaica	0.29	0.26	0.35	0.43	0.81	0.85	0.96
Singapore	0.03	0.36	0.7	0.73	0.72	0.73	0.87
Seychelles	0.08	0.29	0.62	0.66	0.75	0.76	0.99
Solomon Islands	0.38	0.68	0.65	0.68	0.7	0.7	0.72
Tonga	0.45	0.22	0.57	0.65	0.83	0.87	0.87

References

Bagehot, W. (1873). *Lombard Street: A description of the money market*. HS King & Company.

Barajas, A., Beck, T., Belhaj, M., Naceur, S. B., Cerra, V., & Qureshi, M. S. (2020). *Financial inclusion: What have we learned so far? What do we have to learn?* IMF Working Papers, 2020(157).

Beck, T., Demirgüç-Kunt, A., & Levine, R. (2007). Finance, inequality, and the poor. *Journal of Economic Growth, 12*(1), 27–49.

Brooks, C. (2019). *Introductory econometrics for finance*. Cambridge University Press.

Chima, M. M., Babajide, A. A., Adegboye, A., Kehinde, S., & Fasheyitan, O. (2021). The relevance of financial inclusion on sustainable economic growth in Sub-Saharan African nations. *Sustainability, 13*(10), 5581.

Claessens, S., & Perotti, E. (2007). Finance and inequality: Channels and evidence. *Journal of Comparative Economics, 35*(4), 748–773.

Cull, R., Demirgüç-Kunt, A., & Lyman, T. (2012). Financial inclusion and stability: What does research show?

Demetriades, P. O., & Hussein, K. A. (1996). Does financial development cause economic growth? Time-series evidence from 16 countries. *Journal of Development Economics, 51*(2), 387–411.

Demirgüç-Kunt, A., & Singer, D. (2017). *Financial inclusion and inclusive growth: A review of recent empirical evidence*, World Bank Policy Research Working Paper (8040).

Goldsmith, R. (1969). *Financial structure and economic development*. Yale University Pres.

Jahan, M., & Mohan, M. (2019). Role of financial inclusion in promoting inclusive growth. *E-Commerce for Future & Trends, 6*(2), 34–40.

Karlan, D., Ratan, A. L., & Zinman, J. (2014). Savings by and for the poor: A research review and agenda. *Review of Income and Wealth, 60*(1), 36–78.

Kim, D. W., Yu, J. S., & Hassan, M. K. (2018). Financial inclusion and economic growth in OIC countries. *Research in International Business and Finance, 43*, 1–14.

Lenka, S. K., & Sharma, R. (2017). Does financial inclusion spur economic growth in India? *The Journal of Developing Areas, 51*(3), 215–228.

Lucas, R. E., Jr. (1988). On the mechanics of economic development. *Journal of Monetary Economics, 22*(1), 3–42.

Makina, D., & Walle, Y. M. (2019). Financial inclusion and economic growth: Evidence from a panel of selected African countries. In *Extending financial inclusion in Africa* (pp. 193–210). Academic Press.

McKinnon, R. I. (1973). *Money and capital in economic development*. Brookings Institution Press.

Nguyen, T. T. H. (2020). Measuring financial inclusion: A composite FI index for the developing countries. *Journal of Economics and Development*.

Nizam, R., Karim, Z. A., Rahman, A. A., & Sarmidi, T. (2020). Financial inclusiveness and economic growth: New evidence using a threshold regression analysis. *Economic Research-Ekonomska istraživanja, 33*(1), 1465–1484.

Nwogu, M. U. (2021). Evaluating the relationship between financial inclusion, remittance, and economic growth: A panel analysis of ECOWAS cross-sectional countries. *International Journal of Social Sciences and Scientific Studies, 1*(1), 1–14.

Park, C. Y., & Mercado, R. (2018). Financial inclusion: New measurement and cross-country impact assessment. Available at SSRN 3199427.

Patrick, H. T. (1966). Financial development and economic growth in underdeveloped countries. *Economic Development and Cultural Change, 14*(2), 174–189.

Pesaran, M. H., Shin, Y., & Smith, R. P. (1999). Pooled mean group estimation of dynamic heterogeneous panels. *Journal of the American Statistical Association, 94*(446), 621–634.

Purwiyanta, P., Pujiharjanto, C. A., & Astuti, R. D. (2020, October). The impact of financial inclusion on economic growth in Indonesia: panel data 34 provinces. In *Proceeding of LPPM UPN "Veteran" Yogyakarta Conference Series 2020–Economic and Business Series* (Vol. 1, No. 1, pp. 236–241).

Robinson, J. (1952). The model of an expanding economy. *The Economic Journal, 62*(245), 42–53.

Romer, P. M. (1986). Increasing returns and long-run growth. *Journal of Political Economy, 94*(5), 1002–1037.

Romer, P. M. (1990). Endogenous technological change. *Journal of Political Economy, 98*(5, Part 2), S71–S102.

Sahay, M. R., Cihak, M., N'Diaye, M. P., Barajas, M. A., Mitra, M. S., Kyobe, M. A., Mooi, M., & Yousefi, M. R. (2015). *Financial inclusion: Can it meet multiple macroeconomic goals?* International Monetary Fund.

Sarma, M. (2008). *Index of financial inclusion* (No. 215). Working paper.

Sarma, M. (2012). Index of Financial Inclusion - A measure of financial sector inclusiveness. *Berlin working papers on Money, Finance, Trade, and Development, Working Paper No. 07/2012.*

Sarma, M., & Pais, J. (2011). Financial inclusion and development. *Journal of International Development, 23*(5), 613–628.

Seetanah, B. (2008). Financial development and economic growth in an ARDL approach. *Applied Economics Letter, 4*(43), 43–50.

Seetanah, B., Ramessur, S. T., & Rojid, S. (2009). Financial development and economic growth: New evidence from a sample of Island economies. *Journal of Economic Studies, 36*, 124–134.

Sethi, D., & Acharya, D. (2018). Financial inclusion and economic growth linkage: Some cross country evidence. *Journal of Financial Economic Policy, 10*, 369.

Sethi, D., & Sethy, S. K. (2019). Financial inclusion matters for economic growth in India: Some evidence from cointegration analysis. *International Journal of Social Economics, 46*, 132.

Suidarma, I. M. (2019). The nexus between financial inclusion and economic growth in ASEAN. *JEJAK: Jurnal Ekonomi dan Kebijakan, 12*(2), 267–281.

Thathsarani, U. S., Wei, J., & Samaraweera, G. R. S. R. C. (2021). Financial Inclusion's role in economic growth and human capital in South Asia: An econometric approach. *Sustainability, 13*(8), 4303.

UNCTAD (2014). Report of the Expert Meeting on the Impact of Access to Financial Services, including by Highlighting Remittances on Development: Economic Empowerment of Women and Youth.

Van, L. T. H., Vo, A. T., Nguyen, N. T., & Vo, D. H. (2021). Financial inclusion and economic growth: An international evidence. *Emerging Markets Finance and Trade, 57*(1), 239–263.

Yang, Y., Loukoianova, M.E., Sodsriwiboon, P., Wu, M.Y., Jahan, M.S., Hunter, M.L., Jamaludin, M.F., Guo, M.S., Schauer, J., & Rawat, U. (2018). *Financial inclusion in Asia-Pacific* (No. 2018/017). International Monetary Fund.

12 Financial Inclusion: Initiatives and Challenges in the BRICS Countries

Debdutta Choudhury, Apoorva Vaishnavi Bhattaram, D. Priya Dhatri, and Roshan Bhut

Abstract

Financial inclusion is one of the key initiatives taken by policymakers in the Global South to include the section of the population in the mainstream banking and financial system. The initiatives taken by Brazil, Russia, India, China, and South Africa to bring their population into the mainstream financial system can be taken as case studies by the rest of the world as models for financial inclusion. This chapter highlights the initiatives and the results of the various policies across five significant economies, the BRICS countries.

Keywords

BRICS · Financial inclusion · Economics · India · China · Russia · Brazil · South Africa · Development · Regulations · Digitalization · Monetary policy

12.1 Introduction

As a significant emerging global block, the BRICS countries have an economic influence on the rest of the world. However, much of their population still struggles with poverty and inequality. The financial inclusion measures taken by these countries are an effort to lift the living standards for this section of the population. Financial inclusion ensures that citizens directly receive the welfare measures of the respective governments and do not get lost in the maze of bureaucracy and corruption. The case studies of these five countries reveal several common and contrasting initiatives the governments and society took in this direction.

12.2 Brazil

A formal bank account is the key indicator of financial inclusion as it enables access to other financial activities. A 2020 study found that "70% of Brazilian individuals have an account at a formal financial institution" (Altarawneh et al., 2020). Brazil has focused on increasing the physical availability of banking channels to facilitate financial inclusion. A report (IMF & World Bank, 2013) states, "Gaining access to bank accounts in Brazil has been greatly facilitated by the correspondent banking network, which numbers approximately 152,000 locations throughout the country. This network complements the much smaller number of traditional bank branches (approximately 20,000)." By increasing geographical access, Brazil successfully increased

D. Choudhury (✉) · A. V. Bhattaram · D. P. Dhatri · R. Bhut
Woxsen University, Hyderabad, Telangana, India
e-mail: debdutta.choudhury@woxsen.edu.in

G. W. Muschert et al. (eds.), *Financial Inclusion*, Sustainable Development Goals Series,
https://doi.org/10.1007/978-3-031-68803-4_12

convenience and reduced the costs of performing basic banking facilities.

In 1998, Brazil framed a new federal constitution named the "Citizen Constitution," which expanded several citizen rights. Two essential pillars for financial inclusion were introduced then—the national financial system (SFN) and consumer protection. Both elements were designed to promote inclusive economic order and financial democracy.

The 1990s marked economic stability in Brazil, previously plagued with hyperinflation. After the stabilization, Banco Central do Brasil—the Central Bank of Brazil—has taken several steps to construct an inclusive financial system. Some of the key highlights are:

1. Improved regulations on credit unions.
2. Introduced "simplified accounts" with easy process, minimal documentation, and no fee collection.
3. Regulations to increase the transparency of the financial services offered.
4. Implemented customer service structures to answer people about various laws and policies.
5. BCB has institutionalized its Financial Education Program (PEF-BCB) to raise financial education.
6. Created the "Citizen Profile," which detailed financial information in simple language to encourage financial literacy.

In 2013, the Central Bank of Brazil launched its famous "Financial Citizenship Program," which aimed to integrate financial inclusion, financial education, and consumer protection. As a result, Brazil has witnessed a dramatic rise in financial inclusion. According to a study by Banco Central do Brasil (2018), around 87% of adults were banked in Brazil, compared to 60.8% in 2005.

The efforts of financial inclusion in Brazil also extended to gender. In 2005, among adults, 50.8% of men were banked, while among women, it was 49.2%. But in 2017, the proportion of women was higher, with 51.3% of women as part of the banking institutions compared to 48.7% of men (Banco Central do Brasil, 2018). One of the main drivers behind gender equity is the financial aid schemes of the government. According to a report by the Reserve Bank of India (2021), the Bolsa Familia, a conditional cash transfer scheme commenced in 2003, "is a powerful tool that looks at empowering the women with the financial decision." Under the scheme, the eligible families receive monthly financial aid through an "Electronic Benefit Card," which is paid directly to the female head of the family. The card could be used at PoS and also to digitally transfer money, thus bringing the women of low-income households into the financial ecosystem.

The past decade has witnessed the geographical expansion of the points of service in Brazil. Today, the financial network reaches all the municipalities of Brazil, even in the areas with low population density.

However, the most significant push toward digital financial inclusion has come through Pix6, an instant payment program initiative by Banco Central do Brasil launched in November 2020. The scheme has laid down all the regulations and procedures regarding instant payment services. Pix6 has now become an alternative for financial transfers, allowing the digital transfer of funds across accounts. According to the World Economic Forum (Vélez, 2022), "Pix allowed over 40 million people to make their first ever bank transfers. Between November 2020 to March 2022, the number of Pix users increased from 41 million to 124 million."

Covid-19 has also enabled the boom in Brazil's financial inclusion. World Economic Forum (David Vélez, 2022) states that "around 16 million people have been enfranchised into the Brazilian financial system."

Brazil launched the Open Banking Project (BOBP) in 2021, based on the principle that "consumers are the true owners of their data, and, therefore, should be able to use it to their advantage" (Reserve Bank of India, 2021). Under the BOBP scheme, third-party app developers can now use consumer data with consent and build services for financial institutions. The data includes registering bank accounts, transactions, and even credit proposals—the program aimed to minimize the information asymmetry contributing to rising financial service costs. Increased transparency should bring standardization and better credit risk assessments. BOBP is also expected to increase the competition and, as a result, better and cheaper financial products for the customers.

With its pro-innovation landscape and regulations, Brazil enabled banking and financial institutions to prioritize consumer interests. The progress achieved results from an inclusive regulatory framework and focusing on creating services that speak directly to Brazilian consumers.

12.3 Russia

Russia has focused on financial awareness and research in its efforts to bring in financial inclusion. In 2011, it launched a nationwide, five-year "Financial Education Programme" under the World Bank and Ministry of Finance collaboration. As part of the program, the Russian government initiated the "Financial Literacy Trust Fund"in cooperation with the OECD and the World Bank to support the research on improving financial literacy among the people. According to the Ministry of Finance of the Russian Federation (2013), the program targets vulnerable groups and youth and aims for positive socioeconomic outcomes, especially among low-income groups.

As a result of these efforts, more individuals were introduced into the formal financial environment. A report by the Bank of Russia (2018) states that the adult population with bank accounts was 89.4% in 2018. This figure had improved significantly from 2014, when the percentage was 48.2% (Gorshkov, 2017).

The Unified Biometric System (UBS) was launched to increase remote access to financial services in June 2018. UBS is the national digital platform of Russia that allows consumers to remotely open bank accounts and receive loans through identification using biometric data like face and voice recognition.

To align the efforts and interests of various stakeholders, the Bank of Russia set up "Permanent Working Groups" and released "The Financial Inclusion Strategy of Russian Federation for 2018–2020 (extended through 2021)." The policy document contains the priorities of financial inclusion. It details the strategies to create a barrier-free financial ecosystem, especially for low-mobility groups like the elderly, people with disabilities, etc. In addition to these policy initiatives, the Bank of Russia also released "Priorities of the Financial Inclusion Programme of the Russian Federation for 2022–2024," wherein it outlines the initiative planned to enable the creation of quality financial products, financial market development, and balance the interests of both individuals and businesses (Bank of Russia, 2021).

For the better availability of financial services, the Bank of Russia highlights the following initiatives in the "Russian Financial Market Development Program for 2022–2024":

1. Improve ease of access through better physical availability of the financial services.
2. Increase the scope of financial services through online channels for individuals and businesses while diminishing the cyber risks and digital inequality.
3. Widen the opportunities for debt and equity for businesses.

Today, noncash payments have become the norm in Russia. Advances have been made in the infrastructure in the face of solid demands for remote access to accounts and funds by the clients of credit institutions. In 2022, the percentage of accounts with remote access to funds was 91%, and the share of noncash sales in total retail and commercial services was 72.6% (Bank of Russia, 2021).

Russia's National Payment Development Strategy 2021–2023 "aims to provide consumers with convenient, safe, and accessible payment services, shaping a competitive and innovative payment market, and constantly improving the payment infrastructure" (Reserve Bank of India, 2021). The Bank of Russia and the National Payment Card System developed "The Faster Payments System" (FPS) to support digital financial inclusion. FPS is an infrastructure project aimed at enhancing the quality of payment services, promoting healthy competition, and making digital payments cheaper. FPS allows instant transfer of funds using mobile phone numbers.

According to an adult consumer survey as part of the program measuring financial inclusion in May 2021, "41.8% of adult consumers who do not regularly access remote financial services are ready to use remote channels regularly" (Bank of Russia, 2021).

Despite this progress, consumers and businesses have yet to embrace digital payment systems fully. A substantial hurdle is a need for more Internet services in remote and underpopulated areas. Even where the Internet is available, it does not meet the needs of low-mobility groups, the elderly, or rural residents who lack the digital literacy needed to access digital financial services. Territorial discrepancies in the distribution of financial institutions persist in Russia. Financial product providers are mainly concentrated in the western region. Governments and institutions must address inequalities in digital literacy and easy access to financial products to achieve total financial inclusion.

12.4 India

The concept of financial inclusion came into India in 2005, when the Reserve Bank of India introduced it. Most of India's population was not in the loop of financial inclusion as most of the population was into agriculture and mainly dealt in cash. In its economic initiatives, the Government of India gave financial inclusion a top priority. By guaranteeing that India's unbanked population has equitable and affordable access to financial services, initiatives like Jan Dhan, Aadhaar-Mobile, and Digital India have altered the game and reduced income disparities. During the years 2014 to 2017, adult Indians having bank accounts climbed to 79.8%, which was a 27% rise (Global Findex, 2017). The government introduced several schemes to support financial inclusion, such as the Pradhan Mantri Jan Dhan Yojana, which aimed to provide every citizen with a bank account. Likewise, there were many schemes like Jeevan Suraksha Bandhan Yojana, Pradhan Mantri Vaya Vandana Yojana, Pradhan Mantri Mudra Yojana, Stand-Up India scheme, Venture Capital Fund for Scheduled Castes under the social-sector initiatives, Pradhan Mantri Suraksha Bima Yojana, Atal Pension Yojana, Varishtha Pension Bima Yojana, Credit Enhancement Guarantee Scheme for Scheduled Castes, and Sukanya Samriddhi Yojana. Under PMJDY, 192.1 million accounts were opened. In the same year, Aadhar, a biometric-based identification project, was also introduced along with digital India, led by the rapid growth of the Internet. By 2017, 80% of the adult population had bank accounts, with 95% having Aadhar. FinTech also played an essential role in advancing and supporting financial inclusion; as the annual report of RBI suggests, the FI index rose to 53.9 in 2021 compared to 43.4 in 2017. The following Table 12.1 suggests key payment statistics for the past 3 years.

Since 2014, there have been three major phases in developing DFS (Digital Financial Services) in India. The first phase, which spanned from around August 2014 to August 2016, was characterized by a consistent increase in volume growth of 2% each month on the top digital platforms. Demonetization played a significant role in the second phase, while the Unified Payments Interface (UPI) drove the third phase (Journey Map Report, RBI data 2014–2018). The following are the key players:

1. The Pradhan Mantri Jan Dhan Yojana: By 2017, 79% of rural Indians had bank accounts, compared to 76% of urban Indians (Global Findex database, 2017).
2. RuPay: Seventy-nine percent of the 330 million people with PMJDY plan bank accounts had received a RuPay card (Ministry of Finance, GOI). According to the Global Findex research, India's debit card adoption increased from 22% in 2014 to 33% in 2017.
3. Aadhar: As the Aadhaar number has evolved into a legally recognized identification for rural Indians looking to open their first bank account, Aadhaar's growth has been intrinsically connected to PMJDY. "In 2016–17 and

Table 12.1 Payment through various modes in India

S.no.	Indicators	2018–19	2019–20	2020–21
1.	RTGS	1356.88	1311.56	1056
2.	NEFT	227.94	229.46	251.31
3.	IMPS	15.80	23.38	29.41
4.	UPI	8.77	21.32	41.04
5.	BHIM pay	815	1303	2580
6.	Debit/credit cards	11.97	14.35	12.94
7.	Total digital payments	1637.13	1602.89	1414.85
8.	Total digital payments (volume in lakh)	2,32,602	3,41,240	4,37,118

Source: Annual report of Reserve Bank of India, FY 2020–21 (all values in INR 100 billion)

2017–18, there were 138 million e-KYC validations with Aadhaar, up from 48 million in 2016–17." (Source: Journey Map Report, March 2019).

4. Demonetization and GST: Demonetization could have been a driving force behind the sharp rise in digital transactions. Instead of demonetization, the growth at the start of 2016 is the result of the development of UPI, which coincided with a better business environment and the implementation of the GST.
5. Licensing to new levels of banking institutions and creation of payment banks.
6. Fintech: Fintech companies include P2P and retail payment processing companies and software companies like Pine Labs and Mswipe. Since India Stack and Aadhaar have supplied the underlying digital infrastructure, many of these emerging fintech companies now have a business case. "From October 2016 to January 2017, pre-paid instrument (PPI) transactions more than doubled (from 127 million to 296 million)" (RBI data).

Covid-19 also played a crucial role in advancing financial inclusion. As a component of Aatmanirbhar Bharat Abhiyan (self-reliant India project), the Emergency Credit Line Guarantee Scheme (ECLGS) was introduced in May 2020 to assist eligible micro, small, and medium enterprises (MSMEs) and business enterprises in meeting their operational liabilities and resuming their operations in the wake of the COVID-19 pandemic's disruption.

Financial inclusion progress since its launch indicates the progress of banks such as:

1. The number of branches opened: 68,681 to 102,343 from March 2006 to March 2013.
2. The number of banking outlets: 67,694 to 268,454 from March 2006 to March 2013.
3. Kisan credit cards: Issued to small farmers for meeting credit requirements. A total of Rs. 2622.98 billion was outstanding up to March 2013.
4. Expansion of ATM network: From March 2010 to March 2013, there was a CAGR of 30.6% for the overall number of ATMs in rural India, from 5196 to 11,564 from March 2010 to March 2013.

With 48 billion real-time payment transactions, India has the most significant percentage of people who have adopted digital payments at close to 40% (2021).

With mobile and Internet banking transactions rising to 13,615 per 1000 adults in 2020 from 183 in 2015 and the number of bank branches edging up to 14.7 per 1 lakh adults in 2020 from 13.6 in 2015, which is higher than Germany, China, and South Africa, according to a report, India has now surpassed China in terms of financial inclusion metrics.

12.5 China

China, the second largest economy in the world with a GDP of $17,458.04 (Statista) in 2021, is an important subject to understand global demand and distribution of resources. Its rich land and people resources led the country to become the world's manufacturing hub.

Financial inclusion is vital to the country's economic growth and well-being.

According to World Bank's Global Findex Database in 2017, 20% of Chinese residents still need a bank account, and 91% still need loans from formal financial institutions. To curb this scenario, in 2013, the Third Plenary Session of the 18th Central Committee of the Communist Party of China (CPC) explicitly announced the initiative to develop financial inclusion. On December 31, 2015, the State Council of China issued a Plan for Advancing the Development of Financial Inclusion (2016–2020) (Chen & Yuan, 2021).

The primary agenda of financial inclusion is to ensure the availability of quality financial products that fetch satisfaction to MSEs, the poor, physically challenged, rural residents, and urban residents who fall under the low-income category.

The financial inclusion measures by the Republic of China can be categorized into three segments: monetary policies, tax and fiscal policies, and regulatory policies.

12.5.1 Monetary Policies

- The Government of the Republic of China decreased the reserve ratio from 14.6% to 9.4%.

- Reserve ratio: The amount commercial banks can withhold rather than lending or investing (Table 12.2).
- They introduced two rights for mortgage loans:

  1. Farmer's housing property rights.
  2. Management rights to contracted rural lands as collaterals encourage farmers and MSEs to take loans.

12.5.2 Tax Policies

1. VAT exemption and reduction in taxable income calculation by 10% for interest in the income of the financial institution on small loans to farmers.
2. Tax deduction for insurance companies' premium income from the aqua and farming industry.
3. Setup of national guarantee funds (owned by the government) to lend more loans to farmers and MSEs.
4. The earmarked funds of financial inclusion to be used for

   - Awarding financial institutions that showed an incremental increase in agri loans.
   - Subsidies for financial institutions in rural areas.
   - Subsidies for guaranteed loans to start-ups.

12.5.3 Regulatory Policies

Regulatory policies deal with achieving objectives of financial inclusion with the help of laws and regulations.

1. Medium and large commercial banks are encouraged to set up inclusive finance departments in their HQ and branches.
2. Commercial banks are evaluated on an annual basis with respect to financial inclusion.

Table 12.2 Reserve ratio in various banks

Type	Reserve ratio percentage (%)
Large commercial banks	11
Medium commercial banks	9
Small commercial banks	6

Source: Financial inclusion in China: an overview

3. A slack of 2% higher than the industry standards is given to nonperforming loans for MSE loans without penalizing the financial institutions.
4. CSRC (China Securities Regulatory Commission) has initiated lowering financial standards to make equity exchange more straightforward and accessible nationwide.
5. Development and promotion of agri insurance to lower risks for farmers.

12.5.4 Progress of Financial Inclusion in China

China's commercial banks are geographically limited and uneven. However, with various factors like increased market size and policy motivation, the number of service outlets has increased to 8.4%, with a significant focus on village and township banks.

Lowering the risk assessment standards and accepting a broader range of collaterals have led to increased use of credit. According to the People's Bank of China (PBC) and China Banking and Insurance Regulatory Commission (CBIRC), by the end of June 2020, outstanding loans to MSEs and agricultural areas in China reached RMB40.7 trillion and RMB37.8 trillion, accounting for 24% and 22% of the total outstanding loan of financial institutions, respectively. As a crucial part of financial reform, financial inclusion significantly promoted poverty alleviation and regional economic growth in China (Chen & Yuan, 2021). The rate of interest on loans granted to MSEs declined by 0.76% from the average level of 5.94%.

12.5.5 Contribution of Fintech to the Financial Inclusion of China

With the introduction of fintech, financial inclusion has improved extensively. Options like digital payment systems and electronic wallets have made it easier for both urban and rural populations to make use of advanced banking options with safety.

Tsai (2011) shows that China witnessed an explosive growth of fintech products and services through increased demand for web-based services. The government's 2016–2020 plan encouraged digital technologies to promote financial inclusion and social stability. (Tsai).

- Online payment options: Many commercial banks have built "all-in-one" platforms that handle the entire financial transaction with a wide variety of financial products.

Example: WeChat Pay & AliPay

- Internet banks: Given China's vast population and the limited number of banks, Internet banks provide access to users nationwide.

Example: MyBank & WeBank

- Internet microfinance: The financial inclusion measures have encouraged banks to offer financial products that cater to high-frequency, small-amount loans on an urgent basis. Few banks also offer instant credit approval without any guarantee. The cutback in the standards for loan approvals has made loans accessible.

12.5.6 Financial Inclusion Challenges in China

The significant challenges of financial inclusion measures are the slow rate of technology adoption and minimal financial literacy.

The use of formal credit in China is relatively lesser than in most other BRICS countries because most of the lending and borrowing occurs among friends and family. However, a trend has appeared where higher education, age, and experience led to the use of formal credit systems.

The 41st China Statistical Report on Internet Development shows that only 16.5% use the Internet for financial management, though 772 million Internet users exist. This is a significant concern in China's attempt to achieve financial inclusion through the Internet. This may be due to the low financial literacy of the users. Only 20.5% of Internet users own tertiary degrees. However, a survey by the IiMedia Research Group reports that 79.3% of China's digital financial products users own tertiary degrees (Shen et al., 2020).

The use and development of digital inclusion can be challenging, as it is vulnerable to money laundering and illegal data usage.

The financial inclusion measures adopted by China are relatively simplistic. Considering China's vast population, introducing targeted products and sustainable business models can help China's financial inclusion.

12.6 South Africa

Financial inclusion is a fundamental key to reducing poverty in South Africa. It is also a measure to eliminate unemployment and reduce inequality. It will enable the portion of the population who are unbanked or have no access to financial services to meet their requirements adequately. They still need to rely on cash as their purchase medium, as many live in remote regions and have lower access to digital channels. Traditionally, the financial sector in South Africa was run by four big banks, and now five. However, as an aftereffect, many new providers offer flexibility to the workforce who did not have access. The current banking sector of South Africa comprises 17 domestic commercial banks, 2 cooperative banks, 14 local branches of foreign banks, as well as offices of 43 foreign banks. The five banks own and control 80% of the banking sector and are worth 4 trillion rand or US$ 285 billion. The banking sector's infrastructure is excellent, with a network of 5000 branches and 30,000 ATMs. Between 2005 and 2013, there was a 60% surge in electronic financial transactions like debit cards, credit cards, and online payments. South Africa has made significant progress compared to other developing countries.

The National Payment System Framework and Strategy Vision 2025 highlighted the need for financial inclusion through cooperation and collaboration among industry stakeholders in providing access to and effective use of formal financial products for all South Africans. It is the topmost priority of the South African Reserve Bank. In October 2015, the National Treasury of South Africa drafted and implemented the inclusion policy, which promoted improved access to financial services. After this, they sought from relevant authorities a monitoring report to keep track of the initiatives.

Age, education, and the total pay used as a proxy for income, race, gender, and marital status are all factors that influence financial inclusion, according to research. According to these findings, financial inclusion positively correlates with age, education, and total pay as a proxy for income, race, and marital status. This suggests that a rise in any of these factors will have a noticeable impact on South Africa's degree of financial inclusion.

However, being a woman in South Africa lessens one's prospects of being financially integrated. As a result, the government should promote women's use of financial services, and goods and services should be created specifically to meet their demands. The disparities in the likelihood of demand among the various races in South Africa demonstrate the need to educate those in the other races, which are less likely to need financial goods, about the advantages of financial inclusion. The government must ensure that education is strengthened for South Africans so that they would feel it beneficial to engage in the financial sector because education generally increases financial inclusion. This will enable the government to achieve the NDP Vision 2030[(2)].

The National Development Plan Vision 2030 of South Africa, published in 2011, aims to eliminate poverty, reduce inequality, and achieve full employment. It acknowledged financial inclusion as the major playing factor in achieving this goal. It has anticipated that the portion of the population with access to banking and financial services will grow from 63% to 90% by 2030. This same vision identifies broadening access to banking services and reducing costs through competition as a priority.

The 2019 new coronavirus epidemic has highlighted socioeconomic injustices in Africa by having a disproportionately negative impact on the most vulnerable demographics, including the underprivileged, the elderly, people who have been forcefully relocated, women, and other disadvantaged groups.

African nations joined the international community in enacting stringent containment measures, including border closures, curfews, stay-at-home recommendations, and social segregation laws, to halt the spread of the Covid-19 pandemic as it proceeded to claim lives and destroy livelihoods. Due to the widespread closure of bank branches and the suspension of the activities of digital financial service agents, these quarantine measures had a detrimental impact on access to money and financial services. These difficulties have raised the concern about financial inclusion even further. The Covid-19 emergency support programs in South Africa were driven by the country's already existing integrated social registries. In order to partially make up for the loss of school lunches caused by the closure, it boosted child allowances and enhanced unemployment compensation. The government has taken several steps to make it easier to employ digital financial inclusion as a strategy for public health to stop the spread of Covid-19 in Africa.

12.7 Conclusion

This chapter has highlighted the contrasting ways these five countries have tried to tackle the challenges of financial inclusion. The Covid-19 pandemic has accelerated the pace of financial inclusion, as policymakers understood the value of transferring direct benefits to the population suffering the maximum due to prolonged lockdowns and subsequent income loss. Significant initiatives like biometric-based identification systems can become case studies for the rest of the world to bring the entire population into one registry protected by biometric security and thereby have targeted benefits through the financial inclusion system.

References

Altarawneh, Y., Al-Nuaimi, M., & Al-Nimri, A. (2020). The determinants of financial inclusion in Latin America and Europe (Brazil and Romania case). *Systematic Reviews in Pharmacy, 11*(12), 192–196.

Banco Central Do Brasil. (2018). *Member series: Financial Inclusion Journey*. Financial Citizenship Journey in Brazil. https://www.afi-global.org/wp-content/uploads/publications/2018-08/AFI_MS_Brazil_AW_digital.pdf

Bank of Russia. (2018). Report on financial inclusion in The Russian Federation 2017. https://www.cbr.ru/content/document/file/47548/rev_fin_20180907_e.pdf

Bank of Russia. (2021). *Priorities of the financial inclusion programme of The Russian Federation for 2022–2024*. https://www.cbr.ru/content/document/file/133266/prioritet_naprav_finuslug_2022-2024.pdf

Chen, W., & Yuan, X. (2021). Financial inclusion in China: an overview. *Frontiers of Business Research in China, 15*(1), 21. Springer Open. https://fbr.springeropen.com/track/pdf/10.1186/s11782-021-00098-6.pdf

COVID-19 Pandemic: Digital Financial Inclusion as Public Health Tool in Africa. *Africa Policy Journal*. Retrieved from https://apj.hkspublications.org/covid-19-pandemic-digital-financial-inclusion-as-public--health-tool-in-africa/

Digital Financial Inclusion In India—An Overview. Retrieved from https://www.researchgate.net/publication/362184318

Financial Inclusion in India. Retrieved from https://vikaspedia.in/social-welfare/financial-inclusion/financial-inclusion-in-india

Gorshkov, V. A. (2017). Financial inclusion in Russia: Challenges for the promotion of long-term savings and investment. *The Journal of Comparative Economic Studies, 12*, 121–133.

How tech is powering financial inclusion in South Africa. Retrieved from https://www.cio.com/article/189650/how-tech-is-powering-financial-inclusion-in-south-africa.html

India ahead of China in financial inclusion metrics now: SBI report. Retrieved from https://www.business-standard.com/article/economy-policy/india-ahead-of-china-in-financial-inclusion-metrics-now-sbi-report-121110801290_1.html

International Monetary Fund & World Bank. (2013). *Financial inclusion in Brazil: Building on success*. World Bank. https://openknowledge.worldbank.org/handle/10986/16739. License: CC BY 3.0 IGO.

Ministry of Finance of the Russian Federation. (2013). Russian Federation: Financial Education and Financial Literacy Project: on the way to a National Strategy. Published in OECD Advancing National Strategies for

Financial Education. A Joint Publication by Russia's G20 Presidency and the OECD, pp. 217–228.
Reserve Bank of India. (2021). BRICS Digital Financial Inclusion Report, 2021. https://rbidocs.rbi.org.in/rdocs/Publications/PDFs/BRICSDIGITAL7330275ABF0C4198B581897DF2C131CA.PDF
Shen, Y., et al. (2020). Using digital technology to improve financial inclusion in China. *Applied Economics Letters,* 27(1), 5. Taylor & Francis, https://www.tandfonline.com/doi/pdf/10.1080/13504851.2019.1606401
Tsai, K. S. (2011). FinTech and Financial Inclusion in China. https://EconPapers.repec.org/RePEc:hku:briefs:201720
Vélez, D. (2022, May 19). Brazilians are adopting digital payments faster than anyone else- what lessons can we learn? *World Economic Forum.* Weforum.org. https://www.weforum.org/agenda/2022/05/brazilians-are-adopting-digital-payments-faster-than-anyone-else-what-lessons-can-we learn/#:~:text=Brazil%27s%20financial%20inclusion%20boom,have%20access%20to%20financial%20services

13 Assessing Financial Inclusion in 41 African Countries

Calvin Habasonda

Abstract

Many countries have recognized financial inclusion (FI) as a priority for policymaking. Financial inclusion generally has positive welfare effects on the financially excluded segments of society. This paper develops a financial inclusion index for 41 African countries to assess their progress in creating inclusive financial systems. The index covers the period from 2010 to 2019 using three financial inclusion dimensions: access, usage, and availability. The proposed index is easy to compute and provides a helpful tool for tracking progress in financial inclusion. The financial inclusion index will assist in the design of appropriate policy measures aimed at minimizing financial exclusion. The paper finds overall mixed progress in financial inclusion trends among the countries, despite the widely held notion that mobile banking has had a notable impact on financial inclusion in many African countries. Nonetheless, results confirm the significant contribution made by mobile banking. Moreover, the drive for inclusive financial systems is motivated by its perceived benefits on growth and human welfare, which are the focus of the sustainable development goals (SDGs) promoted by the United Nations Development Programme (UNDP).

Keywords

Financial inclusion · Sustainable development goals · Financial system · FinTech · Digital financial services · Mobile money

13.1 Introduction

The World Bank's objective to achieve universal access to financial services by 2030 highlights the significance of financial inclusion (FI) for growth in the economy and poverty alleviation. Financial inclusion has attracted increased attention from policymakers around the globe in recent years. An estimated sixty (60) countries have an explicit national strategy for financial inclusion, while a third of central banks have some form of explicit or implicit goals for financial inclusion. Besides, financial inclusion is a crucial pillar for achieving the United Nations Development Programmes' (UNDP) sustainable development goals (SDGs) on poverty reduction as outlined under SDG 8. While financial inclusion does not explicitly target the achievement of sustainable development goals, there is consensus that increased access to formal financial ser-

C. Habasonda (✉)
University of Stellenbosch,
Stellenbosch, South Africa
e-mail: chabasonda@boz.zm

G. W. Muschert et al. (eds.), *Financial Inclusion*, Sustainable Development Goals Series,
https://doi.org/10.1007/978-3-031-68803-4_13

vices expands opportunities and choices for participation in economic activities. According to the World Bank (2019), financial inclusion is an enabler for 7 out of the 17 SDGs. Moreover, finance is generally a crucial element in economic growth, distributional equity, and the deepening of the financial system (Allen et al., 2016; Beck et al., 2009). The increased focus on financial inclusion is mainly motivated by the notion of financial inclusion's perceived positive welfare effects. However, lately, the focus has shifted to innovations in financial technology (FinTech) and its contributions to increasing access and usage of financial services.

Understanding the trends and impact of financial inclusion depends on the availability of granular data on access, availability, and usage. Similarly, targeting and evaluating the contributions of these innovations require credible granular data. Moreover, regulators are becoming aware of the need to develop financial inclusion databases to monitor financial inclusion progress better. Persistent data gaps still exist in many countries. Nonetheless, studies based on available data point to modest growth in financial inclusion. Digital financial services[1] have been credited as the main driver of this growth (Manyinka et al., 2016).

FinTechs may deliver much-needed financial services at lower unit costs than banks. Additionally, Gomber et al. (2016) view digital finance as encompassing new financial products, financial businesses, finance-related software, and novel customer communication and interaction forms. The increase in ownership of smartphones and access to the Internet have created a conducive environment for digital financial services. Further, the introduction of cyber and data protection laws has created a safe environment for conducting digital financial services. Similarly, countries with cyber laws have been included in the digital finance revolution. They have embraced an adopt-and-learn approach where regulators allow the gradual and incremental introduction of innovations.

Expanded access to financial services increases economic opportunities for financially excluded segments of society. Similarly, there is an inherent assumption that access to finance will automatically resolve the issue of existing poverty and income inequalities. However, clarification is necessary regarding what kind of finance is good for growth. Although studies show that financial inclusion and poverty reduction are positively associated, determining a causal relationship has been more challenging. Although the question of whether financial inclusion contributes to poverty reduction and income inequality appears simple, the answer is less evident because of the definitional issues assigned to financial inclusion. Nonetheless, financial inclusion has a significant positive impact on the well-being of individuals.

13.2 Financial Inclusion Dimensions

Studies on inclusive finance often focus on three dimensions: access, usage, and availability, as discussed below and presented in Fig. 13.1.

13.3 Access

Viewed in the context of financial intermediation, an inclusive financial system that is effective in allocating resources should be accessed by many users. Financial services should be readily avail-

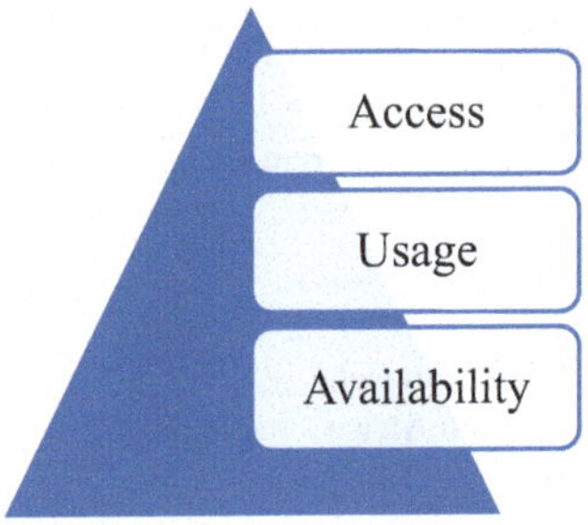

Fig. 13.1 Dimensions of financial inclusion

[1]Digital financial services refer to financial services through mobile phones, personal computers, the Internet, or cards linked to a reliable digital payment system (Manyinka et al., 2016).

able to as many people as possible. Access is the proportion of people with bank accounts versus those without them. Most studies use the number of deposit accounts per 1000 adults. Generally, the number of adults with bank accounts is positively related to that of deposit accounts per 1000 adults, reflecting the banked population. One limitation of using a bank account is that it results in double counting since individuals can have more than one account. This limitation may lead to inaccurate measurement of financial inclusion.

13.4 Availability

This dimension of financial inclusion reflects the number of branches and automated teller machines (ATMs) and the number of mobile money agents per 1000 adult population. Although data on mobile money is still scanty, it is now receiving more attention, considering the leading role of mobile banking in advancing financial inclusion. Further, the availability of Internet banking is included in this dimension, but it constitutes a very insignificant proportion in developing countries. Data is not readily available for this indicator for most countries, but the indicator remains, given available data.

13.5 Usage

Kempson et al. (2004), as cited in Sarma (2012), noted that the "usage dimension is motivated by the notion of under banked or marginally banked." This concept arises from the idea that an individual can have an account or access to a service with minimal usage. The usage is curtailed by distance to the nearest access point, high charges, negative consumer experiences, or inappropriate product design. The leading indicators for this dimension are credit and deposits as a proportion of gross domestic product (GDP), representing the level of private sector credit and deposit liabilities.

Table 13.1 below captures these dimensions and the indicators used to compute the financial inclusion index. The data and definitions derive from the World Bank Development Indicators database and the International Monetary Fund Financial Access Survey (FAS) database.

13.6 Access to Financial Services

Access to formal financial services remains a challenge in most developing countries. Similarly, the cost of accessing financial services is high due to market imperfections and low-income lev-

Table 13.1 Indicators used to compute FI index

Dimensions of financial inclusion		
Availability	Accessibility	Usage
No. of branches per 1000 km sq	No. of deposit accts/1000 km sq	Oustanding deposits(%GDP)
No. of ATMs per 1000 km sq	No. of loan accts/per 1000 km sq	Outstanding loans (%GDP)
No. of active MMA/1000 km sq	No. of registered MM a/cs/1000 km sq	No. of mob transactions (%GDP)
No. of registered MMA/1000 km sq	No. of MM transactions/1000 adults sq	Value of mob transactions (%GDP)
No. NB retail outlets/1000 km sq	No. of credit cards/1000 adults sq	No. of mob and int transactions (%GDP)
	No. of debit cards/1000 adults sq	Value of mob and int transactions (%GDP)
	No. of life policies/1000 adults sq	

Source: Author's computation

els. There are numerous studies on measuring financial inclusion and its impact on welfare. Demirgüç-Kunt and Klapper (2012) compared financial inclusion in 148 countries. The study finds that an estimated 50% of the world's population was unbanked. However, variations occur among countries based on income levels and state of economic development. World Bank (2019) estimated that about 90% of the population in the least developed countries lack access to financial services. As a result, many governments have designed financial inclusion strategies to tackle financial exclusion and achieve universal access to formal financial services.

13.7 Measurement of Financial Inclusion

Most attempts to measure financial inclusion have focused on single supply-side indicators such as the number of ATMs, the percentage of bank accounts and loans, and the number of branches. However, the challenge with this approach is that some applicable informational content from demand-side factors could lead to measurement errors in the usage indicators of financial inclusion. The importance of demand-side factors is emphasized in determining progress in financial inclusion (Kundu (2015). Cámara and Tuesta (2014) developed a multidimensional financial inclusion index based on use, barriers, and access. They find that macroeconomic factors like GDP per capita, schooling, financial system efficiency, and stability are highly correlated.

Some studies have investigated financial inclusion by understanding formal accounts and credit usage patterns. For instance, Demirgüç-Kunt and Klapper (2012) find that under 25% of adults use a formal account in Africa, while access to bank loans by small and medium enterprises (SMEs) is less than 8%. Rhyne (2012) argues that the number of accounts does not reflect the accurate measure of financial inclusion and that the critical factor is using the accounts for transacting. Moreover, using the number of accounts as a sub-indicator is problematic since one person can have more than one account. Most Finscope surveys identify mobile banking as the most significant contributor to growth in financial inclusion (Zambia Finscope Survey, 2020). Owens (2013) also observed that mobile banking promotes quick, sustainable, affordable financial services alongside electronic payments. The Global Findex Database (2014) revealed that financial inclusion had improved in many countries due to innovations in mobile money.

Aduda and Kalunda (2012) argued for the inclusion of informal financial services in the financial inclusion framework in Kenya since these significantly contributed to financial inclusion. However, there is a challenge in obtaining consistent time-series data for informal financial services to enable proper measurement. West (2015) observed that, unlike banks, mobile money offers the most realistic option for the future of financial inclusion. Therefore, studies constructing financial inclusion indices where mobile banking significantly contributes to financial inclusion should include this indicator in the calculation.

13.8 Literature Review

The existence of inequality in society justifies the need for universal financial access. Lack of access to formal financial services limits the economy's growth prospects. The World Bank and the United Nations acknowledge the role of financial inclusion in economic growth, income inequality, and poverty alleviation. Wilkinson and Pickett (2007) note that inequalities in income and assets have adverse implications on socioeconomic deprivation. It also results in unequal opportunities and inefficient resource allocation, inhibiting sound institutional development (World Bank, 2005).

Financial inclusion is a multidimensional concept and has many definitions. Others find it appropriate to define financial exclusion instead of financial inclusion. Ultimately, the choice of the definition determines the nature and focus of the efforts extended toward addressing a particular constraint. Leyshon and Thrift (1995) define financial inclusion as the exclusion of specific segments of society from formal financial systems. Sinclair (2001) focuses on the inaccessibility of financial services due to inappropriate products.

Conroy (2005) defines financial exclusion as preventing poor and disadvantaged social groups from accessing financial systems. Mohan (2006) views financial exclusion as signifying a lack of access to appropriate, low-cost, fair, and safe financial products and services by specific segments of society. Amidžić et al. (2014) and Sarma (2008) have defined it as an economic state where individuals are denied essential financial services. Sarma (2008)[2] extends this definition to view financial inclusion as mechanisms to ensure easy access, availability, and use of the formal financial system. This extended definition clarifies the concept since it delineates the various strands of financial inclusion based on usage, availability, and access, easily measured on a stand-alone basis.

On the other hand, Cámara and Tuesta (2014) define financial inclusion as a financial system that maximizes access and use and minimizes involuntary exclusion. This definition stresses supply- and demand-side factors and associated constraints to financial inclusion. The combination of supply- and demand-side factors creates a lack of conceptual clarity, as it mixes reasons for being financially excluded with the concept of financial inclusion itself.

Demirgüç-Kunt and Klapper (2012) define financial inclusion as those circumstances that feature broad access to financial services without either price or non-price barriers to their use. Chakravarty and Pal (2013) define it as the delivery of an economy's financial system to its members and, thus, consider it synonymous with banking inclusion. In defining financial inclusion, the distinction between voluntary and involuntary exclusion ensures that policy focuses on involuntary exclusion. According to the World Bank (2014), voluntary exclusion involves circumstances where an individual is excluded based on personal experiences or because of cultural or religious reasons. On the other hand, involuntary exclusion emerges from barriers arising from market imperfections such as the design of the financial products, level of bank charges, KYC practices, or distance to access points for financial services.

Developing a broader definition of financial inclusion has implications for a conceptual framework that can identify key factors that lead to an exclusive financial system. Earlier studies focused on social exclusion and the geographical footprint of financial services. However, this is limiting since physical access is one of many factors influencing financial exclusion. Although definitions of financial inclusion vary, some fundamental indicators from both the supply- and demand sides are consistent. A broader definition is more likely to incorporate many aspects of financial inclusion that can have significant implications when estimating the extent of inclusion or exclusion. The diversity of definitions signifies the concept's multidimensionality and helps determine what is measured and how to measure it. [3] The diversity also provides a broader conceptual framework which allows for incorporating country-specific factors in the definition to capture only relevant and significant ones in each country.

The arguments that financial exclusion emanates from the broader issue of social exclusion usually serve as a reason why some segments of the population remain excluded. For example, Dancey (2013) notes that the two concepts are part of the same effort. The argument relies on the idea that socially excluded people cannot meaningfully participate in social and economic activities. As a result of social exclusion, they do

[2]Access: availability of formal, regulated financial services: physical proximity and affordability.

Usage: actual usage of financial services and products: regularity, frequency, duration of time used.

Availability: refers to the penetration of financial services relative to the population.

[3]A multidimensional approach as suggested in Beck, Demirgüç-Kunt, & Martinez-Peria (2007) covers dimensions such as access and actual use of financial services and therefore, acts as a useful indicator of the dimensions to be measured when assessing the state of financial inclusion.

not have sufficient assets, thereby excluding them from the financial system. Many factors can contribute to exclusion from the financial system. In many African countries, cultural norms and gender have been significant barriers for women to acquire assets and access financial services. In addition, financial institutions often require essential identification documents to open an account. In cases where the informational requirements are unavailable, access to the financial system is denied through a process called de-risking. While financial institutions may desire to bank more people, the know-your-customer (KYC) norms based on anti-money laundering requirements become a barrier to financial inclusion. However, this constraint is slowly addressed with technological advancements and the number of financial inclusion products. In cases where a financial inclusion product is low risk, simplified know-your-customer requirements are applied. In any case, the financial action task force (FATF, 2014) has explicit guidance on how to deal with potential impediments to financial inclusion due to the implementation of its standards.

Several studies have used correlation analysis to examine the relationship between financial inclusion and human development (Sarma & Pais, 2011; Nanda & Kaur, 2016; Raichoudhury, 2016). Nanda and Kaur (2016) used both the human development index (HDI) and the modified human development index (MDHI) and found a positive relationship between financial inclusion and human development in 68 countries. Gupta et al. (2014) developed indices for financial inclusion based on penetration, availability, and usage of financial services. They established a positive association between financial inclusion and human development across different states in India. Sarma and Pais (2011) and Dutta and Singh (2019) also used the computed financial inclusion indices to test the relationship between financial inclusion and human development and found that they are positively correlated. Tita and Aziakpono (2017) studied the welfare-enhancing effect of financial inclusion in sub-Saharan African countries using a cross-sectional framework using income inequality as a proxy for welfare. They find that financial inclusion is negatively associated with income inequality, implying that the availability of financial services may only sometimes lead to an increased usage of financial services.

Other studies find that the levels of financial literacy affect the level of inclusion. Customers who are financially literate will likely make informed financial decisions. In addition, financial literacy is a crucial aspect of fostering continued usage of financial services. According to Ramakrishnan (2012), financial literacy is the ability to make informed judgments and effective decisions regarding money use and management. Financial literacy is, therefore, an essential element that has positive spillovers for financial system stability, integrity, and consumer protection. It generates a low-cost base of deposits, enhancing the financial system's resilience while allowing for the scrutiny of transactions for money laundering, or terrorism, or proliferation financing. This study uses supply-side dimensions of financial inclusion to construct a financial inclusion index.

13.9 Methodology

Most studies adopt two widely used approaches in developing financial inclusion indices, namely the nonparametric approach (Sarma, 2016) and the parametric approach adopted by Camara and Tuesta (2018), among others. Under the nonparametric method, assigning weights for the indicators is subjective using judgment. Lockwood (2004) argues that the results are affected in such circumstances since the indices are sensitive to the subjectivity of the weight assigned. This study uses the coefficient variance approach to avoid the problems associated with the arbitrary assignment of weights, which measures the dispersion of data points around the sample mean. Another problem associated with the computation of indices for financial inclusion is multicollinearity, potentially arising from similarities among dimensions of financial inclusion, such as access and availability (Sarma, 2008). Nevertheless, since multicollinearity only affects the coefficients and the p values rather than the model's predictive power, it can be overlooked in many cases, such as in constructing the financial inclusion index. Moreover, the variance inflation factor (VIF) test can still be relied on to determine the extent of multicollinearity among variables.

13.10 Data and Variable Selection

The data used in this study covers the period 2010 to 2019 and includes all African countries. The three dimensions of financial inclusion, namely availability, access, and usage, together with their respective indicators, are used in calculating the index for financial inclusion. However, not all countries have the data required to compute the index. Such countries were dropped leaving 41 countries in the sample. The data used in this study comes from the World Bank Development Indicators and the International Monetary Fund Financial Access Survey (FAS) databases.

The indicators for each dimension of financial inclusion are listed in Table 13.1 below:

13.11 Econometric Estimation

Measurement of financial inclusion presents itself in the many dimensions of the concept. Although much is known about financial inclusion and its importance recognized, approaches to measure financial inclusion still need clarification due to little understanding of how inclusive the financial systems are and what types of financial services are being accessed. In this chapter, we first construct a dimension index for each financial inclusion dimension using the formula below:

$$X_{ij} = \frac{A_{ij} - m_{ij}}{M_{ij} - m_{ij}}$$

Where X_{ij} is the transformed value of indicator j in dimension I, A_{ij} is the actual value, M_{ij} is the maximum of each indicator, and m_{ij} is the minimum of each indicator. The transformed values of each indicator lie between 0 and 1, with 0 indicating complete exclusion while 1 is total inclusion.

We then compute the FII in dimension i using the following identity:

$$FII_i = 1 - \frac{\sqrt{w_{i1}^2\left(1-x_{i1}\right)^2 + w_{i2}^2\left(1-x_{i2}\right)^2 + \ldots + w_{in}^2\left(1-x_{in}\right)^2}}{\sqrt{\left(w_{i1}^2 + w_{i2}^2 + \ldots w_{in}^2\right)}}$$

x_{ij} is the transformed value ($0 \le x_{ij} \le 1$) and w_{ij} is the weight of indicator j in indicator i. This study assigns the weights using the coefficient variation, unlike other studies where the weights are arbitrary and nonscientific. Coefficient variation is the ratio of the standard deviation (σ) to the mean value (μ). Under this approach, the weighting for each indicator is a proportion of its coefficient variation to the numerical sum of all indicators' coefficient variation under the following identity:

$$w_{ij} = \frac{v_{ij}}{\Sigma_j v_{ij}}$$

With w_{ij} representing the weight of indicator j in dimension i, while v_{ij} is the coefficient variation. The next step is then to calculate the FII using the formula below:

$$FII_i = 1 - \frac{\sqrt{w_1^2\left(1-FII_1\right)^2 + w_2^2\left(1-FII_2\right)^2 + w_3^2\left(1-FII_3\right)^2}}{\sqrt{\left(w_1^2 + w_2^2 + w_3^2\right)}}$$

13.12 Analysis and Findings

The financial inclusion index (IFI) was computed for 41 countries in Africa. The summary of inclusiveness (see Table 13.2) shows that only Mauritius and Seychelles had high financial inclusion levels for the entire period except for 2018 when only Mauritius was in that category. No countries had medium financial inclusion in three different years (2010, 2014, and 2017),

Table 13.2 Level of inclusiveness

FI score key	2010	2014	2015	2017	2018	2019
1. $0.7 \leq IFI \leq 1$—High financial inclusion	2	2	2	2	1	2
2. $0.3 \leq IFI < 0.7$—Medium financial inclusion	0	0	1	0	3	1
3. $0.0 \leq IFI < 0.3$—Low financial inclusion	39	39	38	39	37	38
Total	41	41	41	41	41	41

Source: Author's computations

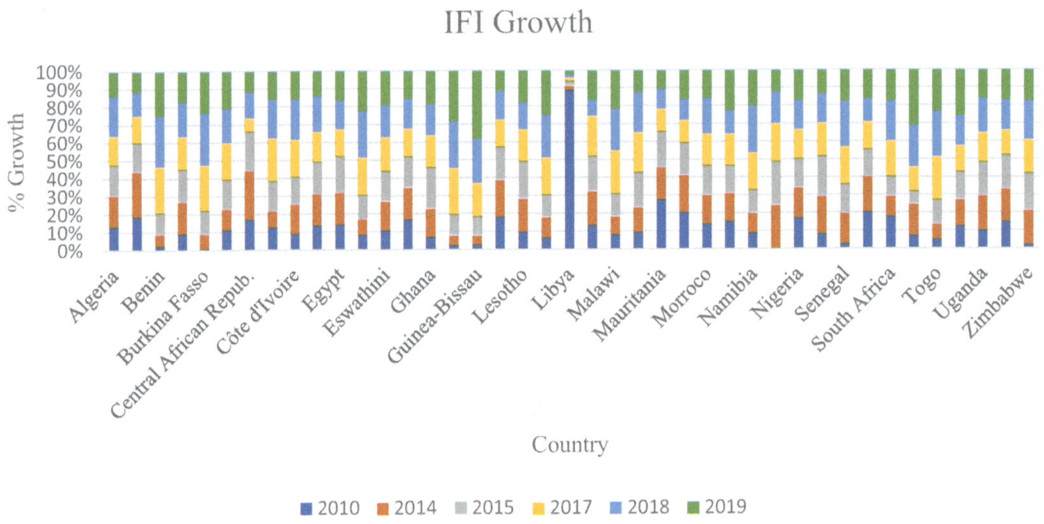

Fig. 13.2 Growth of financial inclusion in Africa. (Source: Author's computation)

while only one (1) country recorded medium financial inclusion in 2015 and 2019. That number was three (3) in 2018. Overall, more than 37 countries fell in the low financial inclusion category. Based on Sarma's index score (2012), 38 countries fell in the lower-ranking category for the entire period from 2010 to 2019. This result underscores the need for continued concerted efforts to expand financial inclusion.

Figure 13.2 below shows the growth levels in financial inclusion based on the calculated financial inclusion index. The graph depicts mixed growth patterns for the countries. However, most countries have recorded higher financial inclusion growth in 2018 and 2019 than in preceding years. In line with empirical studies that suggest a positive correlation between financial inclusion and human development, the mere presence of growth in financial inclusion is desirable. It points to improved welfare effects on the excluded.

The regression analysis results show that availability has the most significant and positive relationship with financial inclusion among the dimensions of financial inclusion, followed by access and usage in that order. These results imply that policies promoting financial services' availability are more likely to increase finances than those related to usage and access. However, based on pairwise correlations, access, and usage are positively associated with availability. This finding implies that policies focusing on increasing the presence of financial services in the community will also increase access and usage (Tables 13.3, 13.4, 13.5, and 13.6).

Table 13.3 Financial inclusion index

Country	Financial inclusion index		Country	Financial inclusion index	
	Value	Ranking		Value	Ranking
Seychelles	0.574508	1	Djibouti	0.148630	10
Morocco	0.554390	2	Eswatini	0.140873	11
Mauritius	0.490470	3	Nigeria	0.137908	12
Egypt	0.352557	4	Tanzania	0.133422	13
South Africa	0.347829	5	Ghana	0.125616	14
Libya	0.248137	6	Togo	0.122365	15
Tunisia	0.208305	7	Mozambique	0.113536	16
Kenya	0.202200	8	Uganda	0.111409	17
Rwanda	0.163220	9	Gambia	0.099492	18
Djibouti	0.148630	10	Algeria	0.071435	27
Eswatini	0.140873	11	Mali	0.064157	28
Nigeria	0.137908	12	Niger	0.060378	29
Tanzania	0.133422	13	Burkina Faso	0.057593	30
Ghana	0.125616	14	Benin	0.056944	31
Togo	0.122365	15	Burundi	0.043037	32
Mozambique	0.113536	16	Madagascar	0.041231	33
Uganda	0.111409	17	Guinea-Bissau	0.039296	34
Gambia	0.099492	18	Angola	0.038952	35
Senegal	0.093575	19	Cameroon	0.037612	36
Lesotho	0.092236	20	Equatorial Guinea	0.032252	37
Botswana	0.085774	21	Zambia	0.031657	38
Namibia	0.085465	22	Guinea	0.020251	39
Zimbabwe	0.083204	23	Chad	0.019609	40
Côte d'Ivoire	0.080202	24	Central African Rep.	0.012824	41
Sudan	0.076168	25			
Malawi	0.071518	26			

Table 13.4 Country index of financial inclusion

		2010	2014	2015	2017	2018	2019
1.	Algeria	0.0556973	0.0713204	0.0701012	0.0659029	0.0928470	0.0542188
2.	Angola	0.0414525	0.0554644	0.0357413	0.0331829	0.0300927	0.0238040
3.	Benin	0.0183753	0.0478703	0.0829074	0.1830128	0.2061759	0.1707780
4.	Botswana	0.0476660	0.0891783	0.0909318	0.0923759	0.0989919	0.0835065
5.	Burkina Faso	0.0070693	0.0543689	0.0862868	0.1687870	0.1965438	0.1500802
6.	Cameroon	0.0315087	0.0313394	0.0446914	0.0557176	0.0535703	0.0546440
7.	Central African	0.0066839	0.0104768	0.0084486	0.0029019	0.0056753	0.0042886
8.	Chad	0.0016899	0.0011889	0.0021908	0.0031502	0.0028723	0.0019774
9.	Côte d'Ivoire	0.0869332	0.1537346	0.1435435	0.1959808	0.2144141	0.1429131
10.	Djibouti	0.0557149	0.0700531	0.0716981	0.0686777	0.0814216	0.0541342
11.	Egypt	0.0989153	0.1251399	0.1400045	0.1073848	0.1147778	0.1103563
12.	Equatorial Guinea	0.0180827	0.0181126	0.0281242	0.0439930	0.0558382	0.0461398
13.	Eswatini	0.0780134	0.1133834	0.1192157	0.1344779	0.1276780	0.1310780
14.	Gambia	0.0908771	0.0942753	0.0912932	0.0846832	0.0904146	0.0792427
15.	Ghana	0.0719917	0.1571914	0.2260390	0.1817945	0.1752437	0.1785191
16.	Guinea	0.0045857	0.0099140	0.0204670	0.0460673	0.0470888	0.0490772
17.	Guinea-Bissau	0.0074747	0.0123825	0.0253850	0.0458115	0.0614572	0.0916617
18.	Kenya	0.1888162	0.2090485	0.1901432	0.1560621	0.1646972	0.1141760
19.	Lesotho	0.0586288	0.1105178	0.1223707	0.1062784	0.0901938	0.1034713

(continued)

Table 13.4 (continued)

		2010	2014	2015	2017	2018	2019
20.	Liberia	0.0110738	0.0187383	0.0202784	0.0340357	0.0393711	0.0395113
21.	Libya	0.0788836	0.0020598	0.0020773	0.0019056	0.0016518	0.0017787
22.	Madagascar	0.0229152	0.0319369	0.0329291	0.0376248	0.0157561	0.0266905
23	Malawi	0.0358068	0.0414830	0.0530691	0.1015984	0.0984723	0.0869482
24.	Mali	0.0502995	0.0720248	0.1013196	0.1171685	0.1184057	0.0604513
25.	Mauritania	0.0576485	0.0370388	0.0415329	0.0262984	0.0233627	0.0209894
26.	Mauritius	0.9732755	0.9746405	0.8615212	0.5943679	0.5664992	0.7392244
27.	Morocco	0.2141342	0.2403737	0.2478074	0.2653902	0.3029246	0.2274920
28.	Mozambique	0.0502002	0.0502489	0.0500237	0.0567800	0.0429926	0.0701806
29.	Namibia	0.0527645	0.0639641	0.0738028	0.1193426	0.1536952	0.1102866
30.	Niger	0.0004962	0.0316285	0.0319748	0.0276806	0.0232916	0.0155586
31.	Nigeria	0.1232669	0.1180828	0.1135588	0.1152302	0.1166073	0.1159188
32.	Rwanda	0.1179332	0.2864735	0.3063595	0.2537413	0.2302130	0.1754611
33.	Senegal	0.0220292	0.1239109	0.1164139	0.1534384	0.1867870	0.1243473
34.	Seychelles	0.7168182	0.6974595	0.5262087	0.5329202	0.4663110	0.5449782
35.	South Africa	0.1942221	0.1216055	0.1172226	0.2152680	0.2396322	0.1829157
36.	Sudan	0.0115736	0.0274108	0.0111023	0.0213548	0.0369768	0.0476542
37.	Togo	0.0347841	0.0555443	0.0877422	0.1593496	0.1673678	0.1494741
38.	Tunisia	0.1618637	0.1871187	0.1990705	0.1876062	0.2169458	0.3172506
39.	Uganda	0.0973892	0.1824609	0.1780809	0.1562550	0.1853399	0.1455444
40.	Zambia	0.0394835	0.0475891	0.0488006	0.0361606	0.0442436	0.0428118
41.	Zimbabwe	0.0214052	0.1596612	0.1789958	0.1591526	0.1871278	0.1439552

Table 13.5 Regression output

. regress main_ifi av_ifi ac_ifi us_ifi

Source	SS	df	MS	Number of obs	=	528
				F(3, 524)	=	13335.34
Model	17.6877686	3	5.89592286	Prob > F	=	0.0000
Residual	0.231674836	524	0.000442128	R-squared	=	0.9871
				Adj R-squared	=	0.9870
Total	17.9194434	527	0.034002739	Root MSE	=	0.02103
main_ifi	Coefficient	Std. err.	t	P>|t|	[95% conf. interval]	
av_ifi	0.6938466	0.0052504	132.15	0.000	0.6835321	0.7041611
ac_ifi	0.1133433	0.00486	23.32	0.000	0.1037959	0.1228908
us_ifi	0.1254112	0.0051203	24.49	0.000	0.1153525	0.13547
_cons	−0.0008679	0.001443	−0.60	0.548	−0.0037026	0.0019668

13.13 Conclusion

This paper constructed a composite financial inclusion index covering 2010 to 2019 for 41 African countries. The data observations were missing for several countries, and the time series available needed to be more consistent—the index aimed at measuring the level of progress made on financial inclusion among African countries. The paper finds mixed progress in levels of financial inclusion, with most countries still falling in the lower category of financial inclusion. Less than 10 countries fall within the medium and high financial inclusion category. The construction of a financial inclusion index is necessary for informing policy responses to promote the expansion of financial inclusion. Inclusive financial systems improve opportunities and choices for individuals to manage their assets and

Table 13.6 Pairwise correlations

. pwcorr main_ifi av_ifi ac_ifi us_ifi, sig

	main_ifi	av_ifi	ac_ifi	us_ifi
main_ifi	1.0000			
av_ifi	0.9673	1.0000		
	0.0000			
ac_ifi	0.6824	0.5527	1.0000	
	0.0000	0.0000		
us_ifi	0.6277	0.4834	0.5459	1.0000
	0.0000	0.0000	0.0000	

money to maximize their welfare. Well-being is an essential outcome of sustainable development goals. Achieving this objective will remain elusive in the presence of low financial inclusion since achieving the SDGs is premised on the positive welfare effects of inclusive financial systems. Though the study presents mixed growth patterns in financial inclusion, the mere fact that financial inclusion is expanding is desirable for poverty alleviation and addressing societal inequalities. Further, growth in financial inclusion is an important signal of how countries are performing toward achieving the goal of universal access to financial services by 2030.

References

Aduda, J. & Kalunda, E. (2012). Financial inclusion and financial sector stability with reference to Kenya: A review of literature. *Journal of Applied Finance & Banking, vol. 2, no. 6, 2012, 95–120*

Allen, F., Demirgüç-Kunt, A., Klapper, L., & Martinez-Peria, M. S. (2016). The foundations of financial inclusion: Understanding ownership and use of formal accounts. *Journal of Financial Intermediation, 27*, 1–30.

Amidžić, G., Massara, A., and Mialou, A. (2014). Assessing countries' financial inclusion standing—A new composite index, IMF Working Paper.

Beck, T., Demirguc-Kunt, A., & Honohan, P. (2009). Access to financial services: measurement, impact, and policies. World Bank Document

Camara, N., & Tuesta, D., (2018). Measuring financial inclusion: a multidimensional index. www.bis.org/ifc/publ/ifcb47p.pdf

Cámara, N., & Tuesta, D. (2014). *Measuring financial inclusion: A multidimensional index*. BBVA Research Working Paper No. 14/26.

Chakravarty, S. R., & Pal, R. (2013). Financial inclusion in India: An axiomatic approach. *Journal of Policy Modeling, 35*(5), 813–837.

Conroy, J. (2005). APEC and financial exclusion: Missed opportunities for collective action? Asia-Pacific Development *Journal, 12*(1), 53–79.

Demirgüç-Kunt, A., & Klapper, L. (2012). *Measuring financial inclusion: The global findex database* (Policy Research Working Paper 6025). The World Bank.

Dutta, S. K., & Singh, K. (2019). Variation and determinants of financial inclusion and association with human development: A cross country analysis. *IIMB Management Review*. https://www.sciencedirect.com/science/article/pii/S0970389616301574

FATF. (2014). *Virtual currencies: Key definitions and potential AML/CFT risks*.

Gomber, P., Koch, J. A., & Siering, M. (2016): Digital finance and Fintech: Current research and future research directions. *Journal of Business Economics*.

Gupta, A., Chotia, V., & Rao, N. V. M. (2014). Financial inclusion and human development: A state-wise analysis from India. *International Journal of Economics, Commerce and Management, II*(5), 1–24.

Kempson, E., Atkinson, A., & Pilley, O. (2004). *Policy level response to financial exclusion in developed economies: lessons for developing countries*. Report of Personal Finance Research Centre, University of Bristol.

Kundu, D. (2015). Addressing the demand side factors of financial inclusion. *Journal of Commerce and Management Thought, 6*, 397.

Leyshon, A., & Thrift, N. (1995). Geographies of financial exclusion: financial abandonment in Britain and the United States. *Transactions of the Institute of British Geographers, New Series, 20*(3), 312–341.

Lockwood, B. (2004), "How robust is the foreign policy-Kearney globalization index?", The World Economy, Vol. 27, pp. 507–523

Manyika, J. Lund, S. Singer, M. White, O. Berry, C. (2016). Digital finance for all: powering inclusive growth in emerging economies. McKinsey Global Institute, USA.

Mohan, R. (2006) Economic growth, financial deepening and financial inclusion. www.bis.org/review/r061121e.pdf

Nanda, K., & Kaur, M. (2016). Financial inclusion and human development: A cross-country evidence. *Management and Labour Studies, 41*(2), 127–153.

Owens, J. (2013). Offering digital financial services to promote financial inclusion: lessons we have learned. MIT Press.

Ramakrishnan, R., (2012). Financial Literacy-The demand side of financial inclusion. (PDF) Financial Literacy and Financial Inclusion (researchgate.net)

Raichoudhury, A. (2016). Financial inclusion and human development: Across country analysis. *Asian Journal of Business Research 6*(1), 34–48.

Rhyne, E. (2012). Money management, financial inclusion and banking the unbanked, Center for Financial Inclusion

Sarma, M., (2016), "Measuring financial inclusion for Asian economies", Financial Inclusion in Asia, Palgrave Macmillan UK, London, pp. 3–34.

Sarma, M. (2008). *Index of financial inclusion*. ICRIER Working Paper 215.
Sarma, M., & Pais, J. (2011). Financial inclusion and development. *Journal of International Development, 23*, 613–628.
Sinclair S. P., (2001). "Financial exclusion: An introductory survey", Report of Centre for Research in Socially Inclusive Services, Heriot-Watt University, Edinburgh
Tita, A., & Aziakpono, M. (2017). The effect of financial inclusion on welfare in sub-Saharan Africa: Evidence from disaggregated; Economic Research Southern Africa (ERSA) is a research programme funded by the National Treasury of South Africa.
Wilkinson, R.G., & Pickett, K.E. (2007). The Enemy between Us: The psychological and social costs of inequality. www.law.nyu.edu/sites/default/files/upload_documents/EJSP%20WilkPick%20final.pdf
World Bank. (2019). World development indicators. http://data.worldbank.org

Part IV

Country Studies

14 The Role of School Banking in Promoting Financial Inclusion: Evidence from Bangladesh

Md. Nur Alam Siddik

Abstract

Financial inclusion has been considered a priority policy by policymakers across the world. Governments and policymakers have adopted several policies and strategies to foster financial inclusion. A school banking program is one of those strategies. School banking aims to acquaint students with current banking services and technologies and create savings behavior among children under 18. The result is that school-going students come under a formal financial system that will allow them to participate in the country's economic activities. Though school banking has gained much attention, it is yet to be empirically investigated as the enabler of promoting financial inclusion. This study fills this gap by examining data from 2011 through 2019. Findings indicate that school banking activities promote access, usage, and depth dimension of financial inclusion. Results suggest that policymakers should make appropriate school banking decisions, leading to full financial inclusion.

Md. N. A. Siddik (✉)
Department of Finance and Banking, Begum Rokeya University, Rangpur, Bangladesh
e-mail: mnas@brur.ac.bd

Keywords

School banking · Savings · Financial inclusion · Sustainable growth · Trend analysis · Bangladesh

14.1 Introduction

Sustainable development goals are one of the most important policies for any country. Economically, having gone from an international basket case to a "laboratory for innovative solutions in the developing world," Bangladesh is a model of development possibilities and a leader in financial innovation. A case in point is the Grameen bank, which innovates the "micro-credit" concept for poor people who do not have access to bank funds due to a lack of sufficient collateral. The country is striving to achieve sustainable development goals (SDGs). Recently, the UN-sponsored Sustainable Development Solutions network (SDSN) has conferred Prime Minister Sheikh Hasina with the "SDG Progress Award" for Bangladesh's steady course in responding to the universal call to action to end poverty, protect the planet, and ensure peace and prosperity for all.[1] One of the main prerequisites

[1] https://www.thedailystar.net/news/bangladesh/news/pm-conferred-uns-sdg-progress-award-2180861

G. W. Muschert et al. (eds.), *Financial Inclusion*, Sustainable Development Goals Series,
https://doi.org/10.1007/978-3-031-68803-4_14

of sustainable development goals is financial inclusion, defined as providing access to and timely use of financial services to all people in society. Hannig and Jansen (2010) stated that financial inclusion aims to bring the underbanked and unbanked people into the official financial system to facilitate access to financial services, including savings, payments, and credit and insurance, in the growth process of a country. This process mainly refers to members of underprivileged groups, whose financial inclusion, in turn, will stimulate inclusive development and growth in the country (Siddik et al., 2015). Accordingly, financial inclusion has been considered a priority by policymakers, governments, and international organizations in many countries worldwide to achieve sustainable development goals (SDGs), in particular, SDG 8.10, sustainable economic growth, which states "sustain per capita economic growth per national circumstances and, in particular, at least 7 percent gross domestic product growth per annum in the least developed countries.".

Governments and policymakers worldwide have enacted numerous policies and strategies to foster financial inclusion. School banking is one of those strategies. School banking advances established banking principles, whose application applies to the welfare of the students. Teaching children essential financial management and saving skills can enable them to adopt responsible banking practices in the future. For this reason, school banking has an influential role in inculcating financial literacy among youth. It teaches essential financial literacy expertise that will bring benefits beyond their school life. This chapter aims to describe how school banking is performing in Bangladesh and examine whether there is a nexus between school banking and financial inclusion. Very few studies appear in this area. In particular, to the researcher's best understanding, there is a relative dearth of empirical studies focused on whether school banking activities promote financial inclusion. Therefore, this study, a descriptive work examining school banking as a financial inclusion strategy, should be significant for governments, policymakers, researchers, and youth.

14.2 Literature Review

14.2.1 Concept of School Banking

Generally, a school banking account is a savings deposit account title with a student's name co-signed by a guardian. School banking is a distinctive kind designed for school-going children, intended to create a practice of savings from an early age. Such accounts are an innovative form of banking that is gradually gaining popularity. School bank accounts have features and advantages for their users. For example, school bank accounts require no fees, free internet banking, relinquishment of the minimum balance prerequisite, use of a debit card at a reduced cost, and others. The school banking program helps students and parents accumulate a generous amount of savings. They also target the creation of savings habits among the students. In short, school banking is one of the steps to expand financial education and service programs that aim to bring school-going children aged 11 to 17 under the umbrella of formal financial channels and teach them about savings and financial management.

14.2.2 Aims and Imperatives of School Banking

Monetary matters are part and parcel of everyone's life. Suppose the practice of handling monetary matters pragmatically is nurtured in an early stage of life. In that case, it sets the stage for doing anything related to finance in a prearranged way in the later years. Without financial literacy, no financial strategy can work. From the inception of Mandell (2005), the working definition of financial literacy is what individuals must know to make vital financial decisions for their best interest. When it comes to school-age children, they learn how to save a tiny amount and see how that amount grows. At one stage of life, that small savings may become an all-important matter. With that end in view, programs in school banking have proliferated.

The aim of this school banking is to get involved with the children to attain financial literacy or how to save money, maintain savings, and calculate the rate of profit, for example. These perspectives will help the children know about modern savings banking. Hasan et al. (2021) found that financial knowledge has a significant impact on getting financial access. Similarly, Lyons and Kass-Hanna (2019) observed that households with higher levels of financial literacy are more likely to be engaged in positive savings behaviors and less likely to be borrowing, especially from informal sources. Shanks et al. (2013) found that recent financial deregulation had prompted the spread of increasingly multifaceted financial products and services, making it challenging for buyers to understand what they were buying. The finding suggested that escalation in consumer distress during good economic times perhaps stemmed from consumers' dearth of financial literacy. Schools can play an essential role in developing this financial knowledge among young children from their early childhood. Oberholtzer (1892) stated that through the suitable inculcation of saving and learning, various undesirable habits that produce disparity could not succeed if people learned self-knowledge, self-dependence, and self-reliance. He argued that if we could inspire the poor to sensible practices, we should exercise a broader and healthier generosity. Such positive behavior would show the middle classes the imprudence of spending all as they get it and the rich people that truthful, systematic care of nature-given blessings results in greater gratification,

For these reasons, school banking aims to enthuse children with financial and economic perceptions from childhood, with the lesson of developing the habit of savings. These small savings will pave the way for showing the benefit of savings for the future. Sherraden and Gilbert (2016) found that lessons learned early are often more effective, and young students have more time for saving and asset-building to enjoy their lives. A youth-based conducive policy may be the best route to a comprehensive and inclusive asset-building policy for the nation (Goldberg, 2005). Bowman (1922) echoed the views and values of the moral reformers of the mid-1800s, which advocated self-discipline, industriousness, wistfulness, thriftiness, practicality, and economy in behavior. Such policies aim to inhibit poverty, criminality, and prodigality while encouraging children to be thrifty, logical, efficient, and judicious in using money. It is essential in building character and preparing youth for future responsibilities. Bowman (1922) argued that good habits and good accounts are necessary assets.

All the students should learn that they should not spend everything in the present, and some should be reserved for the future. Bernheim et al. (2001) have observed a positive influence of financial literacy on peoples' savings behavior. Employing Merrill-Lynch account data for middle-aged investors, the authors found that those who taught financial literacy in their school tend to save a more significant portion of their incomes than those whose schools do not require a financial literacy course (Bernheim et al., 2001). So, school banking can easily be an engine for increasing personal savings, thereby raising national savings. An additional feature of this arrangement is the prospects to draw upon from the evolving area of behavioral economics. Suppose the first aim is to offer every citizen access to a conventional savings account system. In that case, a subsequent step is to make sure that the infrastructure, incentives, and power of inertia can produce that level of savings that can make a remarkable difference in the long run.

Thiry (1890) and Oberholtzer (1892) granted that the success of school banking was not merely about gathering savings but had supercilious goals. According to Thiry (1890), the attainment of the school savings scheme relied on several factors, including the school board's effectiveness, the teachers' enthusiasm, interest, and bank support. While the early contributors to school savings banks replied willingly by saving money, Oberholtzer (1892) argued that this was never the primary goal of school banking. Instead, the main goal was the making kids thrifty, honest, and self-responsible. Ultimately, these goals support the growth of the organic, social, and economic state of affairs under which we live, as well as the financial welfare of the nation—this objective is achievable provided that the purpose and benefits of early saving be an educational priority for youth. With school banking, on the one hand, the government can attain its goal of achieving full financial inclusion On the other, youth would learn savings habits and make improved financial decisions.

14.2.3 History of School Banking in Bangladesh

In Bangladesh, school banking activities started in 1960. Before the liberation war, several banks and Muslim banks first announced the notion of school banking. However, the program ended within a short period with no results. After many years, in 2003, AB Bank Ltd. (previously Arab Bangladesh Bank Ltd.) started school banking activities in the Chattogram district at Sunshine Grammar School. AB Bank Ltd. commenced its school banking program informally and without guidelines from the central Bangladesh Bank. Consequently, their effort was doomed (Al Mamun, 2021; Mondal et al., 2015; Sultana et al., 2020).

In today's world of sustainable development, scholars worldwide agree that financial inclusion plays a vital role in sustainable economic development. Thus, for a brighter future and a sustainable economic system, Bangladesh Bank, the central bank of Bangladesh, took the innovative initiative of introducing the school banking program, an initiative to build the habit of saving among children from a very early age and introduce them to banking services and technology, formally on November 2, 2010.[2] In its recent iteration, school banking is explicitly described as a route for school-going children and their parents to enable low depository obligations, exempted from operational charges, a high-interest rate on deposits with a cutting-edge method of banking transactions and dealings. After taking the initiative of school banking, all commercial banks were instructed to roll out school banking services and asked to open a deposit account for school-going students under 18 years old to make school banking programs more effective.

On October 28, 2013, Bangladesh Bank circulated the policy guidelines for school banking.[3] The guidelines state that students under 18 years are eligible to conduct school banking accounts by depositing only Bangladeshi taka BDT100, equivalent to US$1.18.[4] Parents or guardians can operate the account on behalf of their children. Debit cards with a transaction limit of BDT2000 (US$25) are also allowed. Bangladesh Bank also instructed all scheduled banks to send their quarterly statements of school banking in the designated form.[5] On December 17, 2018, Bangladesh Bank suggested that all the banks convert the school banking accounts into savings accounts with the consent of the account holders over 18.[6]

For all these above initiatives, the school banking program was popular among students from the beginning (Mondal et al., 2015). The response of the young savers, their teachers, and their guardians was highly positive. A total of 55 banks are conducting school banking in Bangladesh, and 2.69 million bank accounts exist under school banking rules with total deposits of BDT19.51 billion, equivalent to US$230.52 million, as of the end of 2020.[7] The grounds for the massive acceptance of school banking among the people is primarily because of its beneficial nature, including secured deposit and insurance facilities, and for assistance in teaching responsible banking activities from an early age. Thus, school banking has many potentials to foster financial inclusion in Bangladesh.

[2] https://www.bb.org.bd/mediaroom/circulars/brpd/nov022010brpdl12.pdf

[3] https://www.bb.org.bd/mediaroom/circulars/gbcrd/oct282013gbcrd07e.pdf

[4] Taka is the name of the Bangladeshi currency, whose short form is BDT.

[5] https://www.bb.org.bd/openpdf.php

[6] https://www.bb.org.bd/mediaroom/circulars/finincld/dec172018fidl02.pdf

[7] https://www.bb.org.bd//pub/quaterly/financial_inclution_report/dec2020.pdf

14.2.4 Scenarios of School Banking in Bangladesh

The annual participation in school banking in Bangladesh appears in Table 14.1.

From Figs. 14.1 and 14.2, it is clear that the total number of school bank accounts and the total amount of deposits are increasing over time. Figure 14.1 shows that school banking accounts have an increased straight-line upward trend, which indicates that school banking is gaining popularity over time. Although the increase in the number of bank accounts has been consistent over time, the number of bank accounts increased significantly in 2014 compared with previous years. This increase may derive from the release of circular no. 7, the guiding principle of school banking provided by the Bangladesh Bank in 2013, which stipulated that students under 18 years would be eligible for conducting school banking accounts by depositing only BDT100, equivalent to US$1.18.

Figure 14.2 depicts the deposits for the same period of 2011–2020. Similar to Figs. 14.1 and 14.2 indicates that school banking deposits have increased over time. On average, we found an increase in the number of accounts and deposits under school banking in Bangladesh, indicating an increase in financial inclusion. This trend suggests that any government aiming to achieve full financial inclusion should undertake conducive policies to expand school banking.

In order to make a robust scholarly contribution to the study, informal conversations have been conducted with parents, teachers, school leaders, students, or bank personnel handling such accounts as follows:

Parents with children having school banking accounts are happy with the school banking program. Parents often permit children to participate because they want their children to know how to save and secure their future financial needs. "My daughter accumulates small money into a school banking account. I encourage her and hope she will be able to meet her higher education expenses from this school banking savings"—mother of a class six student.

Parents like the school banking program; it demonstrates to children the importance of money and teaches them how to save and operate an account. It also explains that there are rewards for saving one's money. A parent of a class eight student said that "the great advantage of school banking accounts is that my kids have financial knowledge, they save money, earn a small profit from the account, and banks charge nothing for operating this account."

Many parents believe school banking programs should be spread, and every parent should participate in them as it has many beneficial features, including no charge for operating the account. "My son receives his scholarship money in the school banking account. He also saves money and now has financial literacy. I think every parent should

Table 14.1 Extent and progress of school banking in Bangladesh

Year	No. of school banking accounts (million)	Deposit amount	
		BDT (million)	US$ (million)
2011	0.03	307.90	3.64
2012	0.13	965.10	11.40
2013	0.30	3057.90	36.13
2014	0.85	7174.90	84.78
2015	1.03	8441.90	99.75
2016	1.26	10207.90	120.62
2017	1.45	13629.60	161.05
2018	1.82	15103.20	178.46
2019	1.99	16256.10	192.09
2020	2.69	19508.60	230.52

Source: Author's own based on financial inclusion report of Bangladesh Bank
Note: US$1 = BDT84.6292; converted from https://www.xe.com/currencyconverter/

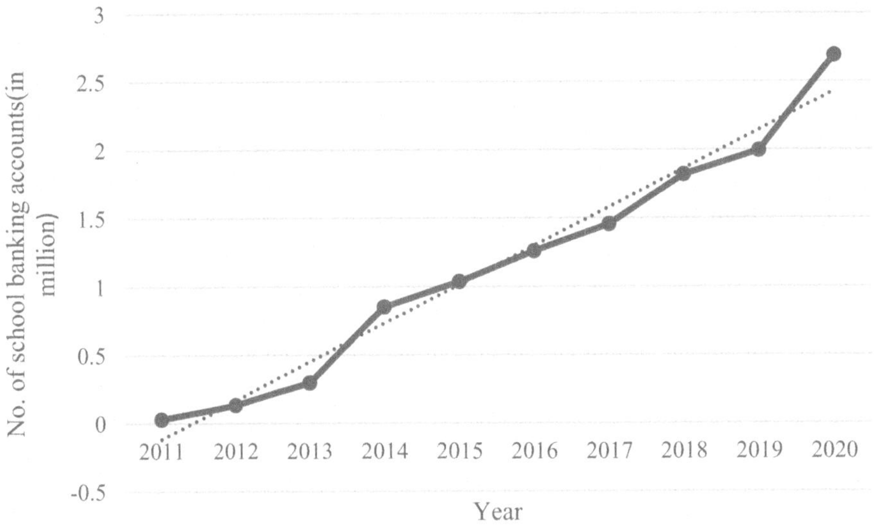

Fig. 14.1 Number of school banking accounts in Bangladesh (2011–2020). (Source: Financial Inclusion Reports of Bangladesh Bank)

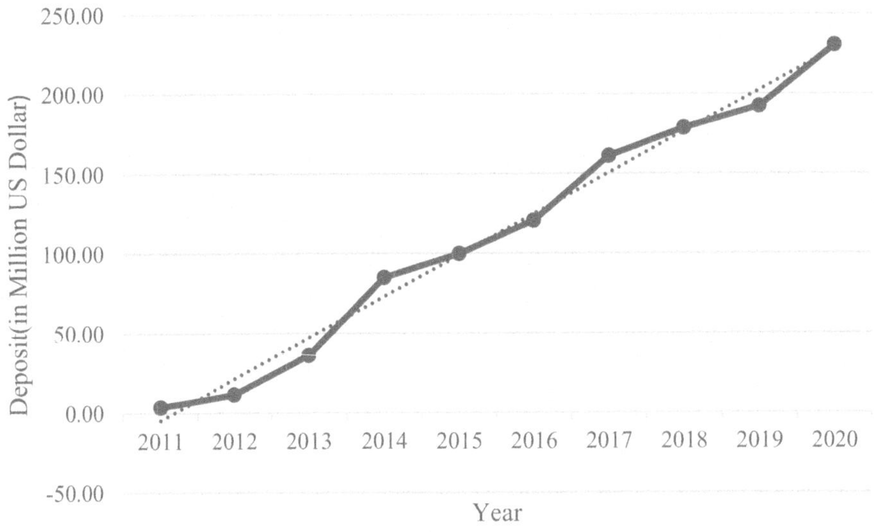

Fig. 14.2 The volume of school banking deposits in Bangladesh (2011–2020). (Source: Financial Inclusion Reports of Bangladesh Bank)

encourage their children to open school banking accounts."—S. Jahan, father of a class ten student.

> We know some students who regularly save money in their school banking accounts. Typically parents and students who are financially sound do not pay much attention to school banking. However, they should understand that, in the long run, school banking may help their children financially. So, we must adopt this school banking program. School banking can certainly help many people handle financial crises, such as meeting financial needs for the education purpose.—Assistant Head Teacher of a high school.

Schools see the programs as trustworthy. School leaders, including principals and administrators, value the future educational opportunity and the banking service they offer to the school community. "School banking creates saving habits among students and should be aimed at high school students as they are more receptive. More seminars and symposiums should be conducted by service providers so that students and their parents know about this program."—A high school leader.

Students love school banking activities as they become acquainted with and learn how to operate bank accounts. "Before, I was afraid of banking activities and the environment. Now, I have a school banking account. I know how to operate this account and have become acquainted with the banking environment."—MST. Sumaiya Nazmin, a class ten student.

School banking creates a savings habit among youngsters that helps them build a strong future. "I want to save money for future education and other expenditures. School banking creates a savings habit of mine. Having school banking accounts, I can save money which will grow over time. I hope I can meet my financial needs."—Sarwar Ahmed, a class nine student.

Among many benefits, school banking programs allow kids to deposit their scholarships and other money directly into their accounts and draw as per their needs. So, there is less opportunity for wastage of money. "I like the school banking account. My scholarship money directly deposits into this account. Sometimes I can save money from my tiffin and deposit it in my account for future expense."—Abrar Shoeb, a class eight student.

Bank personnel handling school banking accounts were asked to comment on their school banking program, its role, and its plan. Most of the school banking program providers state on their websites that the main objective of offering a school banking account is to inspire school-going students to grow decent savings habits and enhance their knowledge of the significance of saving money. Bank personnel state that school banking program creates a savings habit among youngsters. Because the bank does not charge for operating this account, students can save little money, and their scholarship money can be directly deposited in the school banking account, increasing the bank's deposit volume. "The unique features of the school banking program encourage students to conduct financial transactions through banks. To me, I think every student should have a school banking account so that they can know the banking environment and engage in transacting through financial institutions. Such a way can increase the country's financial inclusion, which would positively impact the economic growth."—Mr. Shish Mahmud, branch manager of a privately owned bank.

> We open school banking accounts per the directions of Bangladesh Bank, the Central Bank of Bangladesh. School banking grows financial literacy among students at an early age. It teaches them how and when to save money, enabling them to make better financial decisions. To ensure that every school student gets the chance to open a school banking account, the central bank may closely monitor and supervise the school banking program of banks so that banks can go ahead to achieve targets.—Mr. Sarwar Ahmed, branch manager of a state-owned bank.

These informal conversations show that the school banking program is gradually increasing, fostering financial inclusion in Bangladesh.

14.3 Methodology

To explore whether school banking promotes financial inclusion, this chapter examined data on school banking, the independent variable, derived from school banking reports produced by Bangladesh Bank from 2011 to 2019. In this study, school banking has been measured by the number of school bank accounts on the argument that the more the number of school banking accounts, the greater the volume of school banking activities in the country. Al Mamun (2021) used the number of school banking accounts to proxy the school banking concept.

Three proxies measure financial inclusion. Sarma (2015) used the number of depositors with commercial banks per 1000 adults as an indicator to measure the access dimension, arguing that the more the number of depositors with commercial banks per adult, the greater the access to the for-

Table 14.2 List of variables, measurements, and data sources

Variable	Measurement	Legend	Source
Financial inclusion	Dimension "Access": Number of depositors with commercial banks per 1000 adults	ACCESS	WDI
	Dimension "Use": Outstanding deposits with commercial banks	USE	FAS
	Dimension "Depth": Domestic credit provided to the private sector by domestic money banks as a ratio of GDP	DEPTH	WDI
School banking	No. of school banking account	SB	BB

Source: Author
Note: *WDI* World development indicators; *FAS* Financial Access Survey; *BB* Bangladesh Bank

mal financial system. In this study, similar to Sarma (2015), the "Access" dimension of financial inclusion is measured by the number of depositors with commercial banks per 1000 adults. Data on this variable comes from World Development Indicator.[8] Beck et al. (2007) and Sarma (2015) measured the "Usage" dimension of outstanding deposits with commercial banks, and as such, we measured this dimension by outstanding deposits with commercial banks and collected data from the Financial Access Survey (FAS) of the International Monetary Fund.[9] Arguing similarly with Siddik et al. (2015), the "Depth" of financial inclusion has been measured as domestic credit provided to the private sector by domestic money banks as a ratio of GDP, and data on this variable have been collected from World Development Indicator.[10] A list of variables, measurements, and data sources appears in Table 14.2.

After accumulating data, trend line analysis with graphs examines whether school banking contributes to the country's financial inclusion. The critical advantage of the trend line approach is that it is very modest and economical. Upon availability of historical data, a trend line can rapidly be drawn and variances assessed. While this method is not precise, it does offer a fast valuation of a program's likely outcomes. Thus, we use this approach to examine the role of school banking in expanding financial inclusion.

14.4 Findings and Their Discussion

This section presents the findings and their discussion. Figures 14.3, 14.4, and 14.5 exhibit the results of trend analysis, where the X-axis represents the school banking and the Y-axis represents the financial inclusion proxies.

Figure 14.3 shows a positive trend between the number of depositors with commercial banks per 1000 adults and the number of school banking accounts. This trend means that the more the number of school banking accounts, the more access to financial inclusion, indicating an increase in financial inclusion. The finding suggests that any government aiming to achieve full financial inclusion should undertake conducive policies to expand school banking across the country.

Figure 14.4 presents the role of school banking in accelerating financial inclusion, specifically the usage dimension of financial inclusion. School banking fosters financial inclusion with a positive trend line analysis, as increased school banking accounts positively relate to outstanding deposits with commercial banks. In other words, school banking plays an affirmative role in the usage dimension of financial inclusion. Thus, we observe that there are positive effects of SB on

[8]https://databank.worldbank.org/source/world-development-indicators

[9]https://data.imf.org/?sk=E5DCAB7E-A5CA-4892-A6EA-598B5463A34C

[10]https://databank.worldbank.org/source/world-development-indicators

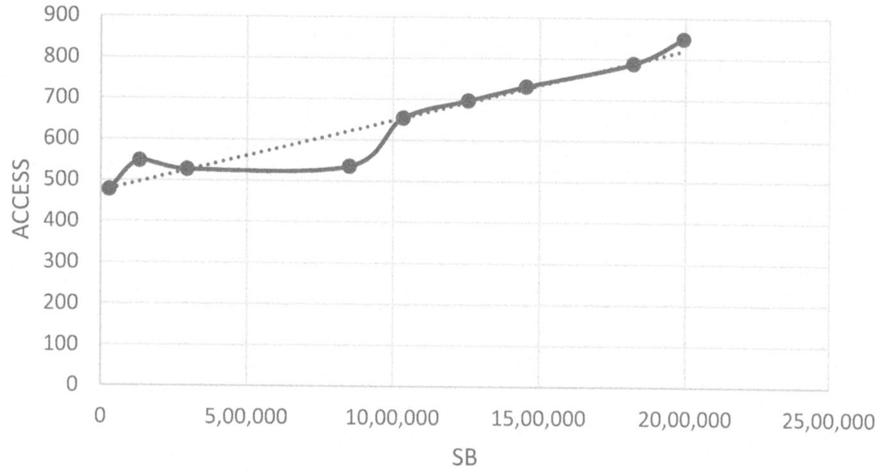

Fig. 14.3 Impact of school banking (SB) on the "access" dimension (ACCESS) of financial inclusion. (Source: Author's analysis based on data accumulated from WDI and BB)

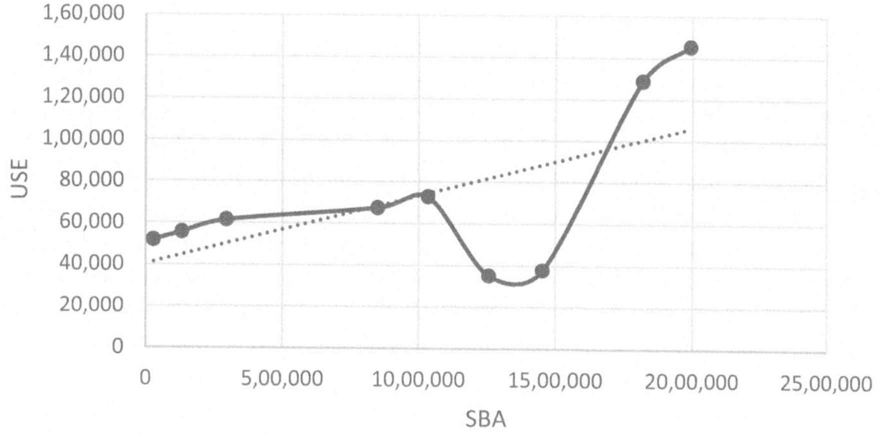

Fig. 14.4 Impact of school banking (SB) on the "usage" dimension (USE) of financial inclusion. (Source: Author's analysis based on data accumulated from FAS and BB)

USE. Therefore, the government should encourage all scheduled banks to conduct more school banking activities to foster financial inclusion.

The third proxy of financial inclusion we employed is domestic money banks' domestic credit provided to the private sector. Outcomes of trend analysis, shown in Fig. 14.5, indicate a positive association of SB with DEPTH. Therefore, it is evident that school banking activities foster financial inclusion in the country, which in turn helps achieve SDG 8.10 sustainable economic growth, which states sustain per capita economic growth per national circumstances.

14.5 Conclusions

School banking is not highly recognized in Bangladesh but is gaining popularity among students and all scheduled banks nowadays. The school banking program educates children on essential skills likely to constructively lead them in financial management. This initiative aims to accustom school-going students to banking products and services. It teaches them banking terminology and practices, aiming to include school-going students in the formal financial system. It is a dynamic, attractive approach for

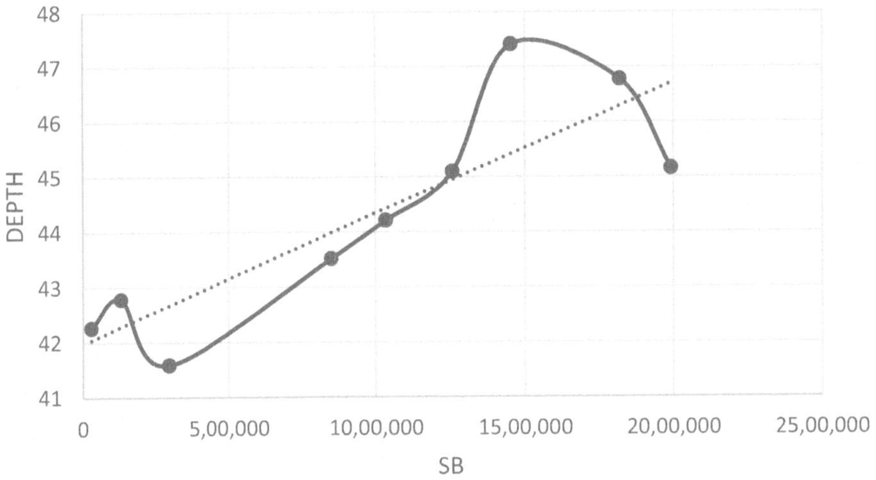

Fig. 14.5 Impact of school banking (SB) on the "depth" dimension (DEPTH) of financial inclusion. (Source: Author's analysis based on data accumulated from WDI and BB)

juvenile students to achieve practical knowledge in lifetime financial management expertise that inspires decent savings behaviors.

Employing data from 2011 to 2019 in Bangladesh, this study performed trend analysis and observed that school banking has substantial encouraging effects on all financial inclusion proxies. Informal conversations have been conducted with stakeholders, specifically parents, teachers, students, and bank personnel handling school banking accounts, to provide further evidence. All these stakeholders have praised the benefits of the school banking program and confirmed that the school banking program could foster financial inclusion, which in turn helps in achieving sustainable development goals, specifically SDG 8.10 sustainable economic growth, which states sustain per capita economic growth per national circumstance. Thus, we observed that the school banking program fosters financial inclusion and helps Bangladesh to achieve SDG 8.10. Based on the findings, we argue that policymakers should adopt conducive policies and strategies to stimulate school banking throughout the country. This will persuade financial inclusion resulting in the country's inclusive and sustainable development goals. Though the analyses indicate that school banking contributes to the financial inclusion campaign, the progress is insufficient to run smooth school banking activities in Bangladesh. Bangladesh Bank has previously issued and constantly provided policy guiding principles to all banks, but that is not everything. Instead, the country has to walk a long way toward the success of school banking. The central bank should closely observe and oversee school banking programs conducted by all banks and instruct banks to make identical school banking reporting systems.

References

Al Mamun, M. A. (2021). School banking, its trend and nexus with profitability in Bangladesh. *International Journal of Finance and Banking Research, 7*(1), 21–28.

Beck, T., Demirguc-Kunt, A., & Peria, M. S. M. (2007). Reaching out: Access to and use of banking services across countries. *Journal of Financial Economics, 85*(1), 234–266.

Bernheim, B. D., Garrett, D. M., & Maki, D. M. (2001). Education and saving: The long-term effects of high school financial curriculum mandates. *Journal of Public Economics, 80*(3), 435–465.

Bowman, M. E. (1922). The school savings bank. *The Elementary School Journal, 23*(1), 56–67.

Goldberg, F. (2005). The universal piggy bank: Designing and implementing a system of savings accounts for children. In *Inclusion in the American dream: Assets, poverty, and public policy* (pp. 303–322).

Hannig, A., & Jansen, S. (2010). *Financial inclusion and financial stability: Current olicy issues*. ADBI Working Paper No. 259. Asian Development Bank Institute.

Hasan, M., Le, T., & Hoque, A. (2021). How does financial literacy impact on inclusive finance? *Financial Innovation, 7*(1), 1–23.

Lyons, A. C., & Kass-Hanna, J. (2019). Financial inclusion, financial literacy and economically vulnerable populations in the Middle East and North Africa. *Emerging Markets Finance and Trade*, 1–40.

Mandell, L. (2005). *Financial literacy: Does it matter*. University of Buffalo.

Mondal, U., Hossain, M. S., & Khan, M. A. M. (2015). School banking: A new idea of banking operation in Bangladesh. *International Journal of Science and Research, 4*(11), 1926–1931.

Oberholtzer, S. L. (1892). School savings banks. *The Annals of the American Academy of Political and Social Science, 3*(1), 14-29.

Sarma, M. (2015). Measuring financial inclusion. *Economics Bulletin, 35*(1), 604–611.

Shanks, T. R. W., Mandell, L., & Adams, D. (2013). Financial education and financial access: Lessons learned from child development account research. *Innovations: Technology, Governance, Globalization, 8*(1–2), 159–175.

Sherraden, M., & Gilbert, N. (2016). *Assets and the Poor: New American welfare policy*. Routledge.

Siddik, M. N. A., Sun, G., & Kabiraj, S. (2015). Financial inclusion and its determinants: A study of Bangladesh. *Indian Journal of Finance, 9*(6), 7–29.

Sultana, M., Hossain, M. J., & Banu, M. L. A. (2020). School banking: Concept, contribution and prospect: The perspective of Dutch Bangla Bank Ltd. (DBBL). *European Journal of Business and Management, 12*(6), 70-81.

Thiry, J. H. (1890). *School savings banks in the United States-A manual for use of teachers*. Anthony Stumpf & Co.

Microenterprises in the Agricultural Sector and (Lack of) Insurance: A Study of Fresh Fruit Farmers and Retailers in Northern Pakistan

15

Ahmad Arslan, Saranjam Baig, and Aftab Ahmed Khan

Abstract

This chapter is one of the first studies to specifically analyze the lack of insurance for agricultural microentrepreneurs in Pakistan's mountainous peripheral Gilgit-Baltistan region. Based on field interviews with 20 fresh fruit farmers and retailers, we found out that even though they are knowledgeable about micro-health insurance, and most have access to banking services, none of the respondents use business or agricultural insurance. The knowledge of agriculture and business insurance (including the benefits and access possibilities) could have been better for microentrepreneurs operating in these sectors. Currently, the sample microentrepreneurs rely on social networks, including family, in case of difficulties, as informal insurance mechanisms. Due to the perishable nature of produce, lack of proper storage facilities, and supply chain problems, the vulnerability of such businesses is high, and lack of insurance further complicates this situation. Based on respondents' feedback and literature review, our chapter offers several policy implications to further financial inclusion, particularly on the insurance aspect in this region.

Keywords

Agriculture · Fresh fruit · Gilgit-Baltistan · Insurance · Microentrepreneurs · Pakistan · Periphery

15.1 Introduction

Insurance (or lack of it in a professional form) is an increasingly important topic associated with financial inclusion by scholars (e.g., Carbo et al., 2005; Cappiello, 2018). However, access to and acceptance of insurance is limited in many developing countries due to various institutional and cultural factors (e.g., Mahul & Stutley, 2010; Hasanabadizadeh et al., 2019). At the same time, prior researchers have established that in the absence of formal insurance

A. Arslan (✉)
Department of Marketing, Management & International Business, University of Oulu, Oulu, Finland

InnoLab, University of Vaasa, Vaasa, Finland
e-mail: ahmad.arslan@oulu.fi

S. Baig
College of Economics and Political Science, Sultan Qaboos University, Muscat, Oman
e-mail: s.baig1@squ.edu.om

A. A. Khan
Karakoram International University, Gilgit, Pakistan
e-mail: aftab.ahmed@kiu.edu.pk

G. W. Muschert et al. (eds.), *Financial Inclusion*, Sustainable Development Goals Series,
https://doi.org/10.1007/978-3-031-68803-4_15

mechanisms and policies, microbusinesses resort to reliance on social networks and ethnic (family) ties in case of business loss or financial difficulties (e.g., Nagler & Naude, 2017; Belissa et al., 2019). Despite these traditional mechanisms used by microbusinesses, there is an increasing trend toward increased professionalization of business services in emerging economies such as Pakistan. Due to the changing socioeconomic realities, reliance solely on microentrepreneur's insurance's informal mechanisms is increasingly becoming unviable. Even though, in recent years, some success has been achieved from the perspective of financial inclusion through the introduction of micro-health insurance, microbusiness insurance (including agricultural insurance) is still either invisible or extremely limited (e.g., Fahad et al., 2018). At the same time, scholars have stressed that despite increasing research on banking-related aspects of financial inclusion in emerging and developing economies, research explicitly focusing on insurance dynamics could be more extensive (e.g., Mohammed et al., 2020). Therefore, there is a visible academic and contextual gap that our chapter aims to fill.

We undertook qualitative exploratory research by conducting interviews with microenterprises' owners active in fresh fruit production (farmers) and selling (retailers) in the peripheral (and under-researched) northern region of Pakistan called Gilgit Baltistan. We focus on a peripheral region because prior studies indicate such regions as lagging or excluded by many financial services players (including insurance firms) due to a range of reasons, such as (unattractive or undetermined) market size, access difficulties, or organizational reasons (Claessens, 2006; Chikalipah, 2017). At the same time, some recent studies focusing on services provision (including financial services) for such ignored population segments have highlighted that there is a significant potential for social value creation as well as tapping the untapped potential for the organizations (e.g., Arslan et al., 2022; Kamara et al., 2022). It is also important to stress that agriculture (farming) and microentrepreneurship related to agriculture tend to be professions for most populations in the peripheral regions (e.g., Simoes & do Rio, 2020). Hence, assessing access (or lack thereof) to critical financial services like insurance in these contexts becomes further critical to offer tangible, practical implications and fill the gap in extant literature.

Taking Gilgit-Baltistan—a mountainous, relatively less accessible, and rural region in the north of Pakistan—as a testbed, our chapter contributes to the literature as one of the first to incorporate the views of microbusiness owners concerning insurance (or the lack thereof). There has been an evolving trend in the region, especially in the last decade when the role of agriculture has shifted from subsistence to cash crops, especially as tourism has gained traction due to physical infrastructure development (e.g., Hussain et al., 2017; Baig et al., 2022). Despite the region being home to excellent organic fresh fruits, hiccups in the supply chains and the potential for losses due to the perishable nature of the agricultural produce (fresh fruits) are the primary financial threats for microentrepreneurs. Since the lack of economies of scale and conditions peculiar to the region limit the scope of large firms to operate, microenterprises, especially in agriculture, play a crucial part in community development. In the absence of formal insurance mechanisms for microenterprises, their sustainability rests on informal mechanisms. Our chapter explores the informal insurance mechanisms, highlights their limitations, and offers suggestions for introducing formal (and accessible) micro-insurance for agricultural sector microentrepreneurs.

The rest of the chapter is organized so that the next section offers a brief literature review linking insurance with financial inclusion, especially in microenterprises. After that contextual overview appears along with data collection details, this discussion precedes a presentation of study findings. The chapter concludes with a presentation of implications and future research directions.

15.2 Literature Review

Financial inclusion refers to the access to and use of financial services by households and firms in a specific country (e.g., Cavoli & Shrestha, 2021). Scholars have stressed that there is no universally accepted definition of financial inclusion or exclusion, as there is a variance in the conceptualization of which financial services fall under the umbrella of this term. According to the definition by the World Bank, *"financial inclusion means that individuals and businesses have access to useful and affordable financial products and services that meet their needs—transactions, payments, savings, credit and insurance—delivered in a responsible and sustainable way"* (World Bank, 2021: 1). This specific term gained the attention of scholars, especially in the emerging economies' context due to the liberalization of the commercial banking system from the 1990s onward (Demirguc-Kunt & Klapper, 2012). Due to limited soft capital (cash) availability and lack of hard capital (collateral), a significant portion of the population, especially in emerging and developing countries, could not access formal financial services, including banking, before financial liberalization. The situation has significantly improved, especially considering banking services during the last two decades. The role of mobile (smart) phone penetration and fintech has further contributed to this aspect (Salampasis & Mention, 2018; Arslan et al., 2022). However, insurance is an essential aspect of financial inclusion. Despite the rise of insuretech and microinsurance, access is still limited, especially in rural communities (Beck, 2020). Hence, the risk exposure of already vulnerable microenterprises in emerging and developing countries remains high, ultimately influencing their survival chances (e.g., Sohns & Diez, 2019).

In many developing countries, traditional insurance providers have yet to actively market insurance services to microenterprises due to high-risk perceptions and other practical challenges. Microentrepreneurs may need more documentation due to the informal nature of business operations in those contexts (e.g., Moore et al., 2008; Chatterjee & Wehrhahn, 2017), making it difficult for insurance providers to assess their business viability and associated risks. The complex nature of the insurance business and the diversity of offers, including variance in premiums and detailed contracts, can sometimes discourage microentrepreneurs from developing countries. In short, the potential consumers of insurance services may need help understanding the offering (Patt et al., 2009; Islam, 2019). Also, as many microentrepreneurs, especially in the agriculture sector, tend (prefer) to operate informally and on a cash basis, hiring seasonal workers as needed, traditional insurance products and services may not precisely fit their needs. Hence, the scholars have stressed the need to undertake a multilevel strategy not only to raise awareness regarding the insurance's importance but also to enhance the value proposition for the clients by offering tailored services with easy-to-understand terms and conditions (e.g., Fahad et al., 2018; Islam, 2019; Hassanabadizadeh et al., 2019). Some examples of this concern are the Association of Kenyan Insurers launched a radio station in 2010 to improve the general populace's risk management and insurance literacy, especially of microentrepreneurs (e.g., Bel and Caicedo, 2013). Scholars have recently started to refer to the PACE (product, access, cost, and experience) tool to improve client value proposition (e.g., De Bock and Gelade, 2012). In Bangladesh, India, and Pakistan, some NGOs focusing poor in urban and rural areas have received praise for developing unique health and disaster insurance offerings (e.g., Rahman et al., 2017). Several other insurance initiatives, primarily focusing on education and health, can be found in other countries, including national health insurance schemes like Sehat Insaf Card in Pakistan (e.g., Rashid & Saeed, 2019).

However, despite these initiatives, access to insurance for agricultural microentrepreneurs remains limited in developing countries, including Pakistan, though Adamjee Insurance has launched some policies in Pakistan, among others, as well as some initiatives at the provincial government level in Punjab and KP provinces (e.g., Fahad et al., 2018). However, the users of these possibilities remain limited to well-connected main areas of the country and prominent agricultural players (farmers and entrepreneurs). Insurance usage by agricultural microentrepreneurs needs to be improved, especially in peripheral regions such as Gilgit-Baltistan in Pakistan. Finally, a lack of awareness and unwillingness to use insurance services, in general, remain visible in the Pakistani context for individuals and businesses (e.g., Cheema et al., 2020). Consequently, this lack of insurance exposes them to higher risks of default and significant financial hardships if something goes wrong with them personally or with their businesses.

15.3 Contextual Overview and Data Collection

15.3.1 Gilgit-Baltistan Region

The Gilgit-Baltistan region has an estimated population of 1.5 million people, spread over 72,496 km^2. This region is located in Pakistan's far north and borders China and India. With the Hindu Kush to the west, the Himalayas to the south, the Karakoram to the east, and the Pamir to the north, the Gilgit-Baltistan region is home to one of the world's most mountainous landscapes. There are 10 administrative districts, with 90% mountainous, 4% forest, and only 4.2% of the land cultivable.

The climate in the Gilgit-Baltistan region is ideal for growing various fruits, particularly apricots. For generations, the region's farmers have used horticulture and agriculture as their livelihood management strategy (ZTBL, 2019). Gilgit-Baltistan has a comparative advantage in fruit farming and agriculture. This region is known worldwide for its almond, apricot, cherry, apples, grapes, and other dry fruit production. The climate in the region is ideal for growing a wide variety of delicious, high-quality fruits and vegetables (Khan et al., 2019). According to Hashmi and Shafiullah (2003: 1), agriculture in Gilgit-Baltistan has peculiar characteristics:

> *However, when one looks at the Northern Areas, agriculture is not a factory or industry. It is not merely a sector of production. Agriculture is a way of life, a cultural practice with all the implications of the word culture. They comprise growing crops with local seeds, caring animals that have adapted to the environment, relishing vegetables and fruits of their own kind and quality. There exists a system of self-reliance and sustainability. It is ecological agriculture in its true sense.*

Agriculture in Gilgit-Baltistan tends to be organic and chemical-free. The region has carved out a position in natural and organic farming, thanks to the utilization of pristine glacial water and the absence of insecticides, pesticides, and herbicides. The natural aspect of regional farming has provided local farmers with an opportunity to cater to health-conscious clientele in both domestic and foreign markets. Gilgit-Baltistan produces a range of fresh fruit. However, most of this fruit goes to waste due to our farmers' antiquated harvesting approach, lack of (cold) storage, lack of marketing skills, packaging limitations, and transportation challenges.

Furthermore, financial exclusion, lack of access to finance, and lack of formal insurance mechanisms have increased the existing challenges that the farmers face. Mountain region agricultural microentrepreneurs, especially farmers, do not receive a fair price for their products due to insufficient governmental and institutional assistance, supply chain problems, the perishable nature of produce, and a lack of professional management skills. These problems can further compound due to the lack of formal insurance for microentrepreneurs, as any disruption as bad weather, may lead to bankruptcy. Some recent studies (e.g., Li et al., 2020) have highlighted the potential for fresh fruit farming as a tool of poverty alleviation in rural mountainous areas similar to Gilgit-Baltistan. However, financial exclusion and particularly lack of (awareness, access, and usage) insurance is one of the significant hurdles for both producers (farmers) and sellers (retailers).

15.3.2 Data Collection

We employed a purposeful sampling strategy (also referred to as purposive sampling by some scholars) in the current study. This study is qualitative, as it is exploratory and qualitative data offers in-depth descriptions (Emmel, 2013) and real-life accounts of respondents experiencing the phenomenon under study (e.g., Guest et al., 2013). Keeping in view the exploratory focus of our chapter, the respondents (interviewees) needed to share specific characteristics (Emmel, 2013; Flick, 2017). We emphasize that judgment and understanding of the research context are critical while using such a sampling strategy (Guest et al., 2013; Flick, 2017). Two coauthors of this chapter are from the Gilgit-Baltistan region, fully understand the context, and have undertaken several academic and policy-focused studies on different topics there already.

In line with the adopted purposeful sampling strategy, one of the authors conducted 20 semi-structured field interviews in March 2021 across four districts of the Gilgit-Baltistan region. The respondents fit the following criteria: they should be microentrepreneurs, represent both producers and retailers, represent variance in the background (both by location within the region and education), and have varied experience. The respondents provided informed consent. Data was transcribed (written) on the spot by the researcher collecting data as the respondents were uneasy with recording. The interview questions focused on the respondents' business and agricultural entrepreneurship experiences. The interviews also focused on their understanding and perception of insurance, along with open-ended discussion on (lack of) microinsurance and its influences on agricultural entrepreneurs, and alternatives used by them. Finally, their perceptions of agricultural entrepreneurship policies in the region were also discussed. All interviews occurred in local languages, and two chapter authors are native to the region and can speak the languages fluently. The findings were translated into English by the same authors, who have significant experience in such research and have PhD degrees from the USA and Germany. Finally, the researchers undertook data analysis manually.

15.4 Findings

Table 15.1 summarizes the key demographic characteristics of our interviewees.

The interviewees' experience running agricultural microbusinesses varied from 2 to 26 years. Notably, four of the interviewed microentrepreneurs were agricultural retailers, while the rest were producers (farmers). Also, the educational background varied from basic education (primary school) to undergraduate degrees. However, despite all this variance in the interviewees' backgrounds, they currently had no business or agricultural insurance for their microenterprises. Half of the respondents were unaware of such mechanisms, while others had some idea. It was further visible that educated (high school and above) respondents understood the importance of agricultural or business insurance. However, they needed to be aware of the possibilities compared to less-educated respondents. This pattern is similar to the findings of financial exclusion in other contexts, where education plays an essential role in awareness of financial services and their overall role in microbusiness survival and competitiveness (e.g., Mohammed et al., 2020; Arslan et al., 2022). We found that reliance on the local family and village networks as informal insurance mechanisms was visible in almost all cases. Hence, we found the situation similar to other developing nations and rural areas, where local ties and family networks serve as informal insurance mechanisms (e.g., Nagler & Naude, 2017; Belissa et al., 2019).

However, interestingly during the discussion with respondents, another element of financial inclusion (or exclusion), access to banking services, was better understood and used. Depending on the policies and initiatives of other financial sector players, other services slowly take root in the target populace (e.g., Cavoli & Shrestha, 2021). Indeed, access to banking services is the first step toward financial inclusion.

Most of the respondents shared their experiences that business losses came from the perish-

Table 15.1 Respondent/interviewee characteristics

District	Education	Experience (years)	Type
Gilgit	High school	4	Farmer
Gilgit	Undergraduate	5	Retailer
Danyore	High school	10	Farmer
Danyore	High school	12	Farmer
Gilgit	High school	20	Farmer
Gilgit	High school	3	Farmer
Ghizer	High school	2	Retailer
Gilgit	High school	15	Retailer
Hunza	High school	18	Farmer
Hunza	High school	26	Famer
Hunza	High school	8	Farmer
Hunza	Higher secondary school	13	Retailer
Hunza	Higher secondary school	15	Farmer
Hunza	Higher secondary school	4	Farmer
Hunza	Higher secondary school	3	Farmer
Hunza	Primary school	18	Farmer
Hunza	Higher secondary school	2	Farmer
Hunza	Primary school	6	Farmer
Hunza	Higher secondary school	7	Farmer

able nature of agricultural products grown by them, given the harsh weather and storage vis-à-vis the areas in Gilgit-Baltistan remains inaccessible for days and weeks due to blockage of the only access road, the Karakoram Highway (KKH). This finding is in line with some prior researchers' findings in other similar mountainous regions (e.g., Orjuela-Castro et al., 2017). Cooperative storage facilities shared by producers in a rural mountainous community and supply chain management skills have been referred to as a solution by some studies focusing on these issues (e.g., Sharangi & Datta, 2015). It is further essential to stress that all respondents were determined to continue their businesses due to high returns and a rapid rise in sales in recent years. Most of them are also planning to expand their businesses soon. This aspect shows the consistency and sustainability of vegetable and fruit businesses in the area, which has benefited from increased tourism due to infrastructure development, as highlighted in recent studies (e.g., Baig & Zehra, 2020; Baig et al., 2022). Almost all the respondents agreed and are interested in availing insurance policies for their microenterprises if provided and given awareness and appropriately guided. The encouraging point about insurance is that most respondents have private health insurance, which shows they have basic knowledge about insurance operations and policies.

Another issue that emerged from the interviews relates to the seasonality of fresh fruit products. The interviewed agricultural microentrepreneurs have more products to sell in the summers; otherwise, in winter, they have minimal products due to the nonavailability of any proper fruit storage in the whole province. Hence, the need for proper storage and warehousing emerged visibly from our field interviews. Some respondents do their businesses only in cash and do not use formal banking methods, which shows their exclusion from formal financial services provided by the government and private sector. However, it also emerged that for those microentrepreneurs whose business has grown considerably, formal banking channels and financial services are increasingly becoming necessary.

Finally, Covid-19 (as it was a significant issue at the time of data collection) and its influences also emerged as a theme during our field interviews. The respondents said their businesses had no significant losses, as the number of cases in this specific region has been minimal, and no lockdown occurred there. Also, there is significant demand for this fresh fruit, and an increasing number of tourists (many tried spending some time away from urban centers of Pakistan where the Covid-19 situation was significantly challenging) helped further in this context.

15.5 Implications, Limitations, and Future Research Directions

Our chapter offers several practical and policy implications based on the findings. Firstly, financial inclusion in contexts such as our research settings requires agricultural microentrepreneurs to access financial services, including microinsurance. However, insurance possibilities must be added or expanded for agricultural microentrepreneurs in these contexts. This exclusion is a potential threat to both the profitability and sustainability of these microentrepreneurs. Introducing microinsurance and providing access to microfinance should be an essential policy imperative for the policymakers in the Gilgit-Baltistan region. The insurance's complexities (including documentation requirements) make it difficult for the agricultural sector microentrepreneurs to insure their products since most of them operate informally. It is vital to introduce a separate insurance mechanism for informal and micro businesses. The public sector can intervene in this concern by offering policy incentives. Also, bringing informal businesses into the formal net can encourage private financial service providers to engage with local entrepreneurs.

Agricultural microentrepreneurs need help with many problems, including the perishability of fresh produce. Financial exclusion further complicates these issues. In this concern, cooperative initiatives supported by large financial institutions, including insurance firms, regarding joint cold storage possibilities for fresh produce (perishable agricultural products) will also benefit farmers as the product remains fresh. Public sector institutions and NGOs can also be a stakeholder in this process by offering practical support, including skills development possibilities in warehousing and supply chain management of perishable agricultural products. Most local microentrepreneurs in the region now have some access to micro-health insurance. It means they are well aware of the potential benefits of the different insurances. Awareness and easiness of accessing financial services require efforts by policymakers, local administrators, and interest by firms eager to tap this new market and contribute socially.

Like any other academic study, our chapter has several limitations. Firstly, field interviews with only 20 agricultural microentrepreneurs can limit the generalizability of our findings. However, our findings are a good representation of agricultural microentrepreneurs in the region due to similarities in fresh fruit farming and retailing dynamics across the region, including the size and market structure. Another limitation is the lack of application of a specific theoretical paradigm in this chapter. However, this limitation also relates to our observation concerning a need for prior research, particularly in peripheral mountainous regions, where microentrepreneurship dynamics differ significantly from urban centers or even rural areas in heartlands of densely populated countries like Pakistan. Hence, context plays a critical role in such studies; an aspect also clearly highlighted in this chapter. We suggest future scholars build on our work and undertake quantitative and qualitative studies in a similar context to develop a further conceptual and theoretical understanding of financial inclusion and microentrepreneurship in the peripheral regions. Finally, future researchers can also undertake comparative studies comparing this region with bordering regions (e.g., in China). Such studies can further highlight best practices, differences, and similarities concerning financial inclusion, warehousing, and supply chain management dynamics of perishable agricultural production (such as fresh fruit) and retailing.

References

Arslan, A., Buchanan, B. G., Kamara, S., & Al Nabulsi, N. (2022). Fintech, base of the pyramid entrepreneurs and social value creation. *Journal of Small Business and Enterprise Development, 29*(3), 335–353.

Baig, S., & Zehra, S. (2020). China-Pakistan economic corridor, governance, and tourism nexus: Evidence from Gilgit-Baltistan, Pakistan. *Current Issues in Tourism, 23*(23), 2884–2889.

Baig, S., Shabbnum, A., & Arslan, A. (2022). Cultural tourism and the wellbeing of local citizens: Examining the mediating role of cultural conservation in Northern Pakistan. In I. Mensah & E. Afenyo-Agbe (Eds.), *Prospects and challenges of community-based tourism and changing demographics* (pp. 130–154). IGI Global.

Beck, T. (2020). *Fintech and financial inclusion: Opportunities and pitfalls* (No. 1165). ADBI working paper series. Available online at https://www.econstor.eu/bitstream/10419/238522/1/adbi-wp1165.pdf

Bel, S., & Caicedo, M. P. (2013). *Audiovisual mass media campaigns for insurance education: Stages and lessons*. Microinsurance Paper, 21. Available online at http://www.impactinsurance.org/publications/mp21

Belissa, T., Bulte, E., Cecchi, F., Gangopadhyay, S., & Lensink, R. (2019). Liquidity constraints, informal institutions, and the adoption of weather insurance: A randomized controlled trial in Ethiopia. *Journal of Development Economics, 140*, 269–278.

Cappiello, A. (2018). *Technology and the insurance industry: Re-configuring the competitive landscape*. Springer.

Carbó, S., Gardner, E., & Molyneux, P. (2005). *Financial exclusion*. Springer.

Cavoli, T., & Shrestha, R. (2021). *Financial inclusion in Asia and beyond: Measurement, development gaps, and economic consequences*. Routledge.

Chatterjee, A. K., & Wehrhahn, R. (2017). *Insurance for micro, small, and medium-sized enterprises*. Available online at https://www.adb.org/publications/insurance-for-msme

Cheema, A. R., Zaidi, S., Najmi, R., Khan, F. A., Kori, S. A., & Shah, N. A. (2020). Availability does not mean utilisation: Analysis of a large micro health insurance Programme in Pakistan. *Global Journal of Health Sciences, 12*(10), 1–4.

Chikalipah, S. (2017). What determines financial inclusion in Sub-Saharan Africa? *African Journal of Economic and Management Studies, 8*(1), 8–18.

Claessens, S. (2006). Access to financial services: A review of the issues and public policy objectives. *The World Bank Research Observer, 21*(2), 207–240.

De Bock, O., & Gelade, W. (2012). *The demand for microinsurance: A literature review*. ILO Microinsurance Innovation Facility Research Paper, Available online at https://www.preventionweb.net/files/29829_29829rp26final1.pdf

Demirgüç-Kunt, A., & Klapper, L. F. (2012). Measuring financial inclusion: The global findex database. *World bank policy research working paper*, (6025). Available online at https://openknowledge.worldbank.org/handle/10986/6042

Emmel, N. (2013). *Sampling and choosing cases in qualitative research: A realist approach*. Sage.

Fahad, S., Wang, J., Hu, G., Wang, H., Yang, X., Shah, A. A., et al. (2018). Empirical analysis of factors influencing farmers' crop insurance decisions in Pakistan: Evidence from Khyber Pakhtunkhwa province. *Land Use Policy, 75*, 459–467.

Flick, U. (2017). *The Sage handbook of qualitative data collection*. Thousand Oaks, CA: Sage.

Guest, G., Namey, E. E., & Mitchell, M. L. (2013). *Collecting qualitative data: A field manual for applied research*. Sage.

Hasanabadizadeh, N., Omidi-Najafabadi, M., Mirdamadi, S. M., & Lashgarara, F. (2019). An agricultural micro-insurance development model for rural areas of Iran. *Eurasian Journal of BioSciences, 13*(2), 2071–2077.

Hashmi, A., & Shafiullah (2003). *Northern areas strategy for sustainable development background paper: Agriculture and food security*. IUCN. Available online at https://portals.iucn.org/library/sites/library/files/documents/2003-095_1.pdf

Hussain, A., Fisher, D., & Espiner, S. (2017). Transport infrastructure and social inclusion: A case study of tourism in the region of Gilgit-Baltistan. *Social Inclusion, 5*(4), 196–208.

Islam, N. (2019). Challenges for increasing insurance awareness among the people of Bangladesh. In *Training Workshop on Raising Awareness Regarding Insurance*, June (pp. 1–16).

Kamara, S., Arslan, A., Kontkanen, M., & Tarba, S. Y. (2022). A bricolage perspective on construction innovation in BOP market: Evidence from a Finnish family-owned micro-enterprise. In D. Dikova & E. Ipsmiller (Eds.), *Research handbook on innovation in international business* (pp. 206–218). Edward Elgar.

Khan, F., Khan, T. U., Ayub, M., & Kabir, R. (2019). Demonstration of technologies and training of growers for handling and value addition of fruits and vegetables in Gilgit-Baltistan. *Advances in Food Technology and Nutrition Science Open Journal, 5*, 67–72.

Li, R., Zheng, H., Zhang, C., Keeler, B., Samberg, L. H., Li, C., et al. (2020). Rural household livelihood and tree plantation dependence in the central mountainous region of Hainan Island, China: Implications for poverty alleviation. *Forests, 11*(2), 248.

Mahul, O., & Stutley, C. J. (2010). *Government support to agricultural insurance: Challenges and options for developing countries*. The World Bank. Available online at https://openknowledge.worldbank.org/handle/10986/2432

Mohammed, F., Barrowclough, M. J., Kibler, M. L., & Boerngen, M. A. (2020). Measuring usage of formal financial services as a proxy for financial inclusion: A case of agricultural households in Ghana. *Agricultural Finance Review, 80*(4), 471–489.

Moore, C. W., Petty, J. W., Palich, L. E., & Longenecker, J. G. (2008). *Managing small business: An entrepreneurial emphasis*. South-Western Cengage.

Nagler, P., & Naudé, W. (2017). Non-farm entrepreneurship in rural sub-Saharan Africa: New empirical evidence. *Food Policy, 67*, 175–191.

Orjuela-Castro, J. A., Sanabria-Coronado, L. A., & Peralta-Lozano, A. M. (2017). Coupling facility location models in the supply chain of perishable fruits. *Research in Transportation Business & Management, 24*, 73–80.

Patt, A., Peterson, N., Carter, M., Velez, M., Hess, U., & Suarez, P. (2009). Making index insurance attractive to farmers. *Mitigation and Adaptation Strategies for Global Change, 14*, 737–753.

Rahman, M. M., Karan, A., Rahman, M. S., Parsons, A., Abe, S. K., Bilano, V., et al. (2017). Progress toward universal health coverage: A comparative analysis in 5 South Asian countries. *JAMA Internal Medicine, 177*(9), 1297–1305.

Rashid, Y., & Saeed, S. (2019). Transforming health in one hundred days: Key achievements. *Annals of King Edward Medical University, 25*(S).

Salampasis, D., & Mention, A. L. (2018). FinTech: Harnessing innovation for financial inclusion. In *Handbook of Blockchain, digital finance, and inclusion* (Vol. 2, pp. 451–461). Academic.

Sharangi, A. B., & Datta, S. (2015). *Value addition of horticultural crops: Recent trends and future directions*. Springer India.

Simões, F., & do Rio, N. B. (2020). How to increase rural NEETs professional involvement in agriculture? The roles of youth representations and vocational training packages improvement. *Journal of Rural Studies, 75*, 9–19.

Sohns, F., & Diez, J. R. (2019). Explaining microenterprise survival in rural Vietnam: A multilevel analysis. *Spatial Economic Analysis, 14*(1), 5–25.

World Bank. (2021). *Financial inclusion overview*. Available online at https://www.worldbank.org/en/topic/financialinclusion/overview#1

ZTBL. (2019). *Promotion of agriculture in GILGIT Baltistan*. Available online at https://www.ztbl.com.pk/wp-content/uploads/Documents/Publications/Research-Studies/AgriPromotionGilgit.pdf

16 Financial Inclusion for Climate Resilience in Rural South Africa: A Qualitative Analysis of Opportunities and Barriers

Maheshvari Appavoo

Abstract

South Africa has become increasingly exposed to climatic extremes that often lead to economic losses. With a growing emphasis on the importance of financial inclusion for risk reduction, financial services have an essential role in addressing the impact of climate-related disasters on household finances. Evidence suggests that access to formal financial services can help smoothen consumption, secure livelihoods, and reduce climate vulnerability after a climate-related disaster. However, studies show that vulnerable rural populations, who bear a disproportionate burden of climate stress, are ill-equipped to deal with shocks, and face significant financial barriers when trying to cope with the financial impacts of climate-related disasters. This study sought to better understand the extent to which financial inclusion contributes to climate resilience in the context of the sustainable development goal, target 8.10, and to identify critical barriers and opportunities faced by survivors of climate-related disasters in rural South Africa. The study involves original field research conducted in South Africa. Semi-structured, in-depth interviews were conducted with experts for the local financial services market, experts in the aid provisions process, and experts on the impact of adverse weather events. The interviewees identified more barriers than opportunities, and were particularly concerned with the extent to which the formal financial sector adequately addresses the needs and circumstances of vulnerable groups. The experts were of the opinion that the best opportunities lie in the informal banking system, and that most barriers can be overcome through digital literacy and financial training.

Keywords

Climate change · Climate-related disasters · Disasters · Financial inclusion · Informal financial sector · Resilience · Rural households

16.1 Introduction

Extreme weather events exacerbate poverty cycles and threaten socioeconomic development. In southern Africa, floods are the most common type of extreme climatic events, droughts have caused the highest economic cost of damage, and storms have been responsible for the displacement of a large majority of people (Davis-Reddy et al., 2017, p. 38). The conse-

M. Appavoo (✉)
Stellenbosch Business School, Stellenbosch University, Stellenbosch, South Africa

G. W. Muschert et al. (eds.), *Financial Inclusion*, Sustainable Development Goals Series,
https://doi.org/10.1007/978-3-031-68803-4_16

quences are devastating, as the damages include the destruction of property, infrastructure, crops, and, in some cases, loss of life (Blamey et al., 2017, p. 2161). Resilience is the ability to anticipate, absorb, manage, or recover from the impacts of a potentially hazardous event (IPCC, 2014) without compromising future welfare. As the region is increasingly exposed to climatic extremes, building resilience to weather and disaster risks has become even more critical for poverty reduction.

Climate change is a development challenge and an indiscriminate risk. The world's poorest and most vulnerable populations are of particular concern, as they contribute the least to climate change but bear a disproportionate burden of climate stress (Islam & Winkel, 2017, p. 152), (Hallegatte et al., 2017, p. 25). Vulnerable populations are more likely to live in areas susceptible to climate disasters, are more dependent on the physical environment for their food and livelihoods, and are likely to have a more significant proportion of their wealth in physical assets, which are often destroyed during extreme weather events (IPCC, 2014, p. 9), (Zhou et al., 2016, p. 349).

The good news is that there are solutions that can help those at risk prepare for, anticipate and respond to climate and disaster risks when they occur. Financial inclusion is the access to and use of affordable and valuable formal financial products and services in a sustainable manner (Demirgüc-Kunt et al., 2018). Rigorous research has found that financial inclusion helps to increase the welfare and resilience of low-income households to adverse events, as they are better able to smooth their consumption when well-designed financial products and services help reduce their vulnerability to shocks (Mawejje, 2019, p. 287; Moore et al., 2019, p. 3; Akotey & Adjasi, 2018, p. 160; Riley, 2018, p. 57; Lensink et al., 2017, p. 335; El-Zoghbi et al., 2017).

Over the years, the South African financial system has become more sophisticated and better regulated. A large part of the formally excluded population has gradually integrated into the financial system and mainstream economy in a continuous effort to redress the uneven development of the past. As a result, financial service providers have actively expanded access to financial services and improved the use of financial services by all South Africans. Nevertheless, rural households remain least prepared for adverse events due to low savings, fluctuating incomes, imperfect credit and insurance markets, and inefficient money and transfer mechanisms, often offset by excessive lending. The problem, however, is that financial solutions are not reaching low-income households effectively and quickly enough.

The South African Financial Services Charter (FSC) was launched in 2003 in a joint effort by major private banks and a state financial institution to deepen financial intermediation and improve access to financial services (Kostov et al., 2014, p. 124). In 2004, the FSC introduced financial inclusion strategies and initiatives, along with the notion of "effective access" to financial services (Financial Sector Charter Council, 2004). This broadened the definition of financial inclusion to define geographical access more precisely (e.g., 20 km distance to the nearest service point) and defined the target group as living standard measure (LSM) 1–5 (i.e., households with specific low-income characteristics). The FSC was replaced in 2012 by the Financial Sector Code, which was amended in 2017 to include a comprehensive measure covering access to electronic customer infrastructure, banking access via densification, and product-based access to geographic access (Department of Trade and Industry, 2017, pp. 283–287).

Despite these policy developments, as well as the National Credit Act and the Consumer Protection Act shaping the financial sector (Arun & Kamath, 2015, p. 281), challenges to financial inclusion persist, with the vulnerable and poor more likely to be excluded from formal financial services (Chitimira & Ncube, 2020, p. 355), (Carter & Tyson, 2017, p. 44). South African banking fees are prohibitive for the low-income classes, who generally do not trust banks because of historical exploitation. They also tend to prefer cash because of a lack of knowledge, excessive documentation required by banks, fear of fraud, and widespread financial illiteracy, reinforcing mistrust of the sector (Chitimira & Ncube, 2020, p. 337). As a result, people are reluctant to engage in convenient cashless transactions, and many tend to engage in informal, trust-based models such as *stokvels* (Chitimira & Ncube, 2020, p. 351).

The National Development Plan (NDP) is South Africa's vision for 2030. In response to the country's shortcomings following the political transition in 1994, the NDP promotes a united society and an inclusive economy, as well as improving the quality of economic management for sustainability and inclusion (Statistics SA, 2019). The goals are to eradicate poverty and reduce inequality and unemployment by 2030 by improving the quality of education and skills and the efficient allocation of resources (Fourie, 2018). About 74% of the sustainable development goal (SDG) targets relate directly to the NDP, and 19% of the remaining targets appear in sectoral programs (Statistics SA, 2019). Notwithstanding the gaps and obstacles in the financial system, the support for financial inclusion is overwhelming, and the country's financial sector and sophisticated infrastructure have the potential to play a proactive role in achieving the country's development goals.

The SDGs address a complex and interconnected set of challenges that require meaningful engagement by all stakeholders. Financial inclusion contributes to SDG 8, which aims to "promote sustained, inclusive and sustainable economic growth, full and productive employment and decent work for all" (Inter-Agency and Expert Group on SDG Indicators, 2020, p. 8). SDG target 8.10 aims to promote universal access to banking, insurance, and financial services, and can offer enormous improvements in inclusivity. Increasing the participation of marginalized individuals within the mainstream economy will play an essential role in strengthening resilience to climate disasters and supporting socioeconomic growth.

This chapter addresses the attainment of SDG 8.10 amidst the systemic risk of climate change. Financial inclusion can bring social and economic benefits to all, and quality financial services are essential tools to improve the resilience of people in fragile and conflict-affected situations. Evidence shows that financial inclusion models can promote economic growth and achieve broader development goals (Klapper et al., 2016). As such, in recent years, financial inclusion has gained much attention from most governments and policymakers for being a facilitator of financial development and economic growth (Barajas et al., 2020, p. 34; Arun & Kamath, 2015, p. 267). Climate-related disasters warrant longer-term solutions to a financially resilient, sustainable recovery. When properly designed, financial services provide displaced communities with the tools they need to save and keep money safe, send or receive money transfers, conduct daily business, and rebuild their economic livelihoods (Martin, 2019, p. 5). The chapter will explore views on whether or not a greater emphasis needs to be placed on the role of financial services in managing crises, minimizing risk exposure, strengthening social protection, and stimulating economic activity.

16.2 Methods

Despite its importance, the literature on financial resilience in climate change adaptation needs a deeper understanding of how climate change affects the lives of low-income households. Little research addresses how survivors cope with the financial shock of weather disasters. The role that well-designed formal financial products and services can play in building the resilience of rural households in South Africa has yet to be fully understood and explored. It is also unclear how these financial products and services can be made more accessible to rural households, especially during times of crisis.

This research applies an evidence-based approach to explore the design, accessibility, and use of various formal financial products and services as a tool to strengthen the resilience of crisis-prone populations. Furthermore, there is a need to further explore the sustained engagement and awareness of financial service providers in promoting financial inclusion.

Given the exploratory nature of the study, a qualitative research paradigm was deemed appropriate. Primary qualitative data derives from experts for the local financial services market, experts in providing local aid, and experts on the impact of adverse weather events. Secondary data derives from an extensive review of academic journal articles and books, and industry reports. To align the literature with industry practices, we interviewed various stakeholders with unique capacities to reflect on the occurrence of financial losses due to weather-related disasters. Through in-depth, semi-structured interviews, personal perspectives and professional viewpoints were explored to elicit diverse and comprehensive data. The objective of the interviews was to gain a better understanding of the extent to which financial inclusion contributes to climate resilience. The aim of the interviews was also to supplement the factors identified in the literature, that is, to identify factors not considered in the literature, and to detect missing variables that could help explain the ability of disaster victims to recover.

When recruiting the sample, there was neither a usable population nor a sampling frame available, listing stakeholders interested in financial losses due to weather-related disasters. Therefore, it was necessary to construct a suitable sampling frame, which was informed by academic sources (e.g., Asbi et al., 2020; AL-Fazari & Kasim, 2019; Golnaraghi & Khalil, 2017). The study used purposive sampling techniques to identify eligible participants. This sampling technique allows for an accurate sampling frame that appears comprehensive, and prevents participants from misrepresenting themselves. This also eliminates sampling bias. This sampling method focuses on the depth of information rather than breadth, and sample sizes are typically small. A more detailed description of the participants appears in Table 16.1.

Participants qualified for inclusion based on their expertise in financial planning, weather disaster management, or their involvement in rural communities. Furthermore, a participant had to be an executive decision-maker or a person in a managerial role who has a special interest in, and has facilitated with the resilience of vulnerable populations.

An in-depth, semi-structured interview guide was developed, based on the literature and refined through pilot interviews and discussions with peers. In addition to requesting biographical details, open-ended questions elicited the participant's views on opportunities and barriers specific to financial inclusion for resilience of crisis-prone communities. Interviews were conducted in person, audio-recorded, and transcribed. Due to resource constraints, logistical challenges, and the preferences of busy professionals, email interviews were also facilitated and conducted directly through the secure REDCap survey software. Follow-up questions of the email interviews were conducted telephonically. Data collection continued until adequate data saturation of the main themes was achieved.

An essential objective of the interview was to find out how participants assess the ability of a rural household to recover financially from a shock. The schedule included questions on various coping strategies concerning financial recovery mechanisms, such as, government and humanitarian assistance, savings and credit, informal risk sharing and insurance, and barriers to financial recovery. Finally, alluding to various behavioral biases and heuristics, participants were asked their opinion on the financial behavior of survivors of climate crises.

Qualitative data analysis software ATLAS.ti was used to conduct a thematic content analysis of the transcripts. Textual data was coded into key themes to identify patterned meanings across participants' perceptions, opinions, views, and experiences. The aim was to code and synthesize common and contrasting themes.

While both email and face-to-face interviews have their strengths and weaknesses, interviews are often criticized for lacking objectivity and not being a reliable method for qualitative data collection. Therefore, the researcher has taken care to address the evaluation criteria by ensuring data credibility by validating interviewees, collecting data to saturation point, peer review, taking careful notes, triangulating thoughts and ideas with

Table 16.1 List of interviewees

	Interviewee	Type of organization	Position/job description
Experts for the local financial services market	1	Credit and financial service provider	CEO
	2	Mobile telecommunication company	CEO
	3	Financial advisory services	Principal and financial advisor
Experts on the local impact of adverse weather events	4	Agriculture	Head of procurement
	5	Agriculture	Proprietor[a]
	6	Traditional leadership	Village chief[a]
	7	NGO	Senior project manager
	8	NGO	Senior social worker[a]
	9	Academic department	Head of department
	10	Government	Ward Councillor
	11	Medical	Community doctor
Experts in the process of providing local aid	12	Development bank	Research manager
	13	NGO	Executive director
	14	NGO	Executive board member
	15	Government	Senior disaster management officer[a]

[a]Survivor of a climate-related disaster

the literature, and securing audio recordings of the face-to-face interviews. Additionally, the trustworthiness and reliability of the data are ensured by underpinning the themes with quotes from the participants to focus on the opinions of the participants rather than the researcher.

16.3 Findings and Discussion

Although the participants confirmed the existence of several barriers and opportunities which were identified in the literature, some new ones were also revealed.

Most interviewees agreed that government transfers such as social grants play a central role in the rural household economy. To improve standards of living and alleviate social inequality, the South African government issued the Social Assistance Act in 2004, mandating the South African Social Security Agency (SASSA) to administer seven different welfare grants and further provide immediate, temporary assistance in the form of food parcels for families most in need (Steinert et al., 2018, pp. 4–5). However, participants were of the view that apart from grants for old age, social transfers can encourage a dependency syndrome, can be misused, and that governments are persisting with what is unsustainable to ensure votes. Nevertheless, social transfers are essential in times of crisis and in maintaining activity within the formal and informal financial systems.

The literature supports the view that transfers such as child support grants and remittances improve food security, particularly among women in rural communities (Aphane et al., 2010, p. 66). Participants believe that remittances play an important role in times of crisis, and are an effective form of support for rural households. A study looking at the microlevel impact of climate change in female-headed rural households consisting of the elderly, suggests that with less capital, these households are most vulnerable, and that receiving remittances or participating in formal savings schemes will reduce their vulnerability (Ncube et al., 2016). In support of this view, Mohapatra et al. (2012) finds evidence that suggests a positive role of remittances in preparing households against natural disasters and coping with the aftermath of loss and damage. Respondents caveat the literature by pointing out that, in some cases, the remittances are irregular and are used to pay loan sharks the moment the household receives them. Still, a FinScope (2018) study suggests that more South Africans depend on money remittances, and informal remittances were higher than bank remittances in 2018.

Low literacy levels influence most of the mentioned barriers to accessing formal financial services. In particular, respondents mentioned a lack of financial literacy and the need to provide literacy courses in rural communities. Financial literacy is a form of human capital and may provide some protection against limitations in cognitive capital. Financial literacy includes the behaviors and attitudes toward finance, knowledge and skill of finance, and the awareness of financial products. The literature on financial literacy in developing countries supports these views and points to the constrained cognitive capacity to optimally process financial information during decision-making processes. In their study, Nanziri and Leibbrandt (2018) construct a financial literacy index to collect data on attitudes toward and access to the use of financial services over the period 2005–2009. They conclude that demographic characteristics contributed up to 10% of the financial literacy differences among individuals in South Africa.

Financial literacy is vital for a nation's economic well-being, and has an apparent beneficial effect on financial inclusion, and on the financial infrastructure (Refera et al., 2016; Grohmann et al., 2018). Financial literacy is low in developing and developed countries, but particularly low in Africa. These are important observations because the adaptive capacity of poor rural households is proving insufficient to deal with the aftermath of recurring, severe natural disasters (Baez et al., 2013). Low levels of financial literacy are obstacles to personal financial decision-making capabilities. Financial education programs will enhance the demand and proper use of beneficial financial services such as savings, microcredit, and insurance (Refera et al., 2016). Enhancing financial literacy and financial inclusion is very important in this region. A contribution to creating awareness of the need for savings and emergency funds will empower survivors to assist themselves when public disaster response is too slow or absent.

It is widely recognized that the effects of climate change lead to inequality in health, gender, and income (Pérez-Peña et al., 2021, p. 10). Participant responses highlighted that women have higher levels of illiteracy in rural communities. Moreover, most respondents believe that households headed by women are most vulnerable to climate-related disasters, and that their ability to recover is severely limited. Severe weather events magnify the countless disproportionate effects. On gender-differentiated impacts of climate change, women suffer more adversely in terms of their assets and well-being because of social and cultural norms that limit their access to and control of assets. The everyday socioeconomic status of women is a significant factor in explaining large-scale gender differences in mortality rates (Goh, 2012; Neumayer & Plümper, 2007). This suggests that inequality will need to be considered while managing the ongoing risks of climate change.

Although respondents highlight the benefits of mobile money, such as easy transactions, instant transfers, bypassing bank accounts and cost efficiency, digital illiteracy and digital poverty remain the most significant challenges regarding mobile money. Participants are of the view that the use of mobile money remains a challenge to women and vulnerable households who will have limited access. Studies have shown that in countries with significant gender disparities, women are left behind by the digitization of financial services (Burjorjee & Scola, 2015, p. 19). Although the FSC has promoted banking densification and electronic customer infrastructure access, the cultural and social norms and socioeconomic status of women are significant factors in explaining the inability of women to access and use mobile phones. Understanding the unique construct of low-income populations is essential to grasp the reactions to technological innovations.

Research has shown that access to formal finance, informal finance and mobile money facilitates the coping strategies of households facing various adverse shocks. Riley (2018, p. 43) examines and compares the impact of a rainfall shock on village households who both use and do not use mobile money, and concludes that after a shock, only the users of mobile money can prevent a drop in their consumption, with no spillover effects to other village members. However, some respondents have expressed concern, as mobile money has proven to be a significant challenge due to technology illiteracy and the lack of access to, or sharing mobile phones. Here, financial inclusion, through new technologies may have future implications on traditional risk-sharing arrangements.

According to the National Stokvel Association of South Africa, over 11 million South Africans belong to more than 810,000 active *stokvel* groups. These community-based models provide the flexibility and social security needed to help lower-income groups manage their money, plan for their future, and cope with emergencies (Storchi, 2018).

Most respondents believe that informal finance, especially informal rotating savings and credit associations, known as *stokvels*, still plays an important role in rural household finance. Formalizing *stokvels* through regulation, and training on financial management skills may strengthen social protection (Dube & Edwell, 2018). However, this is a contentious issue and ongoing debate. A further motivation for *stokvels* is that they represent a self-help approach to social security, particularly for unemployed women and low-income earners, enabling participants to save, invest, and accumulate assets (Matuku & Kaseke, 2014, p. 514). These views confirm those of Bophela and Khumalo (2019) who suggests that money-saving and investment, business opportunities and job creation, and policy development were the main economic contribution factors of *stokvels* in South Africa. However, concern centers around the sustainability of the informal financial sector, as its' prominence results from a combination of weak financial literacy, poor financial inclusion, limited financial access, and growing costs of transacting.

Nevertheless, *stokvels* are attractive to rural communities because *stokvels* are a cultural phenomenon, and are respected and revered. They are well-understood and quite sophisticated. At an independent level, rural households could aggregate into financial collectives, such as informal savings associations *stokvels*, to increase financial reserves for weather disasters. Effectively managed *stokvels* could contribute to the alleviation of crises. These informal savings groups are agencies of social solidarity and social cohesion for the underprivileged classes to deal with the challenges of poverty. These social connections consolidate social capital, leading to trust and reciprocity, the ability to share knowledge, and making decisions in groups or networks.

When the uncertainty of extreme weather events negatively impacts mental health, social capital and communal coping will play an essential role in recovery (Afifi et al., 2011, p. 16). To this end, Tibesigwa et al. (2016) promotes informal social capital and natural resources as cheaper and more accessible coping strategies than insurance, which remains unaffordable in most rural parts of sub-Saharan Africa. However, although Castillo and Carter (2011, p. 16) find a direct impact between adverse weather shocks and trust and reciprocity; their research suggests that too large of a shock might not foster cooperation. Nevertheless, Velazco and Ballester (2016, p. 527) find that physical assets, human capital, and social capital reduce the likelihood of food shortages among rural households in developing countries experiencing climatic, agricultural and health shocks. Moreover, de Bruijn et al. (2017, p. 25) refers to recovery capacity as a function of social, institutional, and economic capital, and suggest that increasing recovery capacity is closely linked to the level of socioeconomic development.

Households most at risk of financial distress and poverty are ones with too little savings, overextended credit, and dropped income levels when there is a sudden increase in demand on expenditure. It was worth exploring whether consumer credit was the most appropriate response option for resilient reconstruction after a shock, such that the costs of financing do not outweigh the potential welfare gains. However, respondents thought that crisis loans make the situation worse because, in a crisis, the tendency is to reach for the nearest solution, which will, in many circumstances, create further challenges with the repayments of these loans, often at very high interest rates. Respondents suggest that it is unfortunate that a vulnerable household must resort to a loan. With most households mired in poverty and barely able to survive on their meager incomes, a loan puts a massive strain on that household regarding food security and other basic needs. Furthermore, respondents suggest that loans will be expensive, and informal loans will be made through unscrupulous players. The literature supports these views by suggesting that low-income groups tend to smooth volatile income by taking on more credit, most often from predatory lenders, as they have less access to mainstream forms of credit (Gandy et al., 2016).

16.4 Concluding Remarks

A dependence on resources sensitive to climate change, accompanied by unequal policies and development patterns, places poor societies in a very vulnerable situation. In reality, vulnerable households face significant barriers to developing and implementing climate-adaptive financial strategies, including ineffective support programs, a lack of property rights, and poor service delivery. Furthermore, they may not have access to social or financial safety nets. Moreover, disaster management policy in developing countries operates in a fiscally constrained environment, and the impact of extreme weather events disrupts the services they provide (Van Niekerk, 2014, p. 864).

It is vital to assist survivors with financial recovery and promote proactive climate-responsive behavior. Externally administered rescue programs are a long-term recipe for disaster, since the problems within volatile, low-income, and developing countries are not necessarily purely financial but more multifaceted and complex. Institutional weaknesses, policy uncertainty, political instability, corruption, mismanagement of public resources, and overall government failure are added layers of concern.

To better understand how the formal sector can represent the social and economic values that drive rural communities' decisions about financial services, further research could include an evaluation of policies that integrate indigenous and scientific knowledge. This could more meaningfully inform the development of adaptation policies and decision-making processes. In support of this view, Melore and Nel (2020, p. 8) suggest that communities that retained some indigenous knowledge appear to improve their resilience to climate change.

Efforts need to be made to help the excluded and vulnerable communities adapt to climate-induced challenges, because financially prepared households fare better during climate shocks and tend to have better financial outcomes (Edmiston, 2017, p. 36). To this end, financial product design should consider the implications for those living in poverty, for migrants, and for women, while remaining mindful of the importance of indigenous knowledge and the informal financial sector.

SDG target 8.10 aims to promote universal access to banking, insurance, and financial services, and can offer enormous improvements in inclusivity. Evidence suggests that increasing access to financial services increases household welfare and resilience, supporting livelihoods and stimulating economic activity when faced with adverse shocks (El-Zoghbi et al., 2017, p. 15). Climate-related disasters warrant longer-term solutions for financially resilient, sustainable recovery. Quality financial products and services are essential to improve people's resilience in fragile and conflict-affected situations. Therefore, more emphasis should be placed on the role of inclusive financial products and services in managing crises, minimizing risks, strengthening social protection, and boosting economic activity.

This research encourages financial inclusion as a practical policy option, to promote economic participation through access to finance and resilience building with greater urgency. Discovering inclusive and accommodating ways in which financial services and products can address the risks of climate change for the poor, will provide insights into developing appropriate policy to improve financial inclusion for vulnerable populations.

Financial resilience offers the potential to reduce the exposure of the most vulnerable, and strengthen adaptive capacity in extreme contexts. Risk and resilience feature prominently in the sustainable development goals. A cross-policy and cross-disciplinary response could enable different stakeholders and actors to work together and consider new policy options that appropriately address the uncertainties in people's lives. Ultimately, the aim is to take a contextual approach to identify the transformational capacity of financial services for financial resilience, while improving capacity and overall development outcomes.

References

Afifi, W. A., Felix, E. D., & Afifi, T. D. (2011). The impact of uncertainty and communal coping on mental health following natural disasters. Anxiety, Stress and Coping, 1–19. https://doi.org/10.1080/10615806.2011.603048.

Akotey, J. O., & Adjasi, C. K. D. (2018). Microinsurance and consumption smoothing among low-income households in Ghana. *The Journal of Developing Areas, 52*(4), 151–165. https://doi.org/10.1353/jda.2018.0057

AL-Fazari, S., & Kasim, N. (2019). Role of stakeholders in mitigating disaster prevalence: Theoretical perspective. *MATEC Web of Conferences, 266*, 1–6. https://doi.org/10.1051/matecconf/201926603008

Aphane, M., Dzivakwi, R., & Jacobs, P. (2010). Livelihood strategies of rural women in Eastern Cape and Limpopo. *Agenda, 24*(84), 66–74. https://doi.org/10.1080/10130950.2010.9676310

Arun, T., & Kamath, R. (2015). Financial inclusion: Policies and practices. *IIMB Management Review, 27*(4), 267–287. https://doi.org/10.1016/j.iimb.2015.09.004

Asbi, A., Ramiah, V., Yu, X., Wallace, D., Moosa, N., & Reddy, K. (2020). The determinants of recovery from theBlack Saturday bushfire: demographic factors, behavioural characteristics and financial literacy. *Accounting and Finance, 60*(1), 15–46. https://doi.org/10.1111/acfi.12575

Baez, J. E., Kronick, D., & Mason, A. D. (2013). Rural households in a changing climate. *World Bank Research Observer, 28*(2), 267–289. https://doi.org/10.1093/wbro/lks008

Barajas, A., Beck, T., Belhaj, M., & Naceur, S. B. (2020). Financial inclusion: What have we learned so far? What do we have to learn? In *IMF Working Papers* (No. WP/20/157). Retrieved from https://www.imf.org/-/media/Files/Publications/WP/2020/English/wpiea2020157-print-pdf.ashx

Blamey, R. C., Middleton, C., Lennard, C., & Reason, C. J. C. (2017). A climatology of potential severe convective environments across South Africa. *Climate Dynamics, 49*(5–6), 2161–2178. https://doi.org/10.1007/s00382-016-3434-7

Bophela, M. J. K., & Khumalo, N. (2019). The role of stokvels in South Africa: A case of economic transformation of a municipality. *Problems and Perspectives in Management, 17*(4), 26–37. https://doi.org/10.21511/ppm.17(4).2019.03

Burjorjee, D. M., & Scola, B. (2015). A market systems approach to financial inclusion; guidelines for funders. *Consultative Group to Assist the Poor (CGAP)*, (September), 60.

Carter, P., & Tyson, J. (2017). Development financing in Africa. Retrieved from https://www.uneca.org/sites/default/files/PublicationFiles/development_financing_in_africa_eng_2017.pdf

Castillo, M., & Carter, M. (2011). Behavioral responses to natural disasters. Retrieved from https://ices.gmu.edu/wp-content/uploads/2011/07/Behavioral-Responses-to-Natural-Disasters-by-Castillo-and-Carter.pdf

Chitimira, H., & Ncube, M. (2020). Legislative and other selected challenges affecting financial inclusion for the poor and low income earners in South Africa. *Journal of African Law, 64*(3), 337–355. https://doi.org/10.1017/S0021855320000182

de Bruijn, K., Buurman, J., Mens, M., Dahm, R., & Klijn, F. (2017). Resilience in practice: Five principles to enable societies to cope with extreme weather events. *Environmental Science and Policy, 70*, 21–30. https://doi.org/10.1016/j.envsci.2017.02.001

Davis-Reddy, C. L., Vincent, K., & Mambo, J. (2017). Socioeconomic impacts of extreme weather events in southern Africa. In *Climate risk and vulnerability: A handbook for southern Africa* (2nd ed., pp. 1–18).

Demirgüc-Kunt, A., Klapper, L., Singer, D., Ansar, S., & Hess, J. (2018). *The global findex database 2017: Measuring financial inclusion and the fintech revolution*. Retrieved from https://globalfindex.worldbank.org/sites/globalfindex/files/2018-04/2017Findex full report_0.pdf

Department of Trade and Industry. (2017). *Broad-based black economic empowerment act 53 of 2003 as amended by act 46 of 2013:* Amended Financial Sector Code, 9 §.

Dube, N., & Edwell, K. (2018). Strengthening informal social security to provide meaningful social protection: The case of stokvels in Soweto. *Journal of Social Development in Africa, 33*(1), 157–181.

Edmiston, K. D. (2017). Financial vulnerability and personal finance outcomes of natural disasters. In *The Federal Reserve Bank of Kansas City Research Working Papers*. https://doi.org/10.2139/ssrn.3064155

El-Zoghbi, M., Chehade, N., McConaghy, P., & Soursourian, M. (2017). The role of financial ser-

vices in humanitarian crises (12). https://doi.org/10.1596/26511
Financial Sector Charter Council. (2004). *South African Financial Sector Charter*, 1–15. Retrieved from http://banking.org.za/wp-content/uploads/2019/07/financial-sector-charter.pdf
FinMark Trust. (2018). How south Africans are leveraging informal and formal financial markets: FinScope SA 2018 fact sheet South African socio-economics banking. *Finscope*, pp. 2–5.
Fourie, W. (2018). Aligning South Africa's National Development Plan with the 2030 Agenda's Sustainable Development Goals: Guidelines from the Policy Coherence for Development movement. *Sustainable Development, 26*(6), 765–771. https://doi.org/10.1002/sd.1745
Gandy, K., King, K., Streeter Hurle, P., Bustin, C., & Glazebrook, K. (2016). Poverty and decision-making: How behavioural science can improve opportunity in the UK. *Department for Education Research Report*, (October), pp. 1–122. https://doi.org/10.1080/09297040802385400.
Goh, A. (2012). A literature review of the gender-differentiated impacts of climate change in developing countries structure. *CAPRi, 106*, 1–38. https://doi.org/10.2499/CAPRiWP106
Grohmann, A., Klühs, T., & Menkhoff, L. (2018). Does financial literacy improve financial inclusion? Cross country evidence. *World Development, 111*, 84–96. https://doi.org/10.1016/j.worlddev.2018.06.020
Golnaraghi, M., & Khalil, P. (2017). *The stakeholder landscape in extreme events and climate risk management*. Retrieved from https://www.genevaassociation.org/media/956576/stakeholder-landscape-in-eecr.pdf
Hallegatte, S., Vogt-schilb, A., Bangalore, M., & Rozenberg, J. (2017). *Unbreakable: Building the resilience of the poor in the face of natural disasters* (Climate Ch). Retrieved from. https://openknowledge.worldbank.org/bitstream/handle/10986/25335
Inter-Agency and Expert Group on SDG Indicators. (2020). Global indicator framework for the Sustainable Development Goals and targets of the 2030 Agenda for Sustainable Development. In *Work of the Statistical Commission pertaining to the 2030 Agenda for Sustainable Development*. Retrieved from https://unstats.un.org/sdgs/indicators/GlobalIndicatorFrameworkafter2019refinement_Eng.pdf%0A, https://unstats.un.org/sdgs/indicators/GlobalIndicatorFramework_A.RES.71.313Annex.pdf
IPCC. (2014). Climate change 2014: Synthesis report. Contribution of working groups 1, 11, 111 to the 5th assessment report of the Intergovernmental Panel on Climate Change. In L. The Core Writing Team, Pachuari, Rajendra, Meyer (Ed.), *Intergovernmental Panel on Climate Change*. https://doi.org/10.1017/CBO9781139177245.003.
Islam, N., & Winkel, J. (2017). Climate change and social inequality. In *Building Resilience to Climate Change – An Opportunity to Reduce Inequalities*. (No. 152). https://doi.org/10.4324/9781315103358.
Kostov, P., Arun, T., & Annim, S. (2014). Banking the unbanked: The Mzansi intervention in South Africa. *Indian Growth and Development Review, 7*(2), 118–141. https://doi.org/10.1108/IGDR-11-2012-0046
Klapper, L., El-Zoghbi, M., & Hess, J. (2016). Achieving the sustainable development goals: The role of financialinclusion. *In Current Science*. Retrieved from https://www.cgap.org/sites/default/files/Working-Paper-Achieving-Sustainable-Development-Goals-Apr-2016_0.pdf
Lensink, R., Servin, R., & van den Berg, M. (2017). Do Savings and credit institutions reduce vulnerability? New evidence from Mexico. *Review of Income and Wealth, 63*(2), 335–352. https://doi.org/10.1111/roiw.12213
Martin, C. (2019). *Roadmap to the Sustainable and Responsible Financial Inclusion of Forcibly Displaced Persons*. Retrieved from https://www.afi-global.org/wp-content/uploads/2020/07/Roadmap_FI-of-FDPs_122019_0.pdf
Matuku, S., & Kaseke, E. (2014). The role of stokvels in improving people's lives: The case in Orange Farm, Johannesburg, South Africa. *Social Work, 50*(4), 1–14. https://doi.org/10.15270/50-4-388
Mawejje, J. (2019). Financial inclusion, shocks and coping strategies: Survey evidence from Uganda. *African Journal of Economic and Management Studies, 10*(3), 286–298. https://doi.org/10.1108/AJEMS-10-2018-0325
Melore, T. W., & Nel, V. (2020). Resilience of informal settlements to climate change in the mountainous areas of Konso, Ethiopia and QwaQwa, South Africa. *Jamba: Journal of Disaster Risk Studies, 12*(1), a778. https://doi.org/10.4102/JAMBA.V12I1.778
Mohapatra, S., Joseph, G., & Ratha, D. (2012). Remittances and natural disasters: Ex-post response and contribution to ex-ante preparedness. *Environment, Development and Sustainability, 14*(3), 365–387. https://doi.org/10.1007/s10668-011-9330-8
Moore, D., Niazi, Z., Rouse, R., & Kramer, B. (2019). *Building resilience through financial inclusion: A review of existing evidence and knowledge gaps*. Retrieved from https://www.poverty-action.org/sites/default/files/publications/Building-Resilience-Through-Financial-Inclusion-January-2019.pdf
Nanziri, E. L., & Leibbrandt, M. (2018). Measuring and profiling financial literacy in South Africa. *South African Journal of Economic and Management Sciences, 21*(1), 1–17. https://doi.org/10.4102/sajems.v21i1.1645
Ncube, M., Madubula, N., Ngwenya, H., Zinyengere, N., Zhou, L., Francis, J., et al. (2016). Climate change, household vulnerability and smart agriculture: The case of two south African provinces. *Jamba: Journal of Disaster Risk Studies, 8*(2), #182. https://doi.org/10.4102/jamba.v8i2.182
Neumayer, E., & Plümper, T. (2007). The gendered nature of natural disasters: The impact of catastrophic events on the gender gap in life expectancy,

1981-2002. *Annals of the Association of American Geographers, 97*(3), 551–566. https://doi.org/10.1111/j.1467-8306.2007.00563.x

Pérez-Peña, M. D. C., Jiménez-García, M., Ruiz-Chico, J., & Peña-Sánchez, A. R. (2021). Analysis of research on the SDGs: The relationship between climate change, poverty and inequality. *Applied Sciences (Switzerland), 11*(19). https://doi.org/10.3390/app11198947

Refera, M. K., Dhaliwal, N. K., & Kaur, J. (2016). Financial literacy for developing countries in Africa: A review of concept, significance and research opportunities. *Journal of African Studies and Development, 8*(1), 1–12. https://doi.org/10.5897/jasd2015.0331

Riley, E. (2018). Mobile money and risk sharing against village shocks. *Journal of Development Economics, 135*(June), 43–58. https://doi.org/10.1016/j.jdeveco.2018.06.015

Statstics SA. (2019). *Sustainable Development Goals - Country report 2019 South Africa*. Retrieved from http://www.statssa.gov.za/MDG/SDGs_Country_Report_2019_South_Africa.pdf

Steinert, J. I., Cluver, L. D., Meinck, F., Doubt, J., & Vollmer, S. (2018). Household economic strengthening through financial and psychosocial programming: Evidence from a field experiment in South Africa. Journal of Development Economics, 134(0), 443–466. https://doi.org/10.1016/j.jdeveco.2018.06.016.

Storchi, S. (2018). Impact evaluation of savings groups and Stokvels in South Africa the economic and social value of group-based financial inclusion October 2018. In *Finmark Trust*. Retrieved from https://www.samra.co.za/wp-content/uploads/2017/02/SAMRA-Journal-2016-EVersion.pdf

Tibesigwa, B., Visser, M., Collinson, M., & Twine, W. (2016). Investigating the sensitivity of household food security to agriculture-related shocks and the implication of social and natural capital. *Sustainability Science, 11*(2), 193–214. https://doi.org/10.1007/s11625-015-0332-6

Van Niekerk, D. (2014). A critical analysis of the south African disaster management act and policy framework. *Disasters, 38*(4), 858–877. https://doi.org/10.1111/disa.12081

Velazco, J., & Ballester, R. (2016). Food access and shocks in rural households: Evidence from Bangladesh and Ethiopia. *Social Indicators Research, 129*(2), 527–549. https://doi.org/10.1007/s11205-015-1118-z

Zhou, L., Sibanda, M., Musemwa, L., & Ndhleve, S. (2016). Vulnerability to climate change related disasters in the eastern Cape Province: An application of the household vulnerability index (HVI). *Journal of Human Ecology, 56*(3), 335–353. https://doi.org/10.1080/09709274.2016.11907071

17 Gendered Social Norms and Financial Inclusion in Turkey

Nuray Karaman

Abstract

A new area of research has recently emerged on the importance of gender disparities in financial inclusion. The sustainable development goals address improving the capabilities and operations of financial institutions, and this is one of the common topics that is highlighted within the context of SDG subgoal 8.10. Financial inclusion is now viewed by many countries in the Global South as one of the most crucial development initiatives. Turkey has joined the movement in recent years, and officials have developed an agenda for financial inclusion. While Turkey's financial inclusion plans aim to increase the reach of the financial system, they mostly disregard the importance of women in financial inclusion, which is becoming more widely recognized. This chapter tries to address a variety of reasons, including political, cultural, and social issues, as well as financial issues that hinder women's financial participation in Turkey. This study aims to show the importance of understanding gendered social norms in favor of fairness in financial inclusion in order to reduce the gender gap in access to and use of financial services. Additionally, understanding the financial inclusion of women in Turkey offers policymakers a fresh viewpoint for future consideration of the new financial inclusion rules.

Keywords

Financial inclusion · Gender gap · Women · Social norm · Patriarchy · Culture · Turkey

17.1 Introduction

There is gender inequality worldwide, and inequality undercuts or openly violates women's human rights. Many studies highlight the significance of gender equality in meeting all development objectives and strategies (Sen & Grown, 1987; Fareed et al., 2017; Ghosh & Vinod, 2017; Siddik, 2017). Sustainable development goals are expanding freedom, reducing poverty, and increasing outcomes for women and men (Sen & Grown, 1987, p. 23). Women's work is essential to human beings' existence and continued reproduction in all nations. While development projects aim to offer equal access to resources and chances for all genders, races, and ethnicities, many researchers show that women do not have equal rights with men in every aspect of life.

The significance of gender inequalities in financial inclusion has been an emerging field of study in recent years. Increasing the capacities

N. Karaman (✉)
Uşak University, Uşak, Turkey
e-mail: nuray.karaman@usak.edu.tr

G. W. Muschert et al. (eds.), *Financial Inclusion*, Sustainable Development Goals Series,
https://doi.org/10.1007/978-3-031-68803-4_17

and functions of financial institutions is handled within sustainable development goals. It is one common area that comes to the fore within SDG subgoal 8.10. In this context, one of the priority areas suggested for SDG is the establishment of entrepreneurial capital funds and ensuring the continuity of enterprise support. Supports given to women entrepreneurs have an important place among the support for small- and medium-sized enterprises. Access to financial services and resources is also an effective tool for women's employment.

On the other hand, entrepreneurship is one of the essential tools for women's involvement in employment. Nevertheless, the ratio of female entrepreneurs in Turkey is also meager. According to Global Entrepreneurship Monitor (GEM) database, less than 20% of all entrepreneurs in Turkey are women (Okten, 2015). This chapter aims to discuss some reasons behind this lag, including culturally rooted practices and beliefs, context including the patriarch, and gendered social norms in Turkey. Understanding of financial inclusion of women in Turkey provides a new perspective to policymakers for future reference of the new regulations of financial inclusion.

17.2 Development and Financial Inclusion

Before delving into the relationship between gender and financial inclusion, we must first define development. The primary textbook definition of development is enhancing individuals' wealth, self-esteem, and independence to raise the standard of their lives and abilities (Todaro & Smith, 2011, p. 5). The Marxist paradigm focuses on the material and inequality aspects of development. Marx is critical of development because it creates exploitation and alienation of workers, creating inequality for the proletariats who do not own any capital. The Marxist critique of development comes from the idea behind the capitalist mode of production, which refers to organizing the production and distribution of material goods in society. Marx's analysis of capitalism led him to conclude that the system was inherently unstable, destructive, and exploitative (Calhoun, 2007, p. 78). This determination is because Marx sees the root of all profit in capitalism as traced to the extraction of surplus value from human labor; as competition increases, exploitation of workers increases through lower wages for them (Calhoun, 2007, p. 79).

Marx was concerned with the conditions of people in the development process. The Marxist paradigm argues that a country's rising gross domestic product (GDP) is not inherently good for the individuals in that country because the capitalist mode of production increases the exploitation of workers. Therefore, Marxian analysis is critical of the division of labor as a wealth creation tool because it exploits workers' labor. From the Marxist perspective, economic growth breeds societal inequality. He argues that as capitalism expands, wage labor is exploited; similarly, inequalities between class and gender negatively impact individuals. Therefore, as a country becomes highly capitalistic, Marx posits that proletarians will rise in opposition as inequality increases.

Poverty and inequality are closely linked, according to a large body of research. Although many developing countries have increased their per capita GDP, the female disadvantages in the labor market have remained persistent. Women continue to earn less than men and are less likely to benefit effectively from financial services. The gender wage gap severely affects women's health, education, employment opportunities, and even the level of authority they have—or do not have—over their own lives and decisions (Hendriks, 2019). For example, the most significant segments of Turkey's poor are women because of the unequal distribution of wealth and opportunities. According to the TÜİK database, in 2011, 16.1% of the Turkish population lived in poverty. Women, without a doubt, comprise a more considerable proportion of this group. The percentage of women who work for a living is equally low, at only 24% (Acuner, 2013). These rates of women's poverty and participation in paid labor show gender inequality in the labor market and some reasons behind women's financial exclusion.

One of the most challenging issues confronting developing economies is ensuring that the financial sector properly promotes equitable growth (Swamy, 2014). Financial inclusion is a term that refers to the process of connecting individuals with banks and the advantages that come with it. Poor and vulnerable people, disadvantaged locations, and lagging sectors need access to secure, accessible, and inexpensive credit, not to mention other financial services, to accelerate growth and reduce income inequalities and poverty. Economic and socially excluded individuals can compete in business and be a part of growth if they have access to a properly working financial system, creating equitable possibilities and shielding them against economic downturns (Swamy, 2014).

According to the Global Findex database, between 2014 and 2017, 515 million individuals worldwide established a bank or mobile money account (Demirguc-Kunt et al., 2018). In high-income countries, 94% have a bank account, whereas 63% in low-income countries do. Individual economies also have a wide range of account ownership (Demirguc-Kunt et al., 2018). In emerging economies, financial inclusion differs from that in developed economies. Academics also suggest that observing financial exclusion based on gender and class in emerging economies is better for studying financial inclusion (Swamy, 2014).

17.3 Women and Financial Inclusion

In many countries, women's social and economic status has progressively improved because of the secular tendencies that have existed for a long time, such as increases in women's enrolment in higher education and adult literacy, labor force participation, and real GDP per capita (Norris, 2001). Even so, progress in women's empowerment in the social, cultural, and economic environment has lagged worldwide, and gender inequality continues to undercut or openly violate women's human rights. Sen and Grown (1987) advocate this idea for some reasons. First, development is expanding freedom, reducing poverty, and increasing outcomes for women and men (Sen & Grown, 1987, p. 23). Second, although being underpaid and devalued, women's labor is essential to human survival and reproduction in all civilizations (Sen & Grown, 1987, p. 23). Third, in many societies, women's work is integrated into all work areas such as business, politics, services, and industries (Sen & Grown, 1987, p. 24). Thus, women must have equal access to and control of economic and financial services to promote gender equality and empowerment and fair and sustainable economic growth and development (Saha, 2020). While development projects aim to offer equal chances for all genders, races, and ethnicities, many researchers show that women still need equal economic, political, cultural, and social rights compared to men.

Women's empowerment has been affected by the division of labor, the strong patriarchy, and the male-dominated market phenomena (Mogdaham, 2003). Feminism would agree with post-structuralism that gender is a social construct that argues that men are seen as more valuable in development because it is assumed so through their place in the household. Furthermore, feminism takes a critical stance against development as a masculine discourse and argues that women in development (WID) paradigm is necessary to create a concept of development that acknowledges both men and women.

Feminism critiques previous development theories because they ignore women and take a masculine approach. One paradigm that came out of the feminist perspective is women in development (WID). The WID paradigm focuses on an effort to "integrate women in development" through training and educational programs (Beneria, 2001, p. xii). Tinker states that development negatively affects poor women because women in higher social classes can access better resources (Tinker, 2001). Tinker (2001) highlights that the goal of WID was to analyze women's place in "economic development" as well as empowerment, education, employment, and legal equality. The economic inequalities in development programs emerge due to gender inequality

(Tinker, 2001, p. 35). The WID explains that women's exclusion emerged not only from the development model but also from women's place in social life and subordinate conditions.

Financial inclusion also includes the knowledge and skills required to use financial goods. Commonly used financial inclusion indicators have a variety of characteristics. Because financial inclusion is a multifaceted term, it influences various issues, including economic growth (Bozkurt & Karakus, 2020). Increased financial inclusion drives women's financial empowerment (Gammage et al., 2017). While the importance of financial systems that work well in reducing income disparity and supporting economic development is well acknowledged, there is limited information on the gender inequality in financial access or the unique hurdles women face compared to males, especially in low-income and middle-income countries. Previous studies have emphasized the fundamental benefit of equitable financial access for women and its importance as a vehicle for starting a positive cycle of empowerment on a social, economic, and political level (Cheston & Kuhn, 2002).

Women can benefit greatly from financial inclusion. Women actively involved in the financial system are better able to manage risk, regulate consumption, pay household expenses such as schooling and household nutrition, and improve the community's results (Duflo, 2012; Hendriks, 2019; Prina, 2015). Not just for women's empowerment but also to alleviate poverty, low-income women need the proper financial instruments to save and borrow money because more women than men live in poverty and are in charge of impoverished children's care (Cawthorne, 2008). The lack of women's engagement in the financial system is problematic for them. However, nations have trouble increasing overall growth figures since women's engagement in finance is low, lowering the country's average participation rate (Demirgüç-Kunt et al., 2017; Goodman, 2018). Inequalities in financial inclusion make it difficult for individuals to escape poverty by limiting investment in education and career options. Increased occupational and educational opportunities enhance saving and borrowing behavior, resulting in increased economic mobility (Klapper & Singh, 2015).

Women's prospects for further education and work outside the home are limited to controlling property ownership, asset inheritance, and financial issues. Financial literacy can help women in homes make better decisions and influence the use of resources (Demirgüç-Kunt et al., 2017; Saha, 2020). Women's access to individual savings accounts has been demonstrated in studies to build a strong economy, empower women to make financial decisions, and enhance their negotiating power in the home (Duflo, 2012; Dupas & Robinson, 2013; Prina, 2015). Education and income are essential in determining financial access and negotiation power in the home. However, political and social variables are far more salient in determining financial usage and benefiting from financial services to improve empowerment in an economic, social, and cultural environment (Ghosh & Vinod, 2017).

Women are a heterogenous group having different economic, political, social, and cultural backgrounds. Therefore, the diversity of women's interests and needs has yet to receive serious consideration to understand the constraints and challenges women encounter in obtaining financial services and their financial knowledge, preferences, attitudes, and behavior regarding financial inclusion initiatives. Mainly focusing on the diversity of women's interests and needs in financial inclusion policies is helpful to develop insight to understand and solve reasons behind women's financial dependency on men and gender inequalities reproduced in reaching and using financial services.

17.4 Challenges to Women's Financial Inclusion: The Turkish Case

Even though social dynamics are essential, most market systems and women's economic empowerment programs still need to grasp how they impact women's financial inclusion. Most programming focuses on increasing opportunities for women to access and use financial services.

However, these programs ignore the sociocultural and religious factors that affect women's behavior and ability to engage with and benefit from these services (Markel et al., 2016). Many market systems programs examine agency problems in the financial system, such as voice, choice, and decision-making authority. They must dive further into the complex social norms and belief systems that influence women's capacity to fully engage in market systems (Markel et al., 2016). Various situations and cultures have different ways of empowering women, but specific components are universal and essential.

All social, economic, and political factors are interconnected and indivisible regarding human rights rhetoric. While there has been a good debate on agency and empowerment, it is clear that empowerment includes resolving current disparities experienced by women and shifts in the sense of self-awareness and agency that dispute patriarchal norms (Sen & Mukherjee, 2014). Understanding and overcoming challenges in women's financial inclusion in Turkey, which reflects the attachment to the majority Islamic and patriarchal culture, it remains vital to analyze women's experiences with the intersection of religion, culture, and agency.

Women in Turkey have a 38% lower chance of holding an account than men after adjusting for other personal characteristics, such as income, employment, education, and age (Demirguc-Kunt & Klapper, 2012). Account ownership increased significantly for both genders, according to Findex 2017; however, the gender divide has expanded. While 83% of men had an account in 2017 in Turkey, and 54% of women did. Some studies show that there is a significant gender difference in financial inclusion in Turkey, and female employment involvement is an essential element in explaining women's financial inclusion (Boudet & Agar, 2014; Demirgüç-Kunt et al., 2018; Ecevit, 2007; Klapper & Singh, 2015; Schwab et al., 2017). According to the Gender Gap Report (2013), there is a significant gender gap in workforce participation in Turkey and substantial consequences for female employment participation. In Turkey, 98% of wage-earning women are banked, compared to 23% of the majority of women who say they are unemployed (Klapper & Singh, 2015). One of the crucial problems arises from a failure to recognize the labor market structure: women's low labor force participation and high informal and unpaid employment (Boudet & Agar, 2014). During the neoliberal period, gendered disparities in labor force participation solidified, making it more difficult to attract women into financial relationships. According to the International Labor Organization (ILO), the participation of women in the workforce fell dramatically in the early 2000s (İlkkaracan, 2012). Women's labor force participation recovered from 24% in 2008 to 33% in 2019, while men's labor participation rate stands at 71.96% (World Bank, 2021). Women have limited access to financial services due to their low labor force participation. The World Bank database shows that 74% of the workforce has an account, whereas only 59% of the unemployed hold an account (Demirgüç-Kunt et al., 2018). Women suffer when social, cultural, religious, economic, and political implications and norms prevent them from working or confines them to low-wage occupations if employment better provides for financial inclusion (Demirgüç-Kunt et al., 2018).

One of the most critical issues to consider is the effect of the patriarchal family on poverty and the exploitation of women in Turkey. According to Walby, patriarchy is a social practice where men are the central authority symbol, leading to the exploitation of women (Haj-Yahia, 2005). In all patriarchal societies, more than women, men have an opportunity for resources, responsibilities, skills, abilities, knowledge, and job opportunities (Haj-Yahia, 2005). Men play the role of being more dominant and achieving compared to women. These beliefs decrease the number of working women and their financial inclusion (Lusardi & Mitchell, 2008). In this view, women are responsible for taking care of their husbands and children and supporting their husbands in their jobs while their husbands earn money and save for their family's welfare. In Turkey, there are many gendered stereotypes, such as the belief that men should earn more than women and be more responsible for financial matters within their relationships. These stereotypes have caused women to be less likely to be entrepreneurs and face more disadvantages in financial inclusion (Cetindamar et al., 2012; Lusardi & Mitchell, 2008).

In Turkey, men are more than three times as likely as women to report receiving money from work or the government through a bank account. Even though just 16% of women claim to be working, 97% of wage and self-employed women in Turkey are banked, compared to only 23% of women who claim to be unemployed. In comparison, 70% of Turkish males who claim to be unemployed have a bank account (Klapper & Singh, 2015). Gender disparities hamper women's access to financing and more extensive economic possibilities, which harm entrepreneurship. According to research on banking and entrepreneurship, women entrepreneurs face significant hurdles to obtaining start-up funding, entrepreneurship training, and loans to build and grow their firms (Cetindamar et al., 2012; Hung et al., 2012; Lusardi & Mitchell, 2008).

Financial services companies are part of the societies in which they operate, and their decisions can reinforce or extend gender norms that limit or increase women's access to financial services. Gender stereotypes hamper women's capacity to attain financial inclusion. According to several studies, financial inclusion problems include a lack of education, employment, and decision-making authority (Cetindamar et al., 2012; Gammage et al., 2017; Goodman, 2018; Güngen, 2018). Gender stereotypes that restrict women from seeking higher education, acquiring employment, and having the same rights as men may also limit women's ability to acquire basic skills to manage their finances and be a consumer of financial services. The conservative government approach exacerbates this by portraying females as family proxies rather than a user of financial communities, reinforcing gender roles in determining financial access (Güngen, 2018, p. 337).

It is critical to consider the hurdles to financial inclusion and the connection between financial exclusion and gender inequalities. People with a low level of education and little control over managing money are more likely to be financially excluded (Demirgüç-Kunt et al., 2017). Cole, Paulson, and Shastry's study (2014) shows that education impacts financial market participation and decreases the effective use of financial services. Education plays a crucial role in financial inclusion, so it is worth considering how different communities perceive women's pursuit of education. Regarding educational achievement, women lag behind men by 5% worldwide (Blake & Propson, 2018). In 1960, just 1.2% of Turkish women were engaged in higher education; by the 2000s, that number had risen to 19%. As a result of this growth, the gap between men's and women's educational attainment rates in Turkey has narrowed, but there is still a gap. In Turkey, 7% of girls between 8 and 12 never attend school, while just 2% of males do (Caner et al., 2016, p. 122). By 2015, female enrolment is about 20% lower than male (Caner et al., 2016, p. 122). According to many studies, this disparity might be due to the gender norm about the significance of education for women (Blake & Propson, 2018; Caner et al., 2016; Cole et al., 2014; Goodman, 2018). Gender role disparities explain gender variations in educational and labor market results. Several studies' findings also show that gender norms, such as the degree of violence against women, the division of household, and the rate of early marriage among women, play a role in explaining the disparity in financial services use between men and women after taking into account other individual and national factors (Demirgüç-Kunt et al., 2013). Policymakers must improve financial knowledge to promote women's labor force participation and financial inclusion in Turkey. However, public policies may exacerbate existing inequalities due to ignoring the country's significant inequalities and the patriarchal division of work in families (Altan-Olcay, 2016; Yildirim et al., 2017). The belief that financial institutions are capable and adequate to provide services does not help women who do not seek financial advice at critical junctures in their life and are incapable of planning budgets or utilizing financial goods to their benefit. While policymakers and the financial industry emphasize individual consciousness, agency, working status, and education by ignoring culturally determined gender ideologies, they do not provide equal opportunities in financial services. Culturally determined gender ideologies have determined gender roles in the family, society, the labor market, and decision-making processes (Altan-Olcay, 2016; Ilkkaracan, 2012; Yıldırım et al., 2017). The impacts on women's empowerment tend to be generally favorable, although they depend on the financial service and cultural and regional context. In Turkey, financial inclusion services do not have a strategy that explicitly marks women without identifying them regarding their families. Instead, it highlights the tactics and initiatives necessary for raising financial awareness.

17.5 Conclusion

The significance of gender in the drive for financial inclusion has recently gained attention. Developing and developed countries' economies are attempting to produce policy answers to this problem. For many nations in the Global South, financial inclusion has become one of the critical development strategies. Turkey has gotten on board in the last decades, and policymakers have created a financial inclusion agenda. While Turkey's financial inclusion strategies intend to expand the financial system's reach, it mostly ignores the significance of women in financial inclusion is gaining growing public awareness. Turkey needs a policy that analyzes women's financial inclusion without putting them in the context of their families. The study gives an overview of the topic in this setting, building on existing information.

A variety of causes hamper women's financial inclusion, including political, cultural, and social issues, as well as financial issues. There is a need for additional attention to gender inequalities in financial inclusion, topics which in the future need thorough investigation. Policymakers can focus on how job discrimination and inequality in education affect a woman's capacity to access financial services and then benefit from these services. Intensifying this approach by analyzing cultural, religious, and social reasons behind inequalities in employment, education, access to, and use of financial services has the potential to result in more effective policies that can help women become more financially independent. Suppose policy-makers address gendered social norms in support of equity in financial inclusion. This will help identify the reasons behind gender inequality in financial inclusion and offer solutions to close the gender gap in access to and use of financial services.

References

Acuner, S. (2013). Gender and development in Turkey. *Turkish Policy Quarterly, 11*(4), 71–78.

Altan-Olcay, Ö. (2016). The entrepreneurial woman in development programs: Thinking through class differences. *Social Politics: International Studies in Gender, State & Society, 23*(3), 389–414.

Beneria, L. (2001). Introduction. In *Gender and development: Theoretical, empirical and practical approaches* (Vol. 1, pp. xi–xxi). Edward Elgar Publishing, Inc.

Blake, M., & Propson, D. (2018). *Advancing financial inclusion metrics: Shifting from access to economic empowerment*. World Economic Forum.

Boudet, A. M. M., & Agar, M. (2014). *Female entrepreneurship in Turkey*. The World Bank.

Bozkurt, İ., & Karakuş, R. (2020). Provincial financial inclusion in Turkey: Measurement and its spatial determinants. *Ege Academic Review, 20*(2), 101–124.

Calhoun, C. (2007). *Nations matter: Culture, history, and the cosmopolitan dream*. Routledge.

Caner, A., Guven, C., Okten, C., & Sakalli, S. O. (2016). Gender roles and the education gender gap in Turkey. *Social Indicators Research, 129*(3), 1231–1254.

Cawthorne, A. (2008). The straight facts on women in poverty. *Center for American Progress, 8*(1), 1–3.

Cetindamar, D., Gupta, V. K., Karadeniz, E. E., & Egrican, N. (2012). What the numbers tell: The impact of human, family, and financial capital on women and men's entry into entrepreneurship in Turkey. *Entrepreneurship & Regional Development, 24*(1–2), 29–51.

Cheston, S., & Kuhn, L. (2002). Empowering women through microfinance. In S. Daley-Harris (Ed.), *Pathways out of poverty: Innovations in microfinance for the poorest families*. Kumarian Press.

Cole, S., Paulson, A., & Shastry, G. K. (2014). Smart money? The effect of education on financial outcomes. *The Review of Financial Studies, 27*(7), 2022–2051.

Demirgüç-Kunt, A., & Klapper, L. F. (2012). *Measuring financial inclusion: The global Findex database*. World Bank Policy Research Working Paper, (6025).

Demirgüç-Kunt, A., Klapper, L. F., & Singer, D. (2013). *Financial inclusion and legal discrimination against women: Evidence from developing countries*. World Bank Policy Research Working Paper, (6416).

Demirguc-Kunt, A., Klapper, L., Singer, D., & Ansar, S. (2018). *The global Findex database 2017: Measuring financial inclusion and the fintech revolution*. World Bank Publications.

Duflo, E. (2012). Women empowerment and economic development. *Journal of Economic Literature, 50*(4), 1051–1079.

Dupas, P., & Robinson, J. (2013). Why don't the poor save more? Evidence from health savings experiments. *American Economic Review, 103*(4), 1138–1171.

Ecevit, Y. (2007). *A critical approach to women's entrepreneurship in Turkey*. International Labor Office.

Fareed, F., Gabriel, M., Lenain, P., & Reynaud, J. (2017). *Financial inclusion and women entrepreneurship* (Working paper no. ECO/WKP(2017)43).

Gammage, S., Kes, A., Winograd, L., Sultana, N., Hiller, S., & Bourgault, S. (2017). *Gender and digital financial inclusion: What do we know and what do we need to know?* International Center for Research on Women.

Ghosh, S., & Vinod, D. (2017). What constrains financial inclusion for women? Evidence from Indian microdata. *World Development, 92*(1), 60–81.

Goodman, C. (2018). *How gender norms impact women's access to financial inclusion*. Economics Undergraduate Honors Theses. Retrieved from https://scholarworks.uark.edu/econuht/27

Güngen, A. R. (2018). Financial inclusion and policymaking: Strategy, campaigns, and microcredit a la Turca. *New Political Economy, 23*(3), 331–347.

Haj-Yahia, M. M. (2005). On the characteristics of patriarchal societies, gender inequality, and wife abuse: The case of Palestinian society. *Adalah's Newsletter, 20*. http://www.adalah.org/newsletter/eng/nov05/fet.pdf

Hendriks, S. (2019). The role of financial inclusion in driving women's economic empowerment. *Development in Practice, 29*(8), 1029–1038.

Hung, A., Yoong, J., & Brown, E. (2012). *Empowering women through financial awareness and education*. https://www.oecd-ilibrary.org/finance-and-investment/empowering-women-through-financial-awareness-and-education_5k9d5v6kh56g-en

İlkkaracan, İ. (2012). Why are so few women in the labor market in Turkey? *Feminist Economics, 18*(1), 1–37.

Klapper, L., & Singh, S. (2015). *The gender gap in the use of financial services in Turkey*.

Lusardi, A., & Mitchell, O. S. (2008). Planning and financial literacy: How do women fare? *American Economic Review, 98*(2), 413–417.

Markel, E., Gettliffe, E., Jones, L., Miller, E., & Kim, L. (2016). The social norms factor: How gendered social norms influence how we empower women in market systems development. *The BEAM Exchange*.

Mogdaham, V. (2003). *Modernizing women –Gender and social change in the Middle East* (Second ed.). Lynne Rienner Publisher.

Norris, P. (2001). Breaking the barriers: Positive discrimination policies for women. In *Has liberalism failed women?* (pp. 89–110). Palgrave Macmillan.

Okten Hasker, C. (2015). *Female entrepreneurship in Turkey: Patterns, characteristics, and trends* (Vol. No. 102013, pp. 1–36). The World Bank.

Prina, S. (2015). Banking the poor via savings accounts: Evidence from a field experiment. *Journal of Development Economics, 115*(1), 16–31.

Saha, A. (2020). Gender interests and financial inclusion: A feminist analysis on the rural women of Barpeta District. *International Journal of Management (IJM), 11*(6), 1047–1057.

Schwab, K., Samans, R., Zahidi, S., Leopold, T. A., Ratcheva, V., Hausmann, R., & Tyson, L. D. (2017, March). *The global gender gap report 2017*. World Economic Forum.

Sen, G., & Grown, C. (1987). *Development alternatives with women for a new era*. Earthscan.

Sen, G., & Mukherjee, A. (2014). No empowerment without rights, no rights without politics: Gender equality, MDGs, and the post-2015 development agenda. *Journal of Human Development and Capabilities, 15*(2–3), 188–202.

Siddik, M. N. A. (2017). Does financial inclusion promote women's empowerment? Evidence from Bangladesh. *Applied Economics and Finance, 4*(4), 169–177.

Swamy, V. (2014). Financial inclusion, gender dimension, and economic impact on poor households. *World Development, 56*(1), 1–15.

Tinker, I. (2001). The making of a field: Advocates, practitioners, and scholars. *Gender and Development: Theoretical, Empirical and Practical Approaches, 1*(1), 70–101.

Todaro, M. P., & Smith, S. C. (2011). *Economic development*. Addison-Wesley, Pearson, ISBN, 10(1), 0–13.

World Economic Forum. (2013). *The global gender gap report 2013*. World Economic Forum.

Yıldırım, M., Bayram, F., Oğuz, A., & Günay, G. (2017). Financial literacy level of individuals and its relationships to demographic variables. *Mediterranean Journal of Social Sciences, 8*(3), 19–33.

Fintech and Financial Inclusion in the Kingdom of Morocco: Demographics and Socioeconomic Characteristics

18

Abdulkader Aljandali and Amal Laaribi

Abstract

The role of fintech in supporting financial inclusion has gained momentum among international bodies. The UN sustainable development agenda of 2030 and the G20 document on digital financial inclusion key principles both stress the need to use fintech to reduce financial exclusion and to enhance income equality. The debate around financial inclusion has increased in the past 2 years, especially with the crisis of Covid-19. The pandemic accelerated the development of Morocco's digital financial environment, with government and private firms deploying fintech tools for payments and other financial transactions. This chapter has the following purposes: (1) to review the literature that examines fintech as a driver to financial inclusion, (2) to investigate critical barriers that prevent some segments of the Moroccan population from accessing financial services while giving indicators that reflect financial exclusion, and (3) to shed light on fintech advances and how the Kingdom of Morocco intends to leverage the contribution of fintech firms in the implementation of its National Financial Inclusion Strategy (SNIF).

Keywords

Fintech · Financial exclusion · MENA · North Africa · Morocco · Financial inclusion

18.1 Financial Inclusion in Emerging Markets

Financial inclusion means individuals and businesses can access effective and affordable financial products and services that meet their needs—transactions, payments, savings, credit, and insurance—delivered responsibly and sustainably (World Bank, 2022). Financial services include a range of products and services banks and other financial institutions provide. However, the main products usually considered are bank accounts, credit cards, ATM services, loans, and other forms of credit. A society is financially inclusive when most people can access these financial products. A vast body of literature now exists on financial inclusion, defined as the state in which an individual or an

A. Aljandali (✉)
School of Business, Computing and Social Sciences, University of Gloucestershire, Gloucester, UK

ISCAE Casablanca, Casablanca, Morocco
e-mail: aaljandali@glos.ac.uk

A. Laaribi
ENCG (Ecole Nationale de Commerce et de Gestion) Casablanca, Université Hassan II, Casablanca, Morocco

G. W. Muschert et al. (eds.), *Financial Inclusion*, Sustainable Development Goals Series,
https://doi.org/10.1007/978-3-031-68803-4_18

entity holds an account with a formal financial institution allowing the person or the entity to save and borrow using the institution's financial instruments (e.g., Zins & Weill, 2016). Financial inclusion is found to empower women (Swamy, 2014), enhance disadvantaged individuals' incomes and employability (Bruhn & Love, 2014), and improve financial stability (Han & Melecky, 2013). Demirgüç-Kunt et al. (2015) provide a statistical comparison between developed and developing countries regarding financial inclusion. Their assessment shows that formal savings are significantly less in developing countries than in high-income economies. For example, only 9% of adults use formal credit in developing countries instead of 18% in advanced economies.

In the African continent, financial inclusion varies substantially across regions and among countries (Demirgüç-Kunt et al., 2013a, b). This applies to both savings and borrowing. Mobile banking is among the financial instruments contributing to the rise in financial inclusion as more individuals possess mobile banking accounts (Demirgüç-Kunt et al., (2015). For the last decade, mobile network operators (MNOs) have dominated mobile money services in Africa. Several barriers are responsible for the low level of financial inclusion, including the following: (1) absence of documentation(s) for women; (2) relying on another member of the family who holds an account; (3) being more concerned about the immediate financial need, particularly for the poor individuals, rather than accessing formal credit; and (4) religious and distance concerns (see, for instance, Allen et al., 2016; Demirgüç-Kunt et al., 2013a, b). Against this background, it is worth noting that Africa is the world leader in mobile money. Mobile financial services have expanded to include various financial services such as credit, insurance, and cross-border remittances. Previous studies show that 37 African markets have 10 times more registered agents than bank branches, which allows customers to cash in and out (McKinsey, 2017).

18.2 Fintech and Financial Inclusion

The dramatic surge of interest in fintech over the past few years has highlighted the need for a better understanding of the value of technological innovations. In line with this reasoning, fintech has emerged as a mechanism that enables banking and nonbanking institutions to cut costs, enhance the quality of their services, and create a much more resilient and diverse financial environment. This fintech development, supported by artificial intelligence (AI) and automation, allows fintech adopters to enter different segments of the financial services industry (Aljandali & Benamraoui, 2020). Fintech is a set of recently developed digital computing technologies that have been applied—or are likely to be—to financial services (Chen et al., 2019). Over the last 10 years, fintech has emerged as a major driving force, which has reshaped the financial industry. Fintech allows for greater control over personal finances and increases accessibility to financial services. Areas such as robo-advisors, insurtech, digital payments, and consumer lending are challenging the traditional financial services sector. Widespread technological adoption has been the major catalyst for fintech innovation, with solutions developed for global problems.

Fintech companies are businesses providing financial services and products by using new technologies. We can compare them to technology-oriented start-ups with a digital marketing and commercial approach relying on a targeted product offer. Fintech companies directly compete with banks and insurers in almost all areas of the financial sector. The landscape of fintech companies worldwide shows that there are nearly 13,000 fintech companies, mainly based in North America (half of them). Another third are in Europe, with a small percentage based in Africa. The fintech sector is the most active subsector within the African start-up ecosystem. It mainly targets solutions to improve financial inclusion by addressing the unbanked segments of the population.

The development of fintech ecosystems happens at different rates of speed in Africa. Some of the most critical drivers for fintech development are low banking rates, strong mobile/Internet penetration, young tech-savvy demographics, and a significant diaspora with ties to their respective African countries. There are at least eight countries leading the fintech movement and three types of ecosystems/regions: Africa's "Leading" fintech ecosystems in Anglophone Africa (e.g., Nigeria, Ghana, South Africa, and Kenya), Northern Africa's "Emerging" fintech ecosystem (e.g., Morocco, Tunisia), and "Promising" fintech ecosystems in Francophone Africa (e.g., Senegal, Côte d'Ivoire). A large and well-trained workforce is required for the continuing growth of the different Fintech ecosystems in Africa. Nevertheless, improving the education system remains a major headache for many African countries. The young population of Africa is one of its greatest assets. Governments and the members of the various African fintech ecosystems should work hand-in-hand to prioritize this. Fintech companies in Africa continue to emerge and grow, thanks to the increasing interest of banks and telecom companies, which offer collaboration options, access to markets, and deep industry expertise. However, the regulatory environment in Africa is yet to be adapted to these new market players and slows down the development of the fintech ecosystem in certain regions. Addressing the regulatory challenges for fintech companies in Africa would complement the governments' efforts in offering financing options tailored to the needs of fintech companies since access to finance continues to be a daunting issue for their founders. So while the rise of fintech companies in Africa is on a positive trend, close collaboration and complete alignment of initiatives by all parties will be required to accelerate and consolidate this growth.

The role of fintech in supporting financial inclusion has gained momentum among international bodies. The fundamental principles of the UN sustainable development agenda 2030 and the G20 document on digital financial inclusion stress the need to use fintech to reduce financial exclusion and enhance income equality. To further reinforce financial inclusion through digitalization, a group of regulators and policymakers representing 90 developing countries established the alliance for financial inclusion. The alliance is funded by the Bill and Melinda Gates Foundation and has the full backing of the G20 countries. Among the key aims of the alliance is to set strategies involving the private sector to engage a global population estimated at 2.5 billion, which still needs to use formal finance. Its mission states that the alliance creatively applies venture capital investments to achieve financial inclusion (Gabor & Brooks, 2017). Aligning with this recent development, several studies have linked financial inclusion to fintech (e.g., Demir et al., 2020; Gosavi, 2018; Tchamyou et al., 2019). Using a global sample of 140 countries, Demir et al. (2020) examine the causality between fintech, financial inclusion, and income equality. They find that fintech has a direct effect on inequality owing to financial inclusion. Among the fintech areas explored is mobile technology. Research points to a direct causality between financial inclusion and the extent of mobile phone penetration or usage per household (Ghosh, 2016).

Individuals use the technology for general financial purposes including making payments, establishing saving accounts, and conducting other banking transactions (Ouma et al., 2017). The impact of mobile and digital payments has become more evident as organizations and governments work toward developing growth in emerging economies and driving financial inclusion. New technologies create an opportunity for financial services to address emerging and developed markets' needs by creating more customized products and services. The use of mobile technology has a positive effect on SMEs' financial inclusion as it allows more access to banking credit (Gosavi, 2018). Overall, the use of digital finance, including mobile banking, is driving economic development (Demirgüç-Kunt et al., 2020).

Research by Sha'ban et al. (2020) show that financial inclusion is more prominent across the use and access measures. Among the countries experiencing a significant increase in financial inclusion are those in Africa and Asia. They further show that financial inclusion is related to the country's gross domestic product (GDP) per capita, level of banking competition, the extent of human development, Internet usage, employment, and government integrity. Technologies such as blockchain (powered by mobile devices) can assist in solving the problem of providing access to services to those who were previously unserved and underserved in emerging markets and have the potential to drive economic growth. Hasan et al. (2021) identify financial technology as playing a pivotal role in providing financial access to people residing in rural areas, noting how mobile banking can be used as an alternative to brick-and-mortar banks as is the case in African nations such as Nigeria. Using individual characteristics, access to digital technology is low among women, poor, and less educated adults (Demirgüç-Kunt et al., 2017). While those with an account, digital or non-digital, are not necessarily using the account for their transactions and instead continue to make payments in cash. Several recommendations have emerged from the studies conducted on fintech and financial inclusion. Sha'ban et al. (2020) stress the importance of enhancing banking competition and technology integration to promote financial inclusion. Digitalization and banking system conditions are perceived to be complementary in their ability to reduce financial exclusion. They further state that improving the quality of education by developing basic technological skills and fostering digital literacy would reduce the digital gap and ultimately improve the usage of financial services and the extent of financial inclusion. Demirgüç-Kunt et al. (2017) assert that alongside fintech, financial inclusion requires a solid physical infrastructure, a well-developed payment system, a robust consumer protection scheme, and pertinent regulations. Besides, the developed financial products must be tailored to the requirements of first-time users, those with less numeracy and literacy skills, people with low incomes, and women.

EY (2019) conducted a survey that examines fintech adoption in six markets—Australia, Canada, Hong Kong, Singapore, the UK, and the USA—and found that the latter surged from 16% in 2015 to 31% in 2017, and finally to 60% in 2019. Although Africa is absent from this list, the promise of mobile financial services is already evident as the M-Pesa shows excellent potential for the continent. In 2007, it had 32 million active monthly users and processed 6.5 billion transactions in the same year (Frost et al., 2019). A Pew Research Center report showed that three out of ten African mobile phone owners claim they use their phones to send and collect payments. Moreover, mobile payment is only the beginning. From insurance and wealth management to highly underserved agricultural and SME funding sectors, fintech is progressing exponentially through many financial markets (Pew Research Center, 2018). Mobile network operators (MNOs) have dominated mobile money services in Africa for the past decade, making Africa the world leader in mobile money, where mobile financial services have expanded to include a broad array of financial services such as credit, insurance, and cross-border remittances (McKinsey, 2017).

18.3 Fintech and Financial Inclusion in Morocco

Given that, only 29% of Moroccan adults have access to a bank account—lower than the 44% Middle East and North Africa (MENA) average—and agriculture accounts for 39% of total employment (World Bank, ILO 2019a, b), there is an opportunity to bank millions of Moroccan farmers without bank accounts. Farmers, historically, have had less access to banking solutions (BAM, 2017). Increasing financial inclusion among Moroccan farmers, including bank accounts, mobile banking, microcredits, insurance, and real-time crop prices, will have direct benefits such as using savings and loans to accelerate consumption, absorb shocks such as health issues or adverse weather, or make investments that increase productivity (Collins et al., 2009). To solve this challenge, there is a need for a simple and fair mobile banking account accessible through both smart and non-smartphones for Moroccan farmers. Unlike generic financial inclusion products, this one will include crop prices and add-on products to help farmers become more productive and competitive. Morocco is the third largest fintech hub in the Arab world, hosting 13% of the 400 active fintech solutions in the region, about 40 fintech solutions, according to research by the financial inclusion think tank Consultative Group to Assist the Poor (CGAP). Total banking penetration in Morocco stood at just 28.6% in 2017, and the economy remains cash-based, with around 80% of transactions made using cash. Systems are the most developed segment of Morocco's 40+ active fintech solutions, payment, remittance, and point-of-sale (POS). Invyad provides a mobile, end-to-end POS solution for small- and medium-sized enterprises (SMEs); Wafacash is a remittance start-up; and Inwi Money and HmizatePay specialize in mobile payments. HmizatePay is the mobile payment platform of the e-commerce start-up Hmizate. OnePay is a newcomer that enables both offline and online businesses to receive digital payments. Its platform acts as an aggregator for payments and value-added services. OnePay raised US$400,000 in July 2020. Other paytech companies in Morocco include Société Maghrebine de Monetique (S2M) and NAPS, a subsidiary of M2M Group.

Morocco's ambition to position itself as a financial hub in Africa has pushed it toward digitalizing its financial sector. Across the world, fintech ecosystems have promoted technological innovation, made capital markets and systems more effective, and improved the customer experience (Diemers et al., 2015). In order to capitalize on such momentum, the Moroccan government is striving to implement the required regulatory support. They focus on developing telecoms, the energy sector, and the entire infrastructure for fintech investors. Hence, law 103–12 was enacted in March 2015, which introduced the concepts of the Payment Establishment and the Payment Agent that has initiated the fintech revolution in Morocco. This law allows nonbank service providers, including telecom operators, to offer payment solutions that end the banks' monopoly to promote financial inclusion. In its chapter dedicated to Morocco, Pricewaterhouse Coopers (PwC) notes that Morocco is a dynamic market in which the fintech sector is becoming more popular (Segal, 2016). The environment is promising, thanks to mature markets, skill availability, and a clear regulatory framework. Industry analysts suggest that the growing influence of fintech is high in emerging markets such as the MENA (Middle East and North Africa) region where a significant proportion of the population is unbanked and where the accessibility to financing for SMEs is one of the lowest in the world (Diemers et al., 2015). This implies that fintech innovations in these markets are more likely to be disruptive in some customer segments (SMEs, millennials, and the unbanked), and in selected financial services/products including crowdfunding, P2P lending, wealth management, and advisory (Zalan & Toufaily, 2017). The last few years have seen the fintech sector flourish. In Morocco, the e-commerce site Hmizate moved into fintech in 2018 with the introduction of the all-in-one smartphone shopping and payment app HmizatePay following the model of Alipay (launched in 2004) and WeChat Pay (launched in 2011) has surpassed 500 million and 900 million monthly active users, respectively. In January 2017, Outlierz Ventures, an African-focused venture capital company located in Casablanca, was founded to invest in tech-enabled companies in the fintech market. Moroccan banks and incumbents have already entered the fintech market. In November 2018, the Popular Central Bank (BPC) group, one of Morocco's most prominent financial entities, introduced the Fintech Challenge, an initiative intended to find innovative start-ups for investment opportunities, and CFG Bank partnered with US-based digital financial innovation specialist Kony as part of the bank's research (Oxford Business Group, 2020). For its part, the government is making efforts to facilitate the financing of Moroccan start-ups. The World Bank has made $50 million available to create the Innov-invest fund, which aims to support the launch of innovative start-ups. In January 2020, King Mohammed 6 launched a program to finance entrepreneurship with a fund of 6 billion dirhams (Finances, 2018).

This context provides growth opportunities with digitalizing payment systems on a clear route toward expansion. According to Morocco's central bank (BAM, 2018), the number of bank cards in circulation between 2015 and 2017 rose from 11.8 to 14.1 million. While Morocco remains a predominantly cash-based culture, the introduction of electronic forms of payment would enable the creation of new service channels. Electronic, Internet, and mobile transactions have been gaining popularity and will enable Morocco to reach further financial inclusion. There were 360,000 mobile wallets as of September 2019, with the volume of electronic payments expected to reach 50 billion dirhams ($5.2 billion) a year by 2030 (Oxford Business Group, 2020).

As more and more mobile subscribers use the Internet and the average selling prices of smartphones decrease, the smartphone penetration rate becomes an essential factor that deserves consideration. The latter directly paves the way for further development. Frost (2020) and Hau et al. (2018) agree that adopting fintech is driven by an unmet demand for financial services; if banks underserve current businesses or customers, there could be potential for faster growth in fintech development. For instance, according to the most recent data, big-tech mobile payments comprised 16% of the GDP in China, but less than 1% in the United States, India, and Brazil. As more and more mobile subscribers use the Internet in Morocco and the average selling prices of smartphones decrease, the smartphone penetration rate becomes an essential factor for consideration. The latter opens the door for further development in the fintech sector. Robo-advisory grows beyond mainstream markets as tech companies look to exploit the Muslim world's economic potential. The introduction of digital advisory services should draw Sharia-compliant clients, who may have difficulties managing the restrictions in Islamic finance. Companies such as Wahed Invest realized astonishing success among Muslim clients allowing them to make Sharia-compliant investments. Morocco can be a compelling host for this technology as the wealth management sector needs better development. Three reasons make the implementation of robo-advisory Sharia-compliant in Morocco very compelling for investors. Firstly, a study by Al Amine (2016) shows that the Islamic financial market's current size in Africa is far below its real potential. Secondly, results from a study conducted by Fulk et al. (2018) attest that users of robo-advisory services generally had lower income, lower net worth, and received no or less inheritance, which may describe a considerable portion of Morocco's population. Thirdly, a survey by the Islamic Research and Training Institute of the Islamic Development Bank found that 98% of Moroccans are interested in Islamic banking (Reuters, 2014).

Morocco's central bank has promoted and developed electronic means of payment as a top objective under the country's national strategy to improve financial inclusion. The regulator has been helped in its ambitions by a banking law introduced in 2015 that allowed it to incorporate a new category of nonbanking institutions, called payment institutions, to serve the market. The first fintech start-up providing payment services launched in 2017, and in 2018 the central bank announced the approval of four new fintech operators; the total number now stands at more than 20, including the services of three telecom operators. In addition to these providers, about 20 banks and insurers also offer digital payment solutions as part of their services. Morocco intends to leverage the contribution of fintech firms in the implementation of its National Financial Inclusion Strategy (SNIF). The first SNIF report, dating from 2019, underlines that, although various actors implemented many actions over the last decade, Morocco still needs a concerted push to close financial inclusion gaps in the system. The most significant disparities in terms of financial services penetration are between women and men, rural and urban dwellers, and youth and older adults. This national strategy concerning fintech relies on two main pillars: developing alternative models that reach the most excluded populations at a lower cost and adapting to these groups' specificities. The Moroccan National Financial Inclusion Strategy (NFIS) relies on eight strategic levers identified after a review of international practices and the analysis of the impact of different models on the penetration of formal financial services at the level of comparable countries in terms of GDP per capita. These levers center around the

- Acceleration of the development of alternative models.
- Incentives for "classic" models to push financial inclusion further.
- Reinforcement of tools to better understand the risk of excluded populations.
- Implementation of a framework and tools facilitating the financing of microenterprises and individuals.
- Creation of conditions for greater use of financial products.
- Management of financial inclusion over time.

This chapter shows that fintech solutions are gradually gaining momentum in Morocco as the government works toward financial inclusion, digitalization, and strengthening the financial services sector. BAM has played a significant role by liberalizing regulations, facilitating access to financial products for start-ups and small- and medium-sized enterprises, and building a mobile payment ecosystem. Programs and policies designed to promote financial inclusion in Morocco aim to promote access and use of formal financial services by women and older people to educate them on mobile banking. The development of mobile banking eliminates all barriers and sources of involuntary exclusion. It allows the inclusion of rural populations, females, and older people marginalized by the conventional financial system. It also solves the problems related to the remoteness of bank branches. It reduces the costs and fees associated with financial transactions and thus eliminates the barrier of the "high cost" of financial services. To sum up, fintech can be the ultimate gateway to include all segments of Moroccan society in financial and economic growth. However, the central bank and public policies can only regulate that to ensure consumer protection and the sustainability of financial inclusion gains.

The environment in Morocco is conducive to the birth of national fintech champions, at least from a regulatory point of view. The environment is favorable for the emergence of this technology. The idea of regulation is that it does not advance before the object of this regulation but is not a barrier to entry. Furthermore, that is the case! We now have the necessary tools to upgrade start-ups to the stage of national champions, agree to say the speakers. The adoption by Bank Al-Maghrib of Law 103–12 is a concrete example, allowing nonbanking services (notably telecom operators) to offer payment solutions ending bank monopolies to promote financial inclusion. In the same vein, the Kingdom is one of the only countries on the continent to make many efforts to pass a law facilitating crowdfunding.

Conflict of Interest We have no conflicts of interest to disclose.

References

Allen, F., Demirguc-Kunt, A., Klapper, L., & Martinez Peria, M. (2016). The foundations of financial inclusion: Understanding ownership and use of formal accounts. *Journal of Financial Intermediation, 27*, 1–30.

Al Amine, M. A. (2016). *Islamic finance and Africa's economic resurgence: Promoting diverse and localized investment*. Springer.

Aljandali, A., & Benamraoui, A. (2020). FinTech innovations: A review of the recent developments and prospects. *Bancaria, 12*, 85–95.

BAM. (2017). Annual report. Available: http://www.bkam.ma/en/content/view/full/14431

BAM. (2018). Annual report. Available: http://www.bkam.ma/en/content/download/667171/7639427/RA_2018_ENG.pdf

Bruhn, M., & Love, I. (2014). The real impact of improved access to finance: Evidence from Mexico. *Journal of Finance, 69*(3), 1347–1376.

Chen, M. A., Wu, Q., & Yang, B. (2019). How Valuable Is FinTech Innovation?, *The Review of Financial Studies, 32*(5), 2062–2106. https://doi.org/10.1093/rfs/hhy130

Collins, D., Morduch, J., Rutherford, S., & Ruthven, O. (2009). *Portfolios of the poor how the World's poor live on $2 a day*. Princeton University Press.

Demir, A., Pesqué-Cela, V., Altunbas, Y., & Murinde, V. (2020). Fintech, financial inclusion, and income inequality: A quantile regression approach. *The European Journal of Finance*, 1–22.

Demirgüç-Kunt, A., Klapper, L., & Randall, D. (2013a). *Islamic finance and financial inclusion: Measuring the use of and demand for formal financial services among Muslim adults* (Policy Working Research Paper No. 6642). The Word Bank.

Demirgüç-Kunt, A., Klapper, L., & Singer, D. (2013b). *Financial inclusion and legal documentation against women: Evidence from developing countries* (Policy Research Working Paper, No. 6416). The Word Bank.

Demirgüç-Kunt, A., Klapper, L., Singer, D., & Van Oudhensden, P. (2015). *The Global Findex Database 2014: Measuring financial inclusion around the world* (Policy Research Working Paper, No. 7255). The Word Bank.

Demirguc-Kunt, A., Klapper, L., & Singer, D. (2017). *Financial inclusion and inclusive growth: A review of recent empirical evidence* (Policy Research Working Paper Series, 8040). The World Bank.

Demirgüç-Kunt, A., Klapper, L., Singer, D., Ansar, S., & Hess, J. (2020). The Global Findex Database 2017: Measuring financial inclusion and opportunities to expand access to and use of financial services. *The World Bank Economic Review, 34*(Supplement), S2–S8.

Diemers, D., Lamaa, A., Salamat, J., & Steffens, T. (2015). *Developing a FinTech ecosystem in the GCC*. Strategy&,16.

EY. (2019). Global FinTech Adoption Index 2019. Available: EY Global FinTech Adoption Index 2019 – Our latest thinking | EY - Global|EY Global FinTech Adoption Index 2019 – Our latest thinking | EY - Global

Finances.gov.ma. (2018). The "CCG" launches A web site devoted to the "Innov invest fund" – MEF – Kingdom Of Morocco.

Frost, J., Gambacorta, L., Huang, Y., Shin, H. S. and Zbinden, P. (2019). "BigTech and the changing structure of financial intermediation."BIS Working Papers No. 779.

Frost, J. (2020). "The economic forces driving fintech adoption across countries."BIS Working Papers No. 838.

Fulk, M., Grable, J. E., Watkins, K., & Kruger, M. (2018). Who uses robo-advisory services, and who does not? *Financial Services Review, 27*(2).

Gabor, D., & Brooks, S. (2017). The digital revolution in financial inclusion: Institutional development in the Fintech era. *New Political Economy, 22*(4), 423–436.

Ghosh, S. (2016). Does mobile telephony spur growth? Evidence from Indian states. *Telecommunications Policy, 40*(10–11), 1020–1031.

Gosavi, A. (2018). Can mobile money help firms mitigate the problem of access to finance in Eastern Sub-Saharan Africa? *Journal of African Business, 19*(3), 343–360.

Han, R., & Melecky, M. (2013). *Financial inclusion for financial stability: Access to bank deposits and the growth of deposits in the Global Financial Crisis* (Policy Research Working Paper No 6577). The Word Bank.

Hasan, R., Ashfaq, M., & Shao, L. (2021). Evaluating drivers of Fintech adoption in The Netherlands. *Global Business Review*. (In press).

Hau, H., Huang, H., Shan, H., & Sheng, Z. (2018). *Fintech credit, financial inclusion, and entrepreneurial growth*. Working Papers.

McKinsey. (2017). *Mobile financial services in Africa: Winning the battle for the customer*. Available: https://www.mckinsey.com/industries/financial-services/our-insights/mobile-financial-services-in-africa-winning-the-battle-for-the-customer

Ouma, S. A., Odongo, T. M., & Were, M. (2017). Mobile financial services and financial inclusion: Is it a boon for savings mobilization? *Review of Development Finance, 7*(1), 29–35.

Oxford Business Group. (2020). *The report: Morocco 2020*. Available: The Report: Morocco 2020 | Oxford Business Group|The Report: Morocco 2020 | Oxford Business Group.

Pew Research Center. (2018). *Internet connectivity seen as having positive impact on life in Sub-Saharan Africa*. Available: Internet Seen as Having Positive Impact in Sub-Saharan Africa | Pew Research Center|Internet Seen as Having Positive Impact in Sub-Saharan Africa | Pew Research Center

Reuters. (2014). *Morocco Islamic Finance 2014: Unlocking the Kingdom's Potential*. Available: https://ceif.iba.edu.pk/pdf/ThomsonReuters-MoroccoIslamicFinance2014UnlockingtheKingdomsPotential.pdf

Segal et al. (2016). *Disrupting Africa: Riding the wave of the digital revolution*. In PwC report available: Disrupting Africa: Riding the wave of the digital revolution: Africa overview: Publications: PwC

Sha'ban, M., Girardone, C., & Sarkisyan, A. (2020). Cross-country variation in financial inclusion: A global perspective. *The European Journal of Finance, 26*(4–5), 319–340.

Swamy, V. (2014). Financial inclusion, gender dimension, and economic impact on poor households. *World Development, 56*, 1–15.

Tchamyou, V. S., Erreyger, G., & Cassimon, D. (2019). Inequality, ICT and financial access in Africa. *Technological Forecasting and Social Change, 139*, 169–184.

World Bank. (2019a). *Employment in agriculture—Morocco*. International Labour Organization, ILOSTAT Database.

World Bank. (2019b). *Financial inclusion and digital economy development policy financing*. Retrieved from http://pubdocs.worldbank.org/en/616321550490336165/Morocco-DPF2019-Factsheet-eng.pdf

World Bank. (2022). *Financial inclusion: Financial inclusion is a key enabler to reducing poverty and boosting prosperity*. World Bank.

Zalan, T., & Toufaily, E. (2017). The promise of fintech in emerging markets: Not as disruptive. *Contemporary Economics, 11*(4), 415–431.

Zins, A., & Weill, A. (2016). The determinants of financial inclusion in Africa. *Review of Development Finance, 6*, 46–57.

19 Mobile Money Transactions for Financial Inclusion in Socioeconomic Development: An Interpretative Account of Two Ghanaian Rural Communities

Kofi A. Boateng, Rosemary Boateng Coffie, and Mercy A. Boateng

Abstract

Mobile money (MoMo) transactions have become increasingly popular in much of the developing world as means of innovative financial electronic transactions. Thus, the increasing popularity of the mobile phone has found itself extraordinarily significant in conducting financial transactions, of which mobile money transfer is crucial. Part of this phenomenon could be attributable to the growing information communication technology (ICT) use that manifests in the application of portable mobile devices. Drawing on the underlying concepts of interpretive research philosophy undergirded by qualitative orientation to data collection and analysis, this academic effort studies two rural communities in Ghana's second most populated region, Ashanti, to understand how MoMo transactions encourage financial inclusion for socioeconomic development in rural communities. The development literature was a lens to yield the findings suggesting financial inclusion's instrumentality in enhancing socioeconomic development in rural communities. This revelation significantly contributes to knowledge and practice as the development literature has underemphasized this technologically inspired component in the qualitative research community. The study, therefore, contributes to the growing body of literature and theory surrounding the issues driving the increasing popularity of mobile money transactions and its associated contributions to the socioeconomic development of the life of rural communities. Future studies should examine how much security and privacy issues can hamper this development effort. Again, future policy should look into strengthening the infrastructural capabilities of rural communities to ensure sustained development through financial inclusion in reinforcing socioeconomic development in disadvantaged communities.

Keywords

Mobile money · financial inclusion · socioeconomic development · poverty · rural communities

K. A. Boateng (✉) · M. A. Boateng
Kwame Nkrumah University of Science and Technology, Kumasi, Ghana
e-mail: kaboateng.ksb@knust.edu.gh

R. B. Coffie
Department of Human Resource and Organisational Development, Kwame Nkrumah University of Science and Technology, Kumasi, Ghana

G. W. Muschert et al. (eds.), *Financial Inclusion*, Sustainable Development Goals Series,
https://doi.org/10.1007/978-3-031-68803-4_19

19.1 Background Matters

The digital revolution brought about by information technologies (ITs) is driving mobile money (MoMo) transactions for socioeconomic development not only urban centers, but also rural communities (Tobbin, 2012). The socioeconomic development process has the potential to influence and shape the lives of the underprivileged in typical rural communities in much of the emerging and developing world. Drawing on Dong et al. (2018, p. 2), we connect with the idea and practice of MoMo as "…a technological innovation that combines mobile technologies and financial services to provide convenient and affordable access to banking systems and to improve financial inclusion." Consistent with Peric, we take the view that "Financial inclusion is not about charity—it is about designing a single financial system that serves the poor and provides an opportunity for commercial players to create a business by serving the poor" (2015, p. 213 cf. Lal & Sachdev, 2015). Indeed, much of the socioeconomic life of contemporary Ghanaian society appears to be engrossed in the MoMo transactions phenomenon. Given this contextual orientation, the study aspires to understand how MoMo transactions orchestrate financial inclusion as a broader agenda for socioeconomic development in rural communities.

This aspiration of the study is legitimate and significant for two reasons. First, there needs to be more discussion regarding MoMo transactions for financial inclusiveness through socioeconomic development for underprivileged rural communities. Second, this study provides an opportunity to link the literature in this area to the broader conversation in terms of policy formulation that is sympathetic to the cause of the rural communities that continue to struggle for the provision of basic social amenities, including infrastructure that has the potential of facilitating the socioeconomic development in rural communities. Last, but not least, it is the abiding conviction of the researchers that a study such as this can prove instrumental in creating some form of awareness in making the SDG8.10 goals a reality for the rural poor.

MoMo transactions are rising in popularity due to a variety of reasons. The growing availability of the mobile phone and its associated increasing handset culture, reasonably relatively easy to understand and use (at least for the literate), convenience and flexibility, swift process, and minimal complexity in the whole MoMo operation. These factors continue to encourage and enroll more mobile money merchants and subscribers in the whole supply chain process of MoMo transactions. It is fair to argue that the global Covid-19 pandemic and the intensified application of modern technology systems in social networking platforms have all contributed to reinforcing MoMo transactions as a formidable pathway to socioeconomic development (see, for instance, Muthukannan et al., 2020).

How do mobile money transactions inspire financial inclusion for socioeconomic development in rural communities?

MoMo transactions involve the application of mobile devices to transfer money from one person to another via the instrumentality of digital or electronic technologies, usually through mobile phones (Jenkins, 2008). MoMo transactions have been touted in much of the developing world for their socioeconomic advantages (REFS), given the recognition of the phenomenon as broadening the appeal of financial inclusion as the practice involves banking the unbanked.

Scholarly attention to MoMo transactions is as varied as they are detailed in terms of their epistemological orientations and approaches (see, for instance, Tobbin, 2011; Wang et al., 2019; Lee et al., 2023).

Given the research question, the unit of analysis of this study focuses on individual mobile money subscribers and merchants. Focusing on these entities is more likely to provide opportunities for detailed, insightful appreciation of the issues connected with financial inclusiveness and the socioeconomic development of the rural poor. Knowledge of these allows us to discuss the implications aimed at enriching our understanding and contributing to the literature on this emerging technology plus socioeconomic phenomenon.

19.2 Mobile Money Transactions in Perspective

MoMo involves electronic remittances undertaken using a mobile phone or other mobile devices (Etim, 2014). MoMo is frequently used for monetary transfers from one person to another or a group of people (Kusimba et al., 2013). The development of MoMo has enhanced and broadened the appeal over and above interpersonal transactions to involve bill settlement, payment for shopping items, payment from employers to staff, withdrawals on the ATMs, rent payments, and access to interest-gaining savings accounts as well as loans. Porteous (2006) have categorized MoMo into mobile banking (m-banking) and mobile payment (m-payment), respectively. According to Dias and Mckee (2010), mobile subscribers in Kenya and South Africa who do not have bank accounts now use MoMo for bill payments, payroll deposits, international transactions remittances, debt settlement, airtime purchases, and groceries.

Ehrbeck et al. (2012) have observed that in sub-Saharan Africa, one critical application of mobile phones for other operations apart from communication is MoMo transactions. MoMo transactions can encourage significant societal shifts for sustainable growth and economic opportunities for the large unbanked population. This view engages with the notion that M-Pesa is the most popular MoMo mode in Kenya and East Africa.

Two categories of MoMo operations can be accounted for by some research enthusiasts, namely, m-payment and m-banking in the literature (Porteous, 2006). The payment categories are "additive and transformative" models. With the additive models, bank account owners can use mobile phones to access an existing bank account and other services such as checking account balances, transferring funds between accounts, and checking transactions. However, the transformative model allows the typically unbanked to access financial products without an existing bank account, mainly through mobile phones (Dias & Mckee, 2010).

MoMo demonstrates a typical platform function. By this function, the platform orchestrates the coming together of stakeholders (platform ecosystem) such as financial services providers, policymakers and regulators, and customers for some anticipated service benefits. The transformative model holds much promise and prospects for the unbanked through financial inclusion for socioeconomic developmental realities.

19.2.1 Mobile Money for Financial Inclusion

There are numerous challenges to reaching the poor with financial services driven employing technological platforms. These impediments could result from many factors, such as limited network technological infrastructure, the digital divide in terms of limited educational exposure, and inadequate social amenities that occasion some disincentives to participation in the MoMo operations. In a developing country like Ghana, the rural poor loses the natural connection to the financial system as these banks require a deposit, almost always located in prominent urban commercial places. Again, the availability of financial services in rural areas had suffered a setback since the mid-1990s when commercial banks closed down due to unattractive profit in rural communities. The situation resulted in the rural and low-income people suffering from limited financial services opportunities, especially for domestic money transfer practices (UN, 2005; World Bank, 2003).

MoMo has the potential to solve many of these issues by giving banks and other financial service providers a cheap way to outsource cash handling and deposit taking, and withdrawal transactions with minimum cost to them. MoMo allows merchants to serve clients at a lower cost per transaction and with reduced investment in infrastructure.

The popularity of mobile money has widened due to the ever-increasing base of mobile phone users. It provides a platform that can serve the financial needs of rural communities (Ivatury & Mas, 2008; Morawczynsk, 2009). Due to the limited reach of banking infrastructure in the developing world, especially in rural areas, MoMo transfers help reach a more comprehensive number of customers faster and cheaper. It reduces the delivery cost of essential financial services, enabling a more significant proportion of the unbanked into the formal economy. The two fundamental attributes of the mobile phone which have led to its increasing usage are mobility and immediate access (Leung & Wei, 2000). Electronic money might displace paper and face-to-face transactions (Leung & Wei, 2000).

MoMo transfer is gaining near-universal acceptance in contemporary times. The benefits derived from MoMo include ease of use and accessibility, banking for the unbanked (Larivière & Martin, 1999), convenient savings, boost to economic growth, and lower cost of transactions, relatively speaking, among others (Dona et al., 2014). Despite this tremendous achievement, some possible drawbacks challenge the industry regarding adoption and penetration rate (Cobert et al., 2012). The difficulties that characterize MoMo transactions are network connectivity problems, security issues, relatively higher transaction costs, and insufficient education. In a rural culture where conservative estimates suggest up to about 80% of the population have no bank account and cannot access banking services such as deposits and loans, MoMo transactions appear to offer an irresistible appeal. Recently, it has been noted that mobile phone owners have exceeded the number of people who own bank accounts in the country (Ghana). Scholars suggested elsewhere that a healthy connection exists between the use of mobile money and financial inclusion (Bold et al., 2012; Jenkins, 2008; Porteous, 2006; Ehrbeck et al., 2012).

19.2.2 Development Through Financial Inclusion

Based on Jenkins (2008, p. 5), we contend, "It is mobile money's ability to facilitate financial inclusion that gives it its enormous potential for development impact." MoMo transactions enable clients to have a dense network of transaction outlets where they work and live, reducing the cost of clients accessing the financial system and the ability to transact at a low cost with banks and other financial service providers. Mobile money has therefore become a tool that allows individuals to make financial transactions using the technologies of the mobile device.

Financial inclusion (FI) is a process that ensures access to appropriate financial products and services available at an affordable cost to the low-income and underprivileged members of society (Duncombe & Boateng, 2009). FI means to reach out to that part of the demographic unit that is underbanked or unbanked of a country's population and provide them with adequate, affordable, and timely financial services such as accounts, credit, and insurance services (EY, 2009).

This has become integral to development policy in many developing countries (Duncombe & Boateng, 2009).

It is the desire of many people in such an economy to have access to financial services to reduce the harsh realities of poverty. Nevertheless, many people need help to afford to open bank accounts and are financially excluded (EY, 2009). This promotes insecure cash-based economies, exploitative financial opportunists, and loan sharks.

19.3 Theoretical Orientation

19.3.1 Economic Development

The framework serves as a guide for the analysis of the research. To this end, concepts in the development management literature derive the themes that provide the basis for interpreting the data gathered from the field. The objectives of this research drive the rationale for adopting a theory from development management. Again, applying a theory derived from development management literature provides opportunities for better explanatory power to appreciate the socioeconomic directions that provide meaning and a sense of purpose for the research. It also serves to explain the essential themes of the research adequately. Investigations into the use of technology in everyday experience and their probable impact lead to socioeconomic development implications, which depend on a theory in development concepts to shed insights for a broader audience.

According to Kerlinger (1986, p. 9), "theory is a set of interrelated constructs (concepts), definitions, and propositions that present a systematic view of phenomenon by specifying relations among variables, to explain and predicting phenomenon." Given this, the constructs in the development management literature act as a theoretical template to inform the discussion and analysis of this research. In other words, socioeconomic development would serve as a lens through which the issues of mobile money transactions are discussed and analyzed (Higgins, 2004).

Development is "a planned and comprehensive economic, social, cultural and political process, in a defined geographic area, that is rights-based and ecologically oriented and aims to continually improve the well-being of the entire population and all its individuals" (Fritz, 2018, p. 1). Development theory generally emphasizes economic development and national income growth that usually measures the change in per capita gross national product (Fritz, 2018). Development is an idea that "functions as both a product of an optimistic vision and an articulation of social interests, and a tool of analysis" (Chojnicki, 2010, p. 1). Socioeconomic development, however, concerns progress connected with economic and social matters within a specified geographic location. The twin concept, socioeconomic, can separately be explained to bring their areas of emphasis to bear on this research. Economic development assumes the growth of prosperity through improved production, distribution, and consumption of services and goods within a given society (Midgley, 1994).

The essence of economic development is to enhance the social and material welfare of all individuals and social institutions to obtain the maximum degree of human development. Therefore, socioeconomic development demands a combination of social and economic development. It also implies meaningful improvement in the quality of both social and economic lives (Kantor & Jang, 2004). Socioeconomic conditions provide a certain degree of security for the people in society. Security, in this sense, identifies the situation in which such people can depend on themselves without relying on others for their livelihood. Such people can exploit the economic conditions to provide for few extended family relations, as in typical urban and rural community settings. Engaging meaningfully in socioeconomic activities at different levels of society can also offer some financial freedom and support some dependents (US Civil Society Coalition, 2002).

19.3.2 Social Development

Social development accounts for the complex nature of social dynamics in terms of the social structures, relationships, and the general constitution of society (Giddens, 1984). Given this, the social centers on the social aspects of the people in terms of their way of life and people-driven participatory methods of development (US Civil Society Coalition, 2002). People must be willing to offer themselves the opportunities provided by the socioeconomic forces to benefit from it, as it is not automatic that under those conditions, people become well-off. The pointers of social development offer ideas of poverty, employment, education, health, income status, security of employment, and the environment, among others (Estes, 2002). In this sense, even starting in one's small way can make a difference and create avenues for more significant and better opportunities in the future. Given this view, the nature of socioeconomic development is "a set of economic agents. The economic agent is a stable macro-structure assigned by the definite economic function (production, consumption of the goods and resources, trade of the goods, credit and monetary operations, redistribution of the incomes)" (Petrov, 2003, p. 4).

Dissemination of information is crucial in these activities within the socioeconomic system and dynamics. The agent's attitude can be explained by the movement of assets and liabilities it produces and spends. The situation then demands that people adapt their attitude to suit external situations relatively quicker to different conditions. Chojnicki (2010, p. 9) succinctly put this relationship: "The nature of socio-economic development is determined by development processes and/or development targets. Generally, those processes are internally ordered sequences of change in states of affairs, or stages in which some states determine other states that follow them."

The attitude of people and their various interactions drive authorities to develop 'regulation mechanisms' of production and distribution of goods that have actual socio-economic forms. Drawing on this, Chojnicki points to the view that two instances can arise from these interactions that enable socio-economic activity. In the first place, there is a spontaneous process in which case the "socio-economic system is not targeted on any final states as specified development goals resulting from people's intentional and rational activity." The last is target-oriented processes that describe a "socio-economic system are guided by the activity and behavior of people (steering, control) and designed to achieve certain specified final states that can be goals of this activity" (Chojnicki, 2010, p. 9).

19.3.3 Research Philosophy and Methods

The research philosophy conditions and provides meaning such that research practice transforms and takes form from the original design to the completed results the researchers conclude. The interpretive research approach has been implicated in studies connected with ethnography (Geertz, 1983), phenomenology (Natanson, 2012), case studies (Darke et al., 1998; Walsham, 1995), and hermeneutics (Bernstein, 1983). This study believes an interpretive research philosophy can offer a rich contextual understanding of the matters and challenges emerging from a broad spectrum of issues connected with MoMo transactions in financial inclusion in the socio-economic development of the two communities studied here. Doing this permits a reflexive glance to be cast on the process that coincides with Walsham's (1995, pp. 4–5) observation that interpretive research strategies aimed at 'producing an understanding of the context of the information system, and the process whereby the information system influences and is also influenced by the context.

Committed to gaining a more profound appreciation of how MoMo transactions occasion financial inclusion for socioeconomic development, this study holds on to the ideals of interpretive research philosophy driven by a phenomenological process of logical reasoning. Phenomenology engages the idea that awareness of reality is socially informed and constructed through the agency of the researcher and that the determination of truth is subjective (Myers, 2004). Given this, the researcher gets to be indissolubly joined with the events that unfold as "the object…cannot be adequately described apart from the subject, nor can the subject be adequately described apart from the object" (Crotty, 2003, p. 79).

Adopting an interpretive posture endows the researcher with the exploratory understanding and "correct causal interpretation of a concrete course of behaviour" (Weber, 1949, p. 35 and 40). Therefore, subjectivism and constructivism reinforce the interpretive way of thinking that resonates with Goodman's (1978) contention that any social reality can be illustrated in one of many ways. To the interpretivist, the reality is not to be taken for granted as it involves a negotiation process conditioned by an individual's beliefs and practices, created by routinization and individual biases. Understandably, the interpretivist orientation acknowledges the diversity of views toward any identified social reality. The viewpoints of MoMo operators and subscribers on financial inclusion and the resulting potential for socioeconomic development implications indicate their experiences and sensemaking of the practice. The construction of such meaningful sensemaking in the spirit of interpretivist mentality is an endeavor that accepts and integrates general cognitive tools to determine unique social realities (Searle, 1995). It is abundantly apparent that the matters and interests of MoMo and financial inclusion are complicatedly diverse and complex to provide a single, inflexible, and thinly explained perspective. Financial inclusion and socioeconomic development indicate a complex web of unique, group, and societal expectations in our daily operations. Furthermore, this imbrication of socioeconomic aspirations and financial inclusion expectations in MoMo transactions inspires the data collection and analysis.

19.3.4 Sampling, Data Collection, and Analysis

Primary data collection techniques motivated the use of interviews to solicit views from respondents. Interviews are one of the effective forms of gathering qualitative data (Myers & Newman, 2007). Since the world cannot be studied all at once, the study adopts a sampling method that considers factors sympathetic to interpretive research tradition. By sampling, the desire is to produce a genuine representation of data subjects in the studied contexts (Fran J, Alston, 2014). A purposive sampling tool is ideal for producing the relevant data from all 30 subjects marked out for data collection. The structure of the 30 data subjects consists of 15 people each from the two towns considered for the research. The researchers ensured that each individual was a registered member of the country's leading mobile network provider (MTN). The remaining 20 data subjects are MoMo merchants comprising 10 from each of the two rural communities involved with the study. These identified individuals possess knowledge and information pertinent to the needs of the researchers. Among the appropriate information are issues that bear on MoMo transactions, security considerations, network connectivity concerns, ease of use, and the usefulness of financial inclusion for the unbanked.

Open-ended, semi-structured, qualitative interview questioning techniques were adopted while interacting with MoMo merchants and network subscribers for the views on the opportunities financial inclusion presented in their socioeconomic lives. This provided a framework for making it possible "...to see that which is not ordinarily on view and examine that which is often looked at but seldom seen" (Rubin & Rubin, 2005 p. vii). For instance, part of the face-to-face interviews and discussions focus on what the MoMo merchants do with the commissions they gain from MoMo transfers over time. It was through responses along similar lines that it became evident that network subscribers receive remittances from their relatives to cater for some educational and health as well as economic needs.

The motley methods of collection techniques, mainly an unstructured set of interview and documentary evidence from newspaper clippings and Internet sources, inspired the gathering of rich qualitative, primary data involving a comprehensive description of the issues surrounding financial inclusivity and its relationship with socioeconomic development among the town folks. Again, the multiple sources of data collection provide triangulation for achieving, among other things, internal validity.

Applying the interview technique allowed respondents to be more detailed, identifying their emotions, manners, and understanding of the issues. This approach was instrumental in recognizing the respondents' opinions, cues, attitudes, and beliefs (Kendall & Kendall, 1993). Before the resumption of the interviews, the respondents received assurances about protecting their identity in the interest of anonymity. The assurance provided some form of psychological security to the respondents. Some MoMo merchants were willing to share certain "operational secrets" that would otherwise have remained a mystery to the researchers.

Dwelling on Gaskell's suggestions for conducting in-depth qualitative interviewing, the background, age, educational qualification, job experiences, and the manner of operations of the MoMo merchants and subscribers were subjected to different questions. In effect, Gaskell advances the idea that matters considered delicate ahead of the interview that exemplifies the research intentions should guarantee to record the contents of the subsequent conversation. The prior knowledge gained from this experience makes provision for exploring the job description and specification of the respondents within the context of MoMo transactions in financial inclusion. The interview process functioned as a reference point in raising many aspects of socioeconomic development in the rural community in which they operate. The final parts of the interviews focus on accessing any supplementary documents considered necessary to the whole research agenda. The researchers ensured that each interview session ended with a grateful appreciation for their time and participation in the exercise.

Other secondary and miscellaneous documentary sources supplemented knowledge and understanding of the MoMo transaction services between the network merchants and their subscribers. Based on the cumulative stock of knowledge from both the primary and secondary data sources, some sense of the socioeconomic life of the informants. After gathering rich and thick stories from the informants in handwritten and tape-recorded versions, the data was transcribed and developed into themes driven by the study's objectives.

19.4 A Tale of Two Rural Communities for Financial Inclusion

Onwe and Akyawkrom are rural communities within the Ejisu municipal assembly, a district in the Ashanti region, Ghana's second most populous per the recent statistical data in the 2021 national census. The choice of these two rural communities is due to their proximity to the researchers for chiefly data collection purposes. The two communities share strikingly similar characteristics regarding the nature and overview of the general population. Farming is the dominant occupation of the people in these communities, characterized by less-developed infrastructural facilities such as poor road networks and the absence of other social amenities that inspire progress. The capital of these two rural communities is Ejisu, one of the 18 municipal assemblies in the Ashanti region.

In what follows, we reproduce the raw statements attributable to the various respondents, including some MoMo subscribers and merchants. We consolidate all the views rather than treat them as fragmentations. Consequently, we present general discussions to reflect the views in a thematic fashion consistent with our interpretive approach to making sense of the matters connected with MoMo and financial inclusion for socioeconomic development.

19.4.1 Ubiquitous Nature of MoMo Transactions

One mobile money transfer operator said, *"In this town alone, there are many people who are doing MTN mobile money. And more and more people are sending monies to their friends and family people by mobile money. Some of the people in this town also come here to receive money"*—**KK**, an MTN mobile money operator (1).

Another respondent stated, "*at the banks one has to join long queue before money can be withdrawn. Most of the banks have their branches at Ejisu and one has to travel from Onwe to Ejisu before you can withdraw your money but the introduction of the MTN mobile money transfer has reduced the transportation cost, because it is easy to come by and is located almost everywhere"*—**KM.**

19.4.2 Digital Transactions in MoMo: A Mixed Reaction

"When my customers come here they don't spend more time here like when they go to the bank to take their own money"—**AS**, an MTN mobile operator.

According to an MMT agent from Onwe, "*With MTN mobile money any amount can be sent at any given time but when they go to banks the situation is different*"—**AB**.

Such factors as ease of transaction, flexibility, and convenience add to the rising popularity of mobile money transactions. Another respondent suggested, "*when I go to the banks, I have to fill some forms before I can withdraw money. With the MTN mobile money, you do not need to fill in any such forms*"—**IO**.

"*Sometimes when you do the transfer, people complain that they have not got their money in their wallet and it is not good*"—**SA**.

"*These days the thieves are killing and robbing mobile money agents…sometimes they beat you up if you challenge them*"—**RA**.

19.4.3 Socioeconomic Development—Views from Onwe

One woman remarked: "*my husband who works at the urban centre can send money electronically to me in the village. These remittances are used to pay for the children's school fees*"—**MDM**.

"*I have no relative to assist me further my education so I'm saving some money with this job to help me to continue with my education in life*"—**KA**, a mobile money operator.

One mobile operator claimed: "*The little money I make from this business helps me buy drugs that my doctor tells me to buy when I fall ill and go to the hospital*"—**KF**.

"*One of the greatest advantages of using mobile money was that it helped solve financial problems immediately at the hospital when my daughter fell ill*"—**YO**.

One customer of a merchant recounted the following experience: "*When my mother was admitted at the hospital, we called my elder brother in Accra (*a reference to Ghana's capital city*) to send me. It is that money that we used to pay for my mother's treatment at the hospital*"—**AO**.

In a related case, an MMT agent said: "*I save the little, little monies I get from the mobile money transfers. And when I put all together in a month, I am able to do some things with it, my sister.*"—**RA**.

Another MMT agent even went to the extent of suggesting that however small she gets from mobile money transactions, it is better than nothing at all. She said: "*I don't care about what people say that this business does not bring quick money…it is better than going to steal. I don't feel ashamed that I do this business. It can give me something to eat*"—**AF**.

"*My uncle sends money to my grandmother every month by mobile money, and I pay my fees and buy clothes because of that*"—mobile money transfer customer.

*Another also said: "My uncle who works with a bank in Accra sends me money to keep his farm in good condition"—**KO**.*

Access to these financial services promotes agriculture, as most rural people are farmers. Using MTN mobile money allowed rural people to purchase more farm inputs, market a more significant proportion of their output, and have higher revenue.

"*Madam you know without this job, I don't know whom I would depend on to feed me every day… I can provide for my everyday needs because of this job I'm doing*"—JT. Another also recounted, "*Since I can buy what I need when I want to, it helps make life meaningful for me instead of sitting at home doing nothing*"—DK. However, there was a contrary view from an MTN mobile money operator who suggested that "*My sister, you can't depend on this job to do something great in life if you want to prosper well. The money in it is small, but I have no choice now*"—**EO.**

19.4.4 Sampled Views from Akyawkrom on MoMo and Socioeconomic Development

"*I have a friend who does the same mobile money job in Kumasi, he has been able to get more money to look after some of his younger siblings in Kumasi. So, I think if I get serious, I would be able to educate myself with this job I am doing"*—**OM**.

"If you come to this town alone, we have so many people doing this mobile money job, and it helps us small, small in our lives"—**AK**.

"MTN people have given the people here a lot of jobs, because before only farming is our work. But now, even some people do farming, and also do mobile money transfer too"—**OS**.

An exciting thought from an MTN MMT agent thus: *"The monies we collect and pay people through mobile money, the government takes some as tax to develop this country...so I'd say this thing we're doing is good for the country"*—**AA**. It is safe to argue that monies generated through taxes this way go a long way to stimulate economic development in the local community.

19.5 Reflections: Financial Inclusivity and Socioeconomic Development for the Rural Poor

The narrative in this study suggests the crucial significance of MoMo transactions in fostering financial inclusiveness for underprivileged rural communities to be part of the socioeconomic development in contemporary times. This observation is inspired by the very people who believe their lives have received some impetus due to the innovative potentialities of systems of technology. The technology system on this occasion is mainly the mobile phone which, through network connectivity functionality, can undertake financial transactions in pretty much the same way as their urban folks would. The mobile handset provides an affordance that orchestrates the many possibilities of banking the people who otherwise would have remained unbanked were it not for the dynamic application of the embodied predisposition of the technological artifact. Put differently, but with the same emphasis, per the mobile phone configuration, folks in rural communities can benefit from banking services that enable them as meaningful contributors to their socioeconomic upliftment.

Most development literature tenets coincide with some stories narrated by the people involved in the MoMo operations for financial inclusion. As a result, the individual stories, when viewed collectively, attest to how financial inclusion drives socioeconomic development in rural communities. The determination to mobilize savings for raising one's social status through education and passing on information via the mobile phone as means for caring for the health needs of family relations, among others, such as the narrative of this study evinces, goes a long way to give a sense of the critical nature of the MoMo transactions for projecting the socioeconomic status of the rural people. Again, the use of MoMo transactions for overseeing the farming activities of relatives and their continued support provided through financial inclusion mechanisms go a long way to reinforce the socioeconomic development agenda of the rural communities.

The manner and trend of socioeconomic development occasioned by financial inclusion provides some exciting thoughts worthy of consideration. Limited infrastructural support in rural communities cannot continue to offer a strong force against development. Financial inclusion means that the avenues for socioeconomic development are as dynamic as can be imagined. Becoming a MoMo merchant can be a manageable initial capital outlay. According to field observations, even a small table and chair are sure to provide an environment for funds mobilization, the volume of which can easily attract the envy of a whole bank branch. This scenario encourages easy cash flow within the rural communities for various personal and group development endeavors.

Social care and support are indispensable components of adding to the socioeconomic development in financial inclusion through MoMo transactions. It is clear from the interviews in this study that family members rely on MoMo transactions to care for them during hospitalization and sickness. Other forms of support include paying school fees to enable their wards and other family members to benefit from getting some level of formal education. The implication of this social support category through financial inclusion is that school dropout rates go down. In addition, caring for the health matters of relatives has implications for their longevity and life expectancy. A favorable disposition toward each of these has positive economic and social consequences for a robust system of family cohesion.

All in all, MoMo transactions have tremendous opportunities for broadening the financial inclusion of the rural folk with enormous potential for development now than at any previous period in history.

19.6 Concluding Account

Thus far, this study has unearthed the extent to which MoMo transactions enable financial inclusion for socioeconomic development in rural dwellings. Doing this has involved the application of an interpretive technique that teases out the views of people considered as the unit of analysis for this study, namely, MoMo subscribers and merchants in two rural communities in the Ashanti region of Ghana. Overall, financial inclusion in MoMo transactions fosters socioeconomic development in underprivileged communities.

Nonetheless, some contrary indications provide food for thought regarding policy directions to expand the scope of socioeconomic development in rural communities. First, the government should continue to broaden and strengthen its infrastructural development programs to make it possible for socioeconomic development to assume a sustainable character. Fiber-optic networks should be extended to rural destinations as they already exist in urban places. Doing this would reduce the worries and troubles associated with network connectivity. Second, communication network providers should also heighten their sense of security to minimize the incidence of fraud arising from insider complicity. Still, at the policy level, the government efforts at enhancing security at the rural community level should be encouraged by installing more closed-circuit television cameras (CCTV) at vantage points in rural communities. The rationale is to deter daylight robberies and also facilitate the prosecution of criminal attacks against MoMo merchants and subscribers.

Theoretically, future researchers should focus on probing the issues hindering financial inclusion among urban dwellers. Such focus should be interesting because there is a particular taken-for-granted assumption that urban centers have all the amenities for socioeconomic development that rural communities lack. Therefore, a study like that will likely upset long-held beliefs, supplying meaningful lessons.

References

Alston, W. P. (2014). Expressing. In *Philosophy in America* (pp. 15–34). Routledge.

Bernstein, R. (1983). *Beyond objectivism and relativism: Science, hermeneutics, and praxis*. University of Pennsylvania press.

Bold, C., Porteous, D., & Rotman, S. (2012). Social cash transfers and financial inclusion: Evidence from four countries. *Population (in millions), 193*(46), 109.

Chojnicki, Z. (2010). Socio-economic development and its axiological aspects. *Quaestiones Geographicae, 29*(2), 7–17.

Cobert, B., Helms, B., & Parker, D. (2012). Mobile money: Getting to scale in emerging markets. *McKinsey & Company, May*, 1–13.

Crotty, M. (2003). The foundations of social research: Meaning and perspective in the research process.

Darke, P., Shanks, G., & Broadbent, M. (1998). Successfully completing case study research: combining rigour, relevance and pragmatism. *Information Systems Journal, 8*(4), 273–289.

Dias, D., & McKee, K. (2010). *Protecting branchless banking consumers: Policy objectives and regulatory options* (Focus Note 64). CGA.

Dona, P. D., Mouri, S. M., Hasan, M., & Abedin, M. Z. (2014). Significance of exponential uses of mobile financial services (MFS) in Bangladesh. *Global Journal of Management and Business Research, 14*(4), 93–101.

Dong, Y., Song, S., Venkataraman, S., & Yao, Y. O. (2018). Mobile technology and financial service bundling: A structural estimation of Mobile money. *SSRN Electronic Journal*. https://doi.org/10.2139/ssrn.3213744

Duncombe, R., & Boateng, R. (2009). Mobile phones and financial services in developing countries: A review of concepts, methods, issues, evidence and future research directions. *Third World Quarterly, 30*, 1237–1258.

Ehrbeck, T., Pickens, M., & Tarazi, M. (2012). *Financially Inclusive Ecosystems: The roles of government today* (Vol. No. 68069, pp. 1–12). The World Bank.

Ernst & Young. (2009). *Mobile money: An overview for global telecommunications operators*. Ernst & Young's Global Telecommunication Centre.

Estes, R. J. (2002). Toward a social development index for Hong Kong: The process of community engagement. In M. R. Hagerty, J. Vogel, & V. Moller (Eds.), *Assessing quality of life and living conditions to guide national policy: The state of the art*. Kluwer Academic Publishers.

Etim, A. S. (2014). Mobile banking and mobile money adoption for financial inclusion.

Fritz, J. M. (2018). Socio-economic developmental social work. *Social and Economic Development, I*.

Giddens, A. (1984). *The constitution of society: Outline of the theory of structuration*. University of California Press.

Geertz, C. (1983). Local knowledge: fact and law in comparative perspective.. Local knowledge.

Goodman, N. (1978). *Ways of worldmaking*. Hackett Pub Co Inc.

Higgins, E. T. (2004). Making theory useful: Lessons handed down. *Personality and Psychology Review., 8*, 138–145.

Ivatury, G., & Mas, I. (2008, April). *The early experience with branchless banking* (Focus Note 46). CGAP.

Jenkins, B. (2008). *Developing mobile money ecosystems*. IFC and Harvard Kennedy School.

Kendall, J. E., & Kendall, K. E. (1993). Metaphors and methodologies: Living beyond the systems machine. *MIS Quarterly*, 149–171.

Kantor, J., & Jang, Y. S. (2004). Yes, there is a (growing) transnational business community: A study of global interlocking directorates. *International Sociology, 19*(3), 355–368.

Kerlinger, F. N. (1986). *Foundations of behavioural research* (3rd ed.). Holt, Rinehart, & Winston.

Kusimba, et al. (2013). *Social networks of mobile money in Kenya*. Institute of Money, Technology and Financial Transactions, Department of Anthropology Northern Illinois University.

Lal, R., & Sachdev, I. (2015). Mobile money services design and development for financial inclusion.

Larivière, S., & Martin, F. (1999). *Innovations in rural microfinance:the challenge of sustainability and outreach, in: Innovations in Microfinance for the Rural Poor:Exchange of Knowledge and Implications for Policy, published by International Food Policy Research Institute, International Fund for Agricultural Development and Bank of Ghana*. DeutscheStiftung für internationale Entwicklung.

Lee, D., Gopal, A., Lee, D., & Shin, D. (2023). Nudging Private Ryan: Mobile Microgiving under Economic Incentives and Audience Effects. *MIS Quarterly, 47*(3), 1101–1146.

Leung, L., & Wei, R. (2000). More than just talk on the move: Uses and gratification of the cellular phone. *Journalism and Mass Communication Quarterly, 77*, 308–320.

Midgley, J. (1994). Defining social development: Historical trends and conceptual formulations. *Social Development Issues, 16*(3), 3–19.

Morawczynski, O. (2009). Exploring the usage and impact of "transformational" mobile financial services: the case of M-PESA in Kenya. *Journal of Eastern African Studies, 3*(3), 509–525.

Muthukannan, P., Tan, B., Gozman, D., & Johnson, L. (2020). The emergence of a Fintech ecosystem: A case study of the Vizag Fintech Valley in India. *Information and Management, 57, 103385*.

Myers, M. D. (2004). Hermeneutics in information systems research. *Social theory and philosophy for information systems*, 103–128.

Myers, M. D., & Newman, M. (2007). The qualitative interview in IS research: Examining the craft. *Information and Organization, 17*(1), 2–26.

Natanson, M. (Ed.). (2012). *Phenomenology and social reality: Essays in memory of Alfred Schutz*. Springer Science & Business Media.

Peric, K. (2015). Digital Financial Inclusion. *Journal of Payments Strategy and Systems, 9*(3), 212–214.

Petrov, A. A. (2003). Systems analysis and modelling of integrated world systems. In *Models of Socio-economic Development* (Vol. II).

Porteous, D (2006). *The enabling environment for mobile banking in Africa*. BankableFrontiers. http://www.bankablefrontier.com/assets/ee.mobil.banking.report.v3.1.pdf

Rubin, H. J., & Rubin, I. S. (2005). Qualitative interviewing (2nd ed.) SAGE Publications, Inc.

Searle, J. R. (1995). *The construction of social reality*. Allen Lane.

Tobbin, P. (2011). Understanding mobile money ecosystem: Roles, structure and strategies. In *2011 10th International Conference on Mobile Business* (pp. 185–194).

Tobbin, P. (2012). Towards a model of adoption in mobile banking by the unbanked: A qualitative study. *Info, 4*(5), 74–88.

United Nations Development Programme. (2005). Human Development Report 2005. United Nations. Retrieved from https://www.undp.org/publications/human-development-report-2005

US Civil Society Coalition. (2002) Responsible reform of the World Bank: The role of the United States in improving the development of effectiveness of the World Bank operations.

Wang, L., Luo, X., Yang, X., & Qiao, Z. (2019). Easy come or easy go? Empirical evidence on switching behaviors in mobile payment applications. *Information & Management. 56*, 10.

Walsham, G. (1995). Interpretive case studies in research: Nature and method. *European Journal of Information Systems, 4*, 74–81.

Weber, M. (1949). Objectivity in Social Science and Social Policy.

World Bank (2003). Providing financial services in rural areas: A fresh look at financial cooperatives,. Report no. 40043-GIB, Washington, DC 2043.

World Bank. (2014). *Digital finance empowering the poor via new technologies*.

Yin, R. K. (2003). *Case study research: Design and methods*. Sage Publications.

Part V

Driving the Research Forward

Nexus Between Financial Inclusion and Economic Activity: A Study About Traditional and Non-Traditional Financial Service Indicators Determining Financial Outreach

20

Florian Gerth

Abstract

This chapter empirically analyzes the link between financial inclusion (SDG 8.10) and economic activity. Instead of following the past literature and approximating financial inclusion by variables only capturing traditional financial services, this chapter considers non-traditional financial services, including mobile money and nonbranch retail agent outlets. With the help of the normalized inverse of the Euclidian distance and a one-way fixed effects panel model, this chapter documents empirically robust results about the positive link between financial inclusion and the level of economic activity. In addition, a break between poverty and financial inclusion is established by regressing the calculated index of financial inclusion on demographic, socioeconomic, and other variables concerning the health and depth of the financial sector. The implications of this finding in this analysis are twofold. First, it highlights the improvements in low, lower-middle, and upper-middle-income countries regarding outreach to financial services in the last decade. Second, it shows that the level of education and the soundness and depth of the local financial sector are essential in reaching higher levels of financial inclusion. Overall, our results emphasize the importance of targeted policies to increase the accessibility, availability, and usage of the financial sector to attain sustainable and long-lasting economic prosperity.

Keywords

Financial inclusion · Non-Traditional financial services · Economic development · Financial Access Survey (FAS)

JEL Classifications

C23 · E13 · E44 · G20 · O16

An earlier version of this manuscript appears in the Munich Personal RePEc Archive (MPRA) as follows: Gerth, F. (2023). "Nexus between Financial Inclusion and Economic Activity: A Study about Traditional and Non-Traditional Financial Service Indicators Determining Financial Outreach," Munich Personal RePEc Archive, MPRA Paper No. 119265, posted November 30, 2023, online at https://mpra.ub.uni-muenchen.de/119265/.
I thank Paul Lohmann and Alex Parker, the participants of the AIB MENA conference in Dubai, UAE and the IAAE conference in Xiamen, China, and the anonymous referees for their valuable suggestions and recommendations.

F. Gerth (✉)
Asian Institute of Management, Makati, Manila, Philippines
e-mail: fgerth@aim.edu

Center for the Study of Global Economic Future (CSGEF), Dubai, United Arab Emirates

G. W. Muschert et al. (eds.), *Financial Inclusion*, Sustainable Development Goals Series,
https://doi.org/10.1007/978-3-031-68803-4_20

20.1 Introduction

Over the last 30 years, the world has made significant progress in lifting people out of poverty. In 1990, the number of people living below $1.90 per day was 1.9130 billion. By 2019, this number had dropped to 644.7 million. More specifically, poverty decreased by more than 13% from 2015 to 2019 and is now at its lowest level ever. This decline in the number of low-income people includes sustained economic growth. In particular, long-term economic growth creates jobs and raises real wages. Such achievement has the potential to sustainably benefit society as a whole.

The growth literature highlights several essential variables for economic progress: the level of investment, educational attainment, productivity, size and growth of the labor force, capital endowment, and inflation.[1] However, even when these determinants perform at their best, insufficient financial development impedes sustained economic growth, preventing economies from climbing out of poverty (Berthelemyje & Varoudakis, 1996). This is because the financial market is a main artery in the economic ecosystem, whereby it facilitates (1) the channeling of savings into investment, (2) risk diversification, and (3) the alleviation of information asymmetry (Levine, 1997, 2005). Only after these tasks have been fulfilled will the economy allocate resources efficiently, leading to a more sustainable accumulation of physical and human capital and technological growth (Goyal et al., 2004; Gerth, 2017). Even with this, Sarma (2012) argues that specific population segments may not be financially included even when an economy has a well-developed financial system. Demirguc-Kunt and Klapper (2012) explain that this can be because of geographical isolation, costs associated with financial services, and invalid documentation. According to Sarma (2012), this lack of inclusion (presence of exclusion) hinders the efficient allocation of productive resources, worsens day-to-day finances, supports the growth of informal sources of credit, and ultimately hampers economic development.

The overall consensus in the literature suggests a positive relationship between financial inclusion and economic development. For example, a study by Bruhn and Love (2014) reinforces the point mentioned in the above paragraph, whereby they show how the opening of 800 bank branches by the Mexican bank *Banco Azteca* led to (1) a reduction in unemployment by about 1.4%, (2) an increase in income by about 7% over 2 years, and (3) poverty in the affected areas decreased as the bank catered for low- and middle-income customers that were previously excluded from having access to financial services. Turning to the sub-Saharan African context, Allen et al. (2012) find recent innovations in helping banking services overcome infrastructural deficiencies and foster economic development. Using a sample of 150 countries, Cihak et al. (2016) show that financial inclusion can produce synergies and mitigate medium-term instability in the financial sector, consequently supporting economic activity positively.[2] Other studies, like Kim (2016), focus on 40 OECD countries to emphasize the positive relationship between financial inclusion and economic prosperity. Wang and Guan (2017) adopt and adapt the methodology of Sarma (2008) to illustrate the spillover effects of being highly financially inclusive in their neighboring countries. According to Wang and Guan, financial inclusion has a positive economic impact on adjacent countries. Furthermore, Kim et al. (2018) and, more recently, Van et al. (2021) provide additional empirical evidence supporting the positive relationship using 55 countries belonging to the Organization of Islamic Cooperation (OIC) and 152 countries around the world, respectively. These studies show that financial inclusion is imperative for economic development.

This research contributes to and extends the literature mentioned above in three ways. Firstly, instead of relying on demand-side financial inclusion indicators, this study collects data using the

[1] For a deeper and more thorough discussion about the determinants of economic growth, see Barro (2000, 2001, 2003, 2013).

[2] See also Gerth and Otsu (2015, 2018) regarding the empirical relationship between financial stability and economic activity.

Financial Access Survey (FAS) conducted and organized by the IMF. This supply-side dataset is unrivaled in its depth in terms of time, geography, and coverage. It is worth mentioning that this database (1) provides a more extended sample size than that of previous studies (2004–2019), (2) includes other countries not studied before (189 countries), and (3) has over 120 variables capturing financial access, availability, and usage. Secondly, unlike prior studies that focused on traditional measures of financial inclusion (number of bank branches and ATMs), this research uses non-traditional indicators such as the availability and use of mobile money and nonbranch retail agent outlets. Thirdly, Sarma (2008) extended the methodology to empirically assess the significance of the three aspects of financial inclusion (financial access, availability, and usage).

The results shown in this research support the prior literature and provide additional evidence not discussed previously. This study shows that regardless of the extension made to build the Index of Financial Inclusion (IFI), the index exhibits dynamic behavior irrespective of the development stage of the economy. The extended sample period enables us to observe a more realistic view of the trend in the IFI. Our results show a consistent rise for low-, lower-middle-, and upper-middle-income economies and an initial stagnation and subsequent fall for the upper-income economies. Furthermore, a new index confirms that countries with a higher IFI score exhibit higher GDP levels. By contrast, GDP growth is negatively related to the level of IFI due to the lower average growth rate of high-income economies. In addition, when controlling for non-traditional financial services, the financial inclusion index is not statistically related to poverty because the increased adoption of mobile money and nonbranch retail agent outlets is predominantly in underdeveloped countries.

This chapter is structured as follows: Sect. 20.1 introduces the paper and discusses the literature on financial inclusion and economic activity. Section 20.2 constructs an alternative financial inclusion index, the Index of Financial Inclusion (IFI) derivation, and a descriptive representation of the IFI's time-series behavior. Section 20.3 presents the empirical models and findings. The last section discusses the policy implications and conclusion.

20.2 Index of Financial Inclusion (IFI)

20.2.1 Motivation and Construction

In 2004, Kempson et al. recognized that one indicator cannot gauge financial inclusion within an economy. The authors suggest that merely having access to some traditional form of banking, say a bank account, does not determine the degree of its utilization. Sarma (2012) also criticizes using a single indicator as it provides a narrow view of the services offered by the financial sector. These two criticisms of the single indicator motivate building another parameter with a multidimensional nature that better represents the current market situation. Consequently, this research refrains from using a single-indicator approach and constructs a multidimensional index to capture financial inclusion.

Based on the consensus in the literature, one single number index must contain different dimensions, namely accessibility, availability, and usage of financial services. The dimensions are respectively referred to as (1) the "banked" population within an economy, (2) the physical and digital availability of financial service outlets, and (3) information about the level of the "underbanked" population.[3] Following Sarma (2008, 2012) and Wang and Guan (2017), the index is constructed.

Each dimension, in turn, consists of several indicators that capture the contemporary, comprehensive, and unique financial environment in each of the 166 countries within our sample (see Table 20.1).[4]

Several indicators in Table 20.1 are contemporary financial advancements that must be considered when determining a consistent degree of country-specific financial inclusion. Recent devel-

[3]This view is widely accepted in the literature; see Sarma (2008, 2012), Sarma and Pais (2011), Kim (2016), Wang and Guan (2017), and Van et al. (2021).

[4]Note that the sample decreased to 166 countries due to the lack of sufficient data for 23 economies.

Table 20.1 Variables of financial inclusion

Dimension	Indicator
Accessibility:	
	Number of deposit accounts with commercial banks per 1000 adults
	Number of loan accounts with commercial banks per 1000 adults
	Number of registered mobile money accounts per 1000 adults
	Number of active mobile money accounts per 1000 adults
	Number of credit cards per 1000 adults
	Number of debit cards per 1000 adults
	Number of life insurance policies per 1000 adults
Availability:	
	Number of commercial bank branches per 100,000 adults
	Number of ATMs per 100,000 adults
	Mobile money agent outlets: active per 100,000 adults
	Number of registered mobile money agent outlets per 100,000 adults
	Number of nonbranch retail agent outlets of commercial banks per 100,000 adults
Usage:	
	Outstanding deposits with commercial banks (% of GDP)
	Outstanding loans with commercial banks (% of GDP)
	Number of mobile money transactions per 1000 adults
	Value of mobile money transactions (% of GDP)
	Number of mobile and internet banking transactions per 1000 adults
	Value of mobile and internet banking transactions (% of GDP)

opments in developing and underdeveloped nations where non-traditional and digital financial services have been introduced and used have accelerated over the past few years (Adrian & Mancini-Griffoli, 2018; Espinosa-Vega et al., 2020), justifying their presence. This is especially true for sub-Saharan Africa and Asia, where the availability and usage of mobile money have increased by as much as 500% from 2015 to 2019 (FAS, 2020). Consequently, using the above indicators is essential, as it will provide a better picture of how well a country's citizens are included in the financial sector.

The *Index of Financial Inclusion, IFI*, is constructed to lie between 0 and 1, where 0 signifies the complete absence of financial inclusion or complete financial exclusion and 1 signifies complete financial inclusion. To do this, transform all indicators to lie between these two lines, and then compute the normalized inverse of the Euclidean distance for each dimension and the entire nation. The result will be one single number that gives each country's relative degree of financial inclusion within the sample.[5]

[5]This mathematical transformation gives the index the following important and helpful mathematical properties: (i) boundedness, (ii) homogeneity, (iii) unit-free property, and (iv) monotonicity.

To normalize each indicator, Eq. (20.1) is used:

$$\varphi_{i,j,c} = \frac{A_{i,j,c} - m_{i,j}}{M_{i,j} - m_{i,j}} \tag{20.1}$$

Where the variable φ captures the transformed value ($0 \leq \varphi \leq 1$) for indicator i in dimension j for country c, see Table 20.1. For example, "Number of deposit accounts with commercial banks per 1,000 adults" indicates $i = 1$ for dimension $j = 1$ "Accessibility." The variable A is the actual observed value, M is the maximum value of this particular indicator within the respective year, and m is the minimum value within the respective year.[6]

In a second step, the IFI for each of the three dimensions is calculated; see Eq. (20.2):

[6]Due to the possibility of extreme outliers in the maximum value of the data, each indicator is *winsorized*. In other words, the second-highest value replaces the highest. This is sensible, since otherwise the situation might arise where one country has a value of 1 for indicator i and all other countries' indicator values range between 0.05 and 0.1. In this particular dataset, this is the case for San Marino, which, for some reason, has an extremely high number of physical banking outlets and ATMs per 100,000 adults.

$$\mathrm{IFI}_{j,c} = 1 - \frac{\sqrt{\omega_{1,j}^2\left(1-\varphi_{1,j,c}\right)^2 + \omega_{2,j}^2\left(1-\varphi_{2,j,c}\right)^2 + \ldots + \omega_{n,j}^2\left(1-\varphi_{n,j,c}\right)^2}}{\sqrt{\left(\omega_{1,j}^2 + \omega_{2,j}^2 + \ldots + \omega_{n,j}^2\right)}} \quad (20.2)$$

where ω is the weight parameter for each indicator within each dimension. These weights are essential to avoid multicollinearity between the dimensions, according to Sarma and Pais (2011). Instead of choosing ω arbitrarily, we follow Wang and Guan (2017) and empirically determine each weight according to the coefficient of variation (CV), Eq. (20.3):[7]

$$\omega_{i,j} = \frac{\vartheta_{i,j}}{\sum_{i,j}\vartheta_{i,j}} \quad (20.3)$$

where ϑ is the CV, defined as the ratio between the standard deviation and the mean of the frequency distributions for each indicator.

Having calculated the indicator weights and IFIs for each dimension in each country within the sample, we calculate the countrywide IFI as follows: Eq. (20.4):

$$\mathrm{IFI}_c = 1 - \frac{\sqrt{\omega_{\mathrm{ac}}^2\left(1-\mathrm{IFI}_{\mathrm{ac},c}\right)^2 + \omega_{\mathrm{av}}^2\left(1-\mathrm{IFI}_{\mathrm{av},c}\right)^2 + \omega_{\mathrm{us}}^2\left(1-\mathrm{IFI}_{\mathrm{us},c}\right)^2}}{\sqrt{\left(\omega_{\mathrm{ac}}^2 + \omega_{\mathrm{av}}^2 + \omega_{\mathrm{us}}^2\right)}} \quad (20.4)$$

where the weights are being calculated for each of the three dimensions: accessibility (ac), availability (av), and usage (us).

20.2.2 Financial Access Survey (FAS)

The data used to construct the IFI introduced in Sect. 20.2.1 comes from the Financial Access Survey (FAS) published by the IMF. Launched in 2009, this unique supply-side dataset covers 189 countries and jurisdictions from 2004 to 2019 and contains more than 120 indicators of the financial sector's access, availability, and usage. The ultimate aim of the FAS is to monitor developments in financial inclusion and to allow a direct comparison of its level and relative progress with other countries within the dataset. Central banks and financial regulators collect data directly from financial institutions and financial service providers within their jurisdictions.

Since its inception in 2009, the FAS has evolved to keep pace with financial advances. In 2014, the need to consider non-traditional financial services was recognized, and the mobile money module was introduced. In 2018, information regarding nonbranch retail agents was included in the survey. In 2019, extensive information and observations concerning mobile and Internet banking were added.

After ensuring sufficient information is available for each country (see variables in Table 20.1), we are left with 166 economies within the dataset (see Table 20.6). According to the World Bank's classification, the sample of countries consists of 19 low-income, 48 lower-middle-income, 47 upper-middle-income, and 52 high-income economies.[8]

[7] Papers that choose the dimension weights according to some subjective rules are Sarma (2008, 2012), Sarma and Pais (2011), Kim (2016), Kim et al. (2018), and Van et al. (2021).

[8] Using the Atlas method, the World Bank classifies high-income economies as those with a GNI per capita of $12,696 or more, upper-middle-income economies as those with a GNI per capita between $4096 and $12,695, lower-income economies as those with a GNI per capita between $1046 and $4095, and low-income economies as those with a GNI per capita of $1045 or less, according to the World Bank (2021).

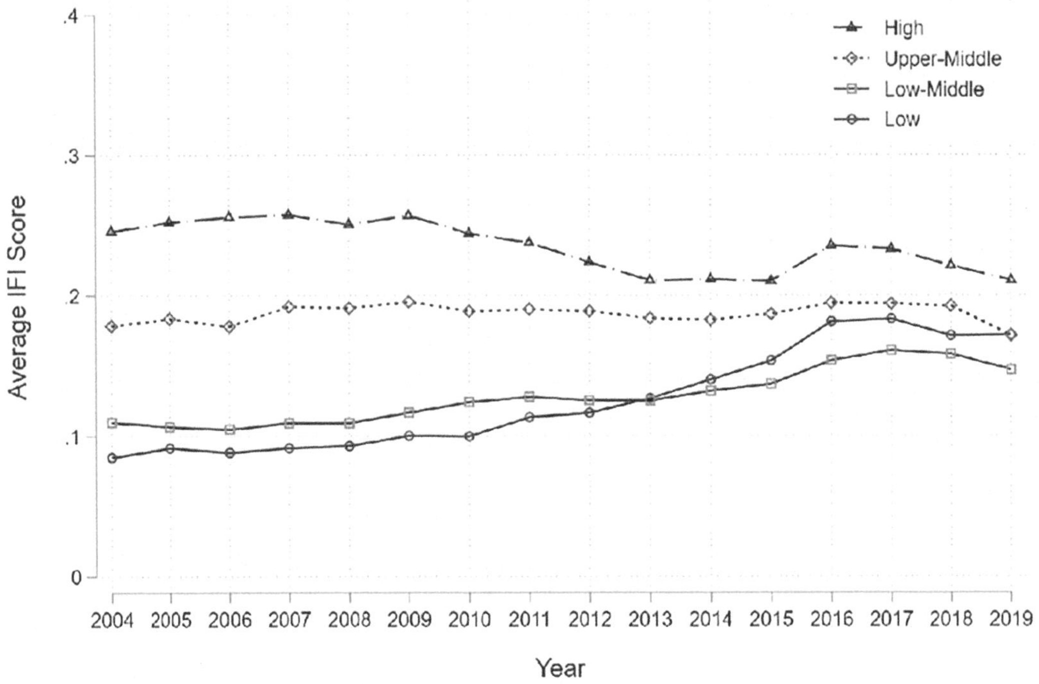

Fig. 20.1 Average IFI scores

20.2.3 Findings

Figure 20.1 shows the average IFI score time series for each development category.

Figure 20.1 (and Table 20.7 in the appendix) present several exciting and important findings. First, the IFI score is dynamic and does not stagnate at a particular level. That is, it fluctuates around a trend. Second, this trend is positive for low- (circle graph symbol), lower-middle- (square graph symbol), and upper-middle-income (diamond graph symbol) countries. Consequently, financial inclusion in these economies improved throughout the sample period. For high-income countries (triangle graph symbol), on the other hand, financial inclusion decreased from 2009 onward. For one, this might be due to the collapse of the economic system during the Global Financial Crisis. This finding might be driven by decreased physical outlets, such as commercial bank branches and ATMs (FAS, 2020).

Furthermore, the adoption of mobile money in these countries is rare. Hence, growth rates in financial inclusion are lower compared to less developed countries that have embraced this kind of technology for the last several years. The third point is that the higher the IFI score increases, the more developed an economy becomes. In 2009, the average IFI for low-income countries was 0.0849, but it increased to 0.2460 for high-income countries, almost three times the size. This gives a descriptive indication of the importance of financial inclusion and the level of economic activity.

20.3 Empirical Analysis

20.3.1 Financial Inclusion and Economic Prosperity

We estimate a one-way fixed effects model to analyze how financial inclusion, as the IFI measures, affects the level of economic activity throughout the 166 countries in our sample. Such a panel model allows us to exploit the longitudinal dimension of the data while accounting for country-level heterogeneity.

$$\mathrm{GDP}_{c,t} = \beta' x_{c,t} + \theta \mathrm{IFI}_{c,t} + \gamma t + \alpha_c + u_{c,t} \qquad (20.5)$$

$$u_{c,t} \sim iid\left(0,\sigma_u^2\right) \quad (20.6)$$

$$\text{corr}\left(\alpha_c, u_{c,t}\right) = 0 \quad (20.7)$$

where GDP in Eq. (20.5) is the level of actual economic activity for time *t*, *(t = 2004,...,2019)*, and for country *c*, *(c = 1,...,166)*. The intercept is the unobservable effect that is unique to each country, β is the vector of coefficients for the control variables contained by *x*, θ is the coefficient for the variable that measures the level of financial inclusion (IFI), γt is a time trend meant to account for possible trend-stationarity in the variables expressed in levels, and u is the error term. Since this model aims to find the statistical relationship between financial inclusion and the level of economic activity, the statistical significance of the coefficient θ is of particular interest. The general-to-specific approach is used to find a statistically robust representation. A general model is being estimated and reduced to its most efficient form.[9] Equations (20.6) and (20.7) are necessary assumptions regarding the independence of the error terms to each other and, throughout time, their homoscedasticity and their uncorrelation with the country-specific intercept.

The final model contains the explanatory variables in Table 20.2.[10]

Barro (2000, 2001, 2003, 2013) inspired the selection of the variables, which aim to capture demand-side factors affecting economic activity through aggregates for private investment (investment), private consumption (consumption), and exports and imports (trade). The variable *schooling* is included to approximate the level of human capital or labor productivity within the economy. The ratio between secondary education and total education was chosen since secondary education comprises primary education and, in addition, more subject- or skill-oriented instruction. The variable population is added to control the potential labor force's size. This variable includes all the people within an economy of working age. Since any variable representing the labor force does not contain discouraged workers, the size of the population is preferred. The variable *unemployment* measures the degree of capacity utilization within the labor market. Finally, following Sarma and Pais (2011), the variable of interest, IFI, is transformed to lie between $-\infty$ and $+\infty$. This is done using Eq. (20.8).

$$\text{IFI}_{\text{trans}} = \ln\left(\frac{\text{IFI}}{1-\text{IFI}}\right) \quad (20.8)$$

This modification makes the variable a logit transformation. The reconfigured variable increases monotonically in IFI and consequently retains its original ordering.[11]

Table 20.3 presents the estimated panel regression model.

When it comes to the control variables, all of them have theoretically consistent signs, except for the ones that measure private consumption and trade. Higher levels of private investment, labor productivity, and population, along with lower levels of unemployment, are statistically significant in explaining higher economic activ-

Table 20.2 Explanatory variables: GDP regression

Variable name	Variable description
Investment	Private gross capital formation (% of GDP)
Consumption	Household final consumption expenditures (% of GDP)
Schooling	The enrollment ratio between secondary and total enrollment
Population	Population (ages 15–64)
Trade	Trade (% of GDP)
Unemployment	Unemployment rate (as a percentage of the labor force)
IFI_trans	For the Index of Financial Inclusion, See Sect. 20.2

[9]For a comprehensive discussion on the general-to-specific approach and its automation, see Krolzig (2001, 2003).

[10]With the exception of the variable IFI, all variables come from the World Bank Development Indicators database.

[11]This step is less important for the regression model presented by Eq. (20.5). Once financial inclusion becomes a dependent variable itself, however, this transformation is necessary to estimate the model via OLS. In the interest of consistency, the transformation is already applied at this step.

Table 20.3 Regression results: GDP regression

Investment	Consumption	Schooling	Population	Trade	Unemployment	IFI_trans
0.0035	−0.0031	0.0028	0.0034	−0.0014	−0.0103	0.421
*(7.22***)*	*(−5.75***)*	*(8.15***)*	*(12.08***)*	*(−9.05***)*	*(−9.91***)*	*(9.57***)*

Note: Values in brackets below the coefficients represent their respective t-statistics. ***, ** and * signify significance at the 1%, 5%, and 10% levels, respectively

ity. The negative sign for private consumption and trade might signify that more developed countries rely proportionally less on consumption and trade to maintain their economic prosperity than developing and underdeveloped countries.[12] This is because more developed countries have higher GDPs than underdeveloped countries. Most importantly, the regression results clearly state that the variable measuring the degree of financial inclusion, IFI_trans, is positively related to the level of economic activity. This implies that an economy where financial inclusion is more advanced enjoys more economic prosperity. This is an important finding since it empirically guides policymakers to strengthen financial literacy to reach a more developed economy.[13]

20.3.2 Financial Inclusion and Its Determinants

While Sect. 20.3.1 empirically shows that financial inclusion is a necessary determinant in explaining the level of economic activity, this section tries to find demographic, socioeconomic, and financial variables essential to describing the degree of financial inclusion. Contrary to the previous section, where only a specific model was presented and discussed, this section estimates a regression model that includes some of the most widely discussed variables in the literature regarding the determinants of financial inclusion.

$$\mathrm{IFI\_trans}_{c,t} = \varphi' z_{c,t} + \delta + \varepsilon_{c,t} \qquad (20.9)$$

the dependent variable, IFI_trans, is the transformed financial inclusion index constructed in Sect. 20.3.1.[14] The score is calculated for country *c*, *(c = 1,…,166)* and time *t*, *(t = 2004,…,2019)*. The coefficient δ is the intercept and φ is the vector of coefficients associated with the independent variables contained by the vector *z*. Lastly, ε is the error term assumed to be independent and identically distributed with a mean of 0 and a variance of σ_{ε}^2.

We use ordinary least squares (OLS) to model Eq. (20.9) for comparison purposes and ease of reproducing our results. We apply adequate care during estimation to comply with the classical linear regression function assumptions. Consequently, the statistical behavior of the residual terms and the parameter values were analyzed and, where necessary, adopted. Furthermore, we calculate the variance as an exponential function of the covariates specified in the model to prevent misleading diagnostic tests. The last step is necessary because OLS explicitly assumes that the residuals of the variables are constant, which does not apply to our data. Concerning the stability of our results, we use the method-of-moments estimation technique as a robustness test. We obtain the same qualitative and quantitative results by assuming predeterminedness and imposing the moment condition of nonstochastic covariates.

[12] As Myint (1958) and his contemporaries argue, underdeveloped countries import manufactured goods and therefore contribute negatively to their economic prosperity. This is because instead of expanding domestic economic activity, they use foreign factors of production that support economic well-being abroad.

[13] The same model is estimated with the *growth rate of GDP* as dependent variable. Not surprisingly, and in line with neoclassical growth theory, the growth rate is negatively related to financial inclusion. This comes from the theoretical argument and empirical finding that highly developed economies have relatively lower growth rates compared to less developed economies. Consequently, richer countries have already experienced their catch-up process and are on its way to experiencing financial inclusion.

[14] See Eq. (20.8) and the corresponding paragraph for the motivation and technical construction of the transformation of the IFI.

Table 20.4 Explanatory variables: IFI regression

Variable name	Variable description
Socioeconomic:	
education	Enrollment ratio between secondary and total enrollment
poverty	Gini coefficient
Demographic:	
urban	Urban population (% of total population)
Financial:	
npl	Bank nonperforming loans (% of total gross loans)
car	Capital-to-asset ratio
credit	Domestic credit to private sector by banks (% of GDP)

Table 20.4 displays the explanatory variables that vector z contains and includes in the regression model.

The first socioeconomic variable, *education*, is the secondary and total education ratio. Its purpose is to approximate the level of education within a country. According to Sarma and Pais (2011), education level is essential for financial inclusion since banks are more likely to welcome well-educated customers. The Gini coefficient roughly measures the possibility of social exclusion and inequality. Kempson and Whyley (1998), Barr (2004), and Buckland et al. (2005) argue that people who belong to low-income groups suffer from income inequality and are consequently less likely to be connected to the financial sector. As a result, the theoretically consistent sign of this variable is negative. The variable *urban* measures the proportion of the population living in urban areas. Leyshon and Thrift (1995) claim that geographical proximity to urban financial centers is expected to positively affect financial inclusion since people living in urban areas can access physical banking outlets more easily. The third category contains variables that control for the soundness and health of the banking system. The variable *npl* measures the ratio of nonperforming loans to total gross loans provided by the banking sector. According to Wang and Guan (2017), the relationship to financial inclusion can be either positive or negative.

Regarding the positive, they contend that NPL will rise as soon as banks include previously excluded customers who are more likely to default on their loans. Consequently, the statistical relationship is expected to be positive. Contrary to that, the relationship can also be damaging. The reason is that a high level of NPL may imply an unhealthy banking sector whereby one needs to be more cautious in lending decisions (excluding potentially risky and low-income customers). The second variable, *car*, represents the capital-to-asset ratio and is the extent of the banks' available capital for their risk-weighted assets. As with the variable capturing the degree of nonperforming loans, its theoretical relationship to financial inclusion can be twofold. The last variable, *credit*, measures domestic credit to the private sector by banks as a percentage of GDP. Therefore, this variable captures financial depth and considers the importance of a well-developed financial system (Beck et al., 2007). Its relationship to financial inclusion is expected to be positive.

Table 20.5 presents the estimated regression model.

Except for *poverty*, all variables are significant at a 1% level. The statistical insignificance of inequality toward financial inclusion is an essential economic finding. It underscores current financial advances in that underdeveloped and developing countries nowadays depend much less on traditional financial services. Instead, they will likely employ non-traditional financial services like mobile money and nonretail agent outlets to access the financial service sector, according to Espinosa-Vega et al. (2020). Consequently, the link between inequality, poverty, and financial inclusion breaks. This is a new and unprecedented finding. While constructing financial inclusion indices, previous literature focuses only on traditional banking services. They do not consider variables like mobile money

Table 20.5 Regression results: IFI regression

education	poverty	urban	npl	car	credit
0.0083	0.0030	0.0092	−0.0118	0.0319	0.0055
*(3.34***)*	*(0.55)*	*(2.91***)*	*(−3.05***)*	*(2.64***)*	*(6.53***)*

Note: Values in brackets below the coefficients represent their respective t-statistics. ***, ** and * signify significance at the 1%, 5%, and 10% levels, respectively

outlets, nonbranch retail agent outlets of commercial banks, registered and active mobile money accounts, and the number and value of mobile money transactions. Consequently, they draw incomplete conclusions (see Sect. 20.2.1).[15]

Economic theory indicates the coefficients for the variables education and urban. Higher education and, proportionally, more people living in urban areas increase financial inclusion. The variable capturing the financial depth of a country, *credit*, also contributes positively to the IFI score. This emphasizes the positive relationship between the size and development of financial institutions and markets and the inclusion of the financial sector. The latter conclusion aligns with Beck et al.'s (2007) analysis. The variable capturing nonperforming loans, *npl*, is negatively correlated to IFI. According to Wang and Guan (2017), this has to do with the health of the financial institution giving out loans; the unhealthier a financial service provider becomes, the less willing it is to issue loans to disadvantaged borrowers. Therefore, financial inclusion will drop, and more credit-worthy customers can access traditional banking services. In line with banks' precautionary motive and institutional risk aversion, the variable capturing the capital-to-assets ratio, *car*, positively correlates to the financial inclusion score. Indeed, this aligns with the previous finding: the healthier and more financially robust financial institutions become, the more willing they are to include previously excluded customers. Should their health deteriorate, they are likely to exclude their riskiest customers.

[15]Some articles that fail to include non-traditional banking services in determining the level of financial inclusion are Barr (2004), Buckland et al. (2005), Honohan (2008), Al-Hussainy et al. (2008), Sarma and Pais (2011), Arora (2014), Kim (2016), Wang and Guan (2017), Kim et al. (2018), and Van et al. (2021).

20.4 Discussion, Policy Recommendation, and Conclusion

Contrary to past literature and the UN's Sustainable Development Goals, incorporating non-traditional financial services into measuring financial inclusion is an exception. Consequently, this chapter refrains from using a single-indicator approach to gauge the level of financial outreach. Instead, it relies on a multidimensional index that takes accessibility, availability, and usage into account. Given that over the past few years, some low-income countries have been experiencing growth of about 500% in adopting mobile money, this seems sensible.

To calculate this augmented Index of Financial Inclusion, the Financial Access Survey (FAS) is used together with further developments of the technique developed by Sarma (2008). While maintaining important mathematical properties, 16 variables are transformed into the normalized inverse of the Euclidean distance. This implies that this augmented index rivals all existing measures and postulates that the proposed measure is more appropriate.

A one-way fixed-effects panel model is estimated to determine the statistical relationship between the newly generated variable's level and economic activity and growth. The results confirm economic theory. The more financially inclusive an economy is, the higher its economic activity. Furthermore, consistent with the neoclassical growth theory, more inclusion is related to less economic growth. The latter finding emphasizes the empirical fact that highly developed countries grow less robustly than their less developed counterparts.

Socioeconomic, demographic, and other variables concerning the health and depth of the financial sector are used to find the statistical determinants of financial inclusion. Results indicate that the empirical link between poverty and financial inclusion breaks when considering non-traditional financial services. This confirms the widely debated issue that mobile money plays a vital role in the financial development of the low- and lower-middle-income world. Furthermore, the health and depth of the financial sector are also crucial. In other words, the deeper and healthier the financial industry, the higher the propensity to cater to lower-income customers. On the other hand, should the solvency of the financial sector deteriorate, riskier customers may be financially excluded.

These findings augment and support the arguments made by Levin (1997) that financial development and economic growth are closely intertwined and that public policy needs to do everything in its power to foster the viability of the financial sector. Consequently, this book chapter recommends that optimal financial sector policies be paramount in lifting the underdeveloped world out of poverty. Increasing financial literacy by improving primary financial education might be beneficial. Conversely, ensuring the financial sector is safe and sound will make banks more open to low-income borrowers. A deep and thorough discussion about particular policies and their calibrations is left for further research.

Appendix

Table 20.6 Countries within sample

Country	Country	Country	Country
Afghanistan	Dominica	Kyrgyz Republic	Poland
Albania	Dominican Republic	Lao PDR	Portugal
Algeria	Ecuador	Latvia	Qatar
Angola	Egypt, Arab Rep.	Lebanon	Romania
Argentina	El Salvador	Lesotho	Russian Federation
Armenia	Equatorial Guinea	Liberia	Rwanda
Australia	Estonia	Libya	Samoa
Austria	Eswatini	Lithuania	San Marino
Azerbaijan	Ethiopia	Luxembourg	Saudi Arabia
Bahamas	Fiji	Macao	Senegal
Bangladesh	Finland	Madagascar	Serbia
Barbados	France	Malawi	Seychelles
Belarus	Gambia	Malaysia	Singapore
Belgium	Georgia	Maldives	Slovak Republic
Belize	Germany	Mali	Slovenia
Benin	Ghana	Malta	South Africa
Bhutan	Greece	Mauritania	Spain
Bolivia	Granada	Mauritius	Sudan
Bosnia and Herzegovina	Guatemala	Mexico	Suriname
Botswana	Guinea	Micronesia	Sweden
Brazil	Guinea-Bissau	Moldova	Switzerland
Brunei Darussalam	Guyana	Mongolia	Tanzania
Bulgaria	Haiti	Montenegro	Thailand
Burkina Faso	Honduras	Morocco	Timor-Leste
Burundi	Hong Kong	Mozambique	Togo
Cabo Verde	Hungary	Myanmar	Tonga
Cambodia	Iceland	Namibia	Trinidad and Tobago
Cameroon	India	Nepal	Tunisia
Canada	Indonesia	Netherlands	Turkey
Central African Rep.	Iran	New Zealand	Uganda

(continued)

Table 20.6 (continued)

Country	Country	Country	Country
Chad	Iraq	Nicaragua	Ukraine
Chile	Ireland	Niger	United Arab Emirates
China	Israel	Nigeria	United Kingdom
Colombia	Italy	North Macedonia	United States
Comoros	Jamaica	Norway	Uruguay
Costa Rica	Japan	Oman	Uzbekistan
Core d'Ivoire	Jordan	Pakistan	Vanuatu
Croatia	Kazakhstan	Panama	Vietnam
Cyprus	Kenya	Papua New Guinea	Zambia
Czech Republic	South Korea	Paraguay	Zimbabwe
Denmark	Kosovo	Peru	
Djibouti	Kuwait	Philippines	

Table 20.7 Average IFI scores

Year	Low	Lower-middle	Upper-middle	High
2004	0.0849	0.1101	0.1785	0.2460
2005	0.0916	0.1068	0.1835	0.2524
2006	0.0883	0.1050	0.1781	0.2559
2007	0.0915	0.1095	0.1921	0.2576
2008	0.0930	0.1095	0.1911	0.2507
2009	0.1002	0.1169	0.1955	0.2572
2010	0.0997	0.1242	0.1888	0.2442
2011	0.1135	0.1279	0.1902	0.2377
2012	0.1166	0.1255	0.1888	0.2237
2013	0.1269	0.1253	0.1837	0.2109
2014	0.1402	0.1321	0.1825	0.2116
2015	0.1536	0.1371	0.1867	0.2104
2016	0.1813	0.1537	0.1946	0.2356
2017	0.1831	0.1609	0.1939	0.2331
2018	0.1714	0.1582	0.1924	0.2212
2019	0.1719	0.1471	0.1713	0.2108

References

Adrian, T., & Mancini-Griffoli, T. (2018). *The rise of digital money*. IMF—Fintech Notes, Note/19/01.

Al-Hussainy, E., Beck, T., Demirguc-Kunt, A., & Zia, B. (2008). *Household use of financial services*. World Bank Development Economics Research Group Working Paper Series.

Allen, F., Demirguc-Kunt, A., Klapper, L., & Peria, M. S. M. (2012). *The foundations of financial inclusion*. World Bank Policy Research Working Paper, No. 6290.

Arora, R. U. (2014). Access to finance: An empirical analysis. *The European Journal of Development Research, 26*(5), 798–814.

Barr, M. (2004). Banking the poor. *Yale Journal on Regulation, 21*, 122–239.

Barro, R. J. (2000). *Education and economic growth* (Working Paper). Harvard University.

Barro, R. J. (2001). *Economic growth in East Asia before and after the financial crisis*. NBER Working Paper Series, Working Paper 8330.

Barro, R. J. (2003). Determinants of economic growth in a panel of countries. *Annals of Economics and Finance, 4*, 231–274.

Barro, R. J. (2013). Inflation and economic growth. *Annals of Economics and Finance, 14*(1), 85–109.

Beck, T., Demirguc-Kunt, A., & Martinez Peria, M. S. (2007). Reaching out: Access to and use of banking services across countries. *Journal of Financial Economics, 85*, 234–266.

Berthelemyje, A.-C. V., & Varoudakis, A. (1996). Economic growth, convergence clubs, and the role of financial development. *Oxford Economics Papers, 48*(2), 300–328.

Bruhn, M., & Love, I. (2014). The real impact of improved access to finance: Evidence from Mexico. *The Journal of Finance, 69*(3), 1347–1376.

Buckland, J., Guenther, B., Boichev, G., Geddie, H., & Mutch, M. (2005). *There are no banks here: Financial and insurance exclusion services in Winnipeg's north end*. Winnipeg Inner-City Research Alliance (WIRA). University of Winnipeg.

Cihak, M., Mare, D. S., & Melecky, M. (2016). *The nexus of financial inclusion and financial stability: A study of trade-offs and synergies*. World Bank Policy Research Working Paper, No. 7722.

Demirguc-Kunt, A., & Klapper, L. (2012). *Measuring financial inclusion: The Global Findex database*. World Bank Policy Research Working Paper, No. 6025.

Espinosa-Vega, M., Shirono, K., Villanova, H. C., Chhabra, E., Das, B., & Fan, Y. (2020). *Measuring financial access: 10 years of the IMF financial access survey*. Departmental Papers/Policy Papers, 20(8).

FAS. (2020). *Financial Access Survey—2020 trends and developments*. https://data.imf.org/?sk=E5DCAB7E-A5CA-4892-A6EA-598B5463A34C

Gerth, F. (2017). *Allocative efficiency of UK firms during the great recession*. School of Economics Discussion Paper Series, KDPE 1714.

Gerth, F. (2023). *Nexus between financial inclusion and economic activity: A study about traditional and non-traditional financial service indicators determining financial outreach*. Munich Personal RePEc Archive, MPRA Paper No. 119265, posted 30 Nov 2023, Online at https://mpra.ub.uni-muenchen.de/119265/

Gerth, F., & Otsu, K. (2015). A neoclassical analysis of the Great Recession: a historical comparison. *Economics Bulletin, 35*(4), 2363–2373.

Gerth, F., & Otsu, K. (2018). The post-crisis slump in Europe: A business cycle accounting analysis. *The B.E. Journal of Macroeconomics, 18*(1), 1–25.

Goyal, R., Creane, S., Mobarak, A. M., & Sab, R. (2004). *Financial sector development in the Middle East and North Africa*. IMF Working Paper, No WP/04/201.

Honohan, P. (2008). Cross-country variation in household access to financial services. *Journal of Banking & Finance, 32*, 2493–2500.

Kempson, E., & Whyley, C. (1998). *Access to current accounts*. British Bankers' Association.

Kempson, E., Atkinson, A., & Pilley, O. (2004). *Policy level response to financial exclusion in developed economies: lessons for developing countries*. Report of Personal Finance Research Centre, University of Bristol.

Kim, J. H. (2016). A study on the effect of financial inclusion on the relationship between income inequality and economic growth. *Emerging Markets Finance and Trade, 52*(2), 498–512.

Kim, D.-W., Yu, J.-S., & Hassan, M. K. (2018). Financial inclusion and economic growth in OIC countries. *Research in International Business and Finance, 43*, 1–14.

Krolzig, H.-M. (2001). General-to-specific reductions in vector autoregressive processes. In R. Friedmann, L. Knüppel, & H. Lütkepohl (Eds.), *Econometric studies – A Festschrift in Honour of Joachim Frohn* (pp. 129–157).

Krolzig, H.-M. (2003). General-to-specific model selection procedures for structural vector autoregressions. *Oxford Bulletin of Economics and Statistics, 65*, 769–801.

Levine, R. (1997). Financial development and economic growth: Views and agenda. *Journal of Economic Literature, 35*(2), 688–726.

Levine, R. (2005). Finance and growth: Theory and evidence. In P. Aghion & S. Durlauf (Eds.), *Handbook of economic growth*. Elsevier Science.

Leyshon, A., & Thrift, N. (1995). Geographies of financial exclusion: financial abandonment in Britain and the United States. *Transactions of the Institute of British Geographers, 20*(3), 312–341.

Myint, H. (1958). The "classical theory" of international trade and the underdeveloped countries. *The Economic Journal, 68*(270), 317–337.

Sarma, M. (2008). *Index of financial inclusion*. ICRIER Working Paper, No.215.

Sarma, M. (2012). *Index of financial inclusion—A measure of financial sector inclusiveness*. Berlin Working Papers on Money, Finance and Trade Development, Working Paper No: 07/2012.

Sarma, M., & Pais, J. (2011). Financial inclusion and development. *Journal of International Development, 23*(5), 613–628.

Van, L. T.-H., Vo, A. T., Nguyen, N. T., & Vo, D. H. (2021). Financial inclusion and economic growth: An international evidence. *Emerging Markets Finance and Trade, 57*(1), 239–263.

Wang, X., & Guan, J. (2017). Financial inclusion: measurement, spatial effects, and influencing factors. *Applied Economics, 49*(18), 1751–1762.

World Bank. (2021). *World Bank country, and lending groups*. https://datahelpdesk.worldbank.org/knowledgebase/articles/906519-world-bank-country-and-lending-groups

21 Micro- and Macro-economic Factors of Mobile Money Adoption in the Middle East and North Africa

Dimitrios Reppas

Abstract

For the past decade, Mobile Money (MM) systems have become very popular in low- and middle-income countries, particularly in sub-Saharan Africa, and have transformed millions of lives by advancing financial inclusion, as required by SDG 8.10. Adoption continues upward within the Middle East and North Africa region (MENA), possibly due to the recent COVID-19 pandemic. Overall, the MM industry has shown remarkable recent growth in MENA, with the fastest global growth in 2021. Nevertheless, adoption rates remain low compared to other regions globally and vary considerably across MENA countries. This chapter sheds novel and much-needed light on a relatively understudied topic: the micro- and macroeconomic determinants for adopting MM systems in the MENA region. To do so, it summarizes the main findings from the existing theoretical and empirical literature, most of which have focused on African countries. This systematic literature review aims to assist researchers in designing future quantitative studies on MM.

D. Reppas (✉)
Department of Economics, University of Birmingham, Dhabi, United Arab Emirates
e-mail: d.reppas@bham.ac.uk

Keywords

Mobile money · Financial inclusion · SDG 8.10 · Economic factors

21.1 Introduction

The Middle East and North Africa region's (MENA) social, economic, and technological development often seems uneven compared with other parts of the world. A case in point is mobile money (MM) systems, which are considered innovative in digital financial services because they lie outside the formal banking system.

MM systems are an electronic form of currency, that is, digital money, which requires an application on a device (such as a tablet or a mobile phone). The way MM systems work, in general, is as follows: Customers first need to register with the provider of the service, typically a telecommunications company (i.e., a Mobile Network Operator [MNO]), by showing some form of government identification to a participating "agent," often small retailer shops (grocery stores or petrol stations) or larger establishments (such as utility companies). Once registered, users can then deposit cash (i.e., official tender) in their MM accounts. What they get, in return, is an equivalent value in the electronic currency (i.e., e-money) to be held within their mobile wallets.

G. W. Muschert et al. (eds.), *Financial Inclusion*, Sustainable Development Goals Series,
https://doi.org/10.1007/978-3-031-68803-4_21

Unlike mobile banking, where customers access their conventional bank account via mobile devices, MM systems enable users to make basic financial transactions—such as transfers, deposits, withdrawals, and savings—without needing a formal bank account.[1] To conduct such cashless transactions, users must send a text message (SMS) through their mobile phone. Although MM systems initially focused on consumer transactions, that is, on person-to-person (P2P) remittances, they may be used for various other purposes, which include person-to-business (P2B) remittances, person-to-government (P2G) remittances (e.g., merchant payments, electricity payments, university, and schooling fees), B2P, B2B, and G2P remittances (e.g., salary payments). Aggregate data provided by the *Global System for Mobile Communications Association*[2] (GSMA, 2022) reveals, for instance, that the share of merchant payments (in low- and middle-income countries) has doubled in 2021 to 21% of their circulating value (compared to 10% in the previous years); the number of employees receiving salaries via MM grew globally by 150% (during the period from September 2020 to June 2021); and the number of bill payments processed via MM increased in 2021 by 37%. The observed diversification in the uses of MM shows clearly how these systems play an essential role in people's daily lives. It also verifies their potential to promote SDG subgoal 8.10, that is, "to encourage and expand access to banking, insurance, and financial services for all."

Although sub-Saharan Africa (particularly Kenya, Tanzania, Uganda, Ethiopia, Nigeria, and Ghana) has traditionally been the epicenter of MM (for instance, in 2021, it counted a total of 161 live services and accounted for about half of all active MM accounts worldwide [GSMA, 2022]), over the past 5 years, MM systems have also gained popularity in other parts of the world, such as in South Asia (e.g., Afghanistan, Pakistan, Indonesia, and Bangladesh) and also in the Latin America and Caribbean (e.g., Colombia and Paraguay). However, in the MENA region, MM systems have been less widespread. For instance, in 2021, MENA counted only 28 live services, accounting for about 4% of all active accounts worldwide (GSMA, 2022). More specifically, Jordan and Tunisia are the two MENA countries in which MM systems seem to be more popular (with 7 and 5 systems deployed, respectively); Morocco, Egypt, Turkey, Iran, Iraq, and the UAE have also had their systems; nevertheless, several other countries (such as Saudi Arabia, Oman, Qatar, Kuwait, Bahrain, Syria, Lebanon, Algeria, and Israel) have not yet developed any. Overall, there is considerable heterogeneity in the adoption of MM systems within MENA countries.

For this reason, a systematic investigation explaining MM systems' lower levels of market penetration for parts of the MENA region seems to be missing. This chapter tries to fill this gap in the research literature by identifying possible socioeconomic determinants for adopting MM in the MENA region. Section 21.2 briefly compares selected metrics for MM usage between MENA countries and other parts of the world. Section 21.3 reviews some of the existing theoretical and empirical studies, providing insights to researchers in designing future quantitative studies for MM adoption in MENA. Section 21.4 concludes.

21.2 Mobile Money in MENA: A Comparison with Other Countries

All figures in this section compare metrics for MM usage from 2015 to 2020 between MENA countries and selected "Rest of the World" countries with a diversified and expanded MM industry (data for all figures are from the IMF's Financial Access Survey).[3]

[1] Nevertheless, contrary to the practice of the formal banking system, interest is paid on deposits or savings of e-money. Also, payment services linked to a traditional banking product or credit card (such as Apple Pay, Google Pay, and Samsung Pay) are not considered MM transactions.

[2] GSMA (https://www.gsma.com/) is an industry organization that represents the interests of MNOs worldwide.

[3] Accessible from https://data.imf.org/?sk=E5DCAB7E-A5CA-4892-A6EA-598B5463A34C

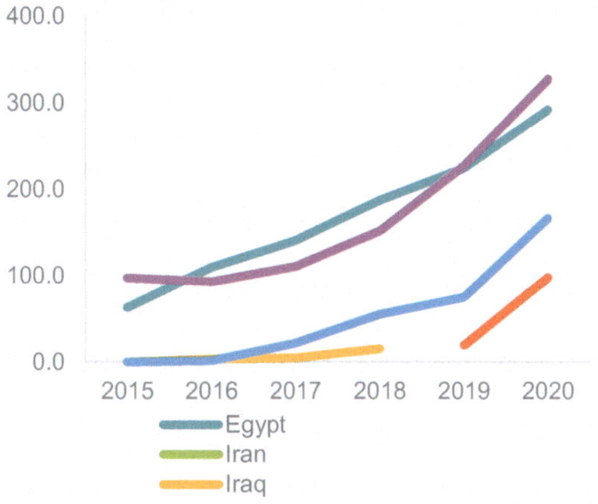

Fig. 21.1 Number of registered MM accounts per 1000 adults in selected MENA countries (2015–2020)

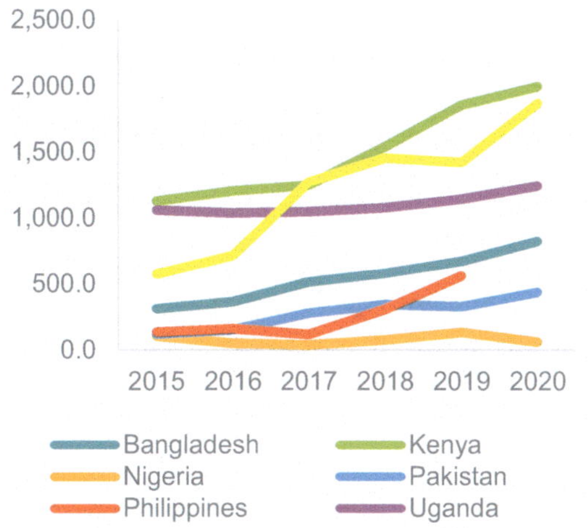

Fig. 21.2 Number of registered MM accounts per 1000 adults in selected Rest of the World countries (2015–2020)

A comparison of Fig. 21.1 (for selected MENA countries) with Fig. 21.2 (for the Rest of the World countries) shows that the number of registered MM accounts per 1000 adults in the MENA region has been at least 10 times smaller than elsewhere (for instance, Egypt and Qatar started with around 100 registered MM accounts in 2015, whereas, at the same time, Kenya and Uganda were in the region of 1000).[4] Likewise, Fig. 21.3 shows that the number of MM transactions per 1000 adults in selected MENA countries has been 5–10 times smaller than the rest of the world (see Fig. 21.4).[5] Moreover, Figs. 21.5 and 21.6 depict the value (rather than the numbers) of MM transactions as part of those countries GDPs. Once again, there is an apparent disparity between the numbers for the MENA region (around 0.5% of their GDPs, in Fig. 21.5) and the Rest of the World (around 10% for most of the considered countries in Fig. 21.6, while Uganda and Kenya are clearly "exceptions," with 50% of their GDPs)!

A comparison of the above three sets of figures reveals the following two main points:

(a) MM systems have lower levels of market penetration in the MENA region than elsewhere. This is likely related to the different degrees of development of the formal banking sector across MENA countries. Preliminary evidence from the IMF suggests that MM is more prevalent in countries with lower access to traditional banking services; therefore, MM systems are virtually nonexistent in economies with robust banking systems (IMF, 2020). This dynamic explains why, within the MENA region, MM has spread particularly into geographical areas not extensively reached by bank branch networks (such as in Jordan, Egypt, Morocco, Iran, and Iraq). In contrast, MM systems have grown significantly smaller in areas where the population has had access to traditional banking services (such as in the UAE, Qatar, and Saudi Arabia). For instance, Fig. 21.7 reveals that Jordan, the MENA country with the most significant number of MM networks, has only had about half the number of deposit accounts with commercial

[4]Also, a similar picture emerges by comparing IMF's data referring to "active" accounts rather than "registered" accounts ("Active" means that the account has been used to perform at least one transaction during the 90 days prior to the end of December for the year indicated).

[5]Note that Fig. 21.4 does not even include Africa's "big players" (such as Kenya, Uganda, and Zambia). If the latter were included, the differences (with the MENA countries) would have been even more significant: Kenya and Zambia fall in the region of 50 K transactions, while Uganda is well above 100 K!

Fig. 21.3 Number of MM transactions per 1000 adults in selected MENA countries (2015–2020)

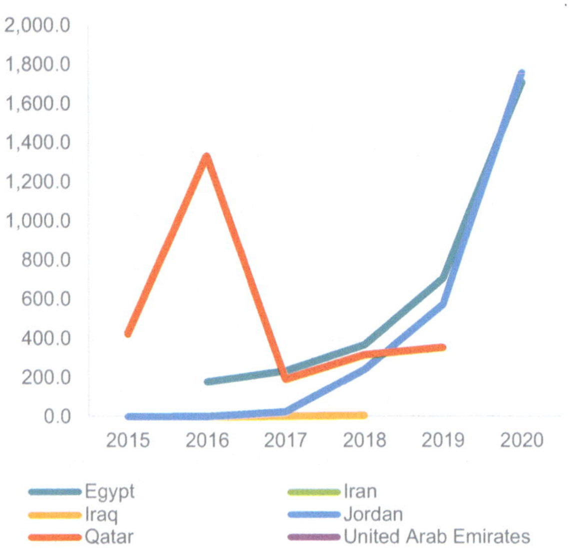

Fig. 21.4 Number of MM transactions per 1000 adults in selected Rest of the World countries (2015–2020)

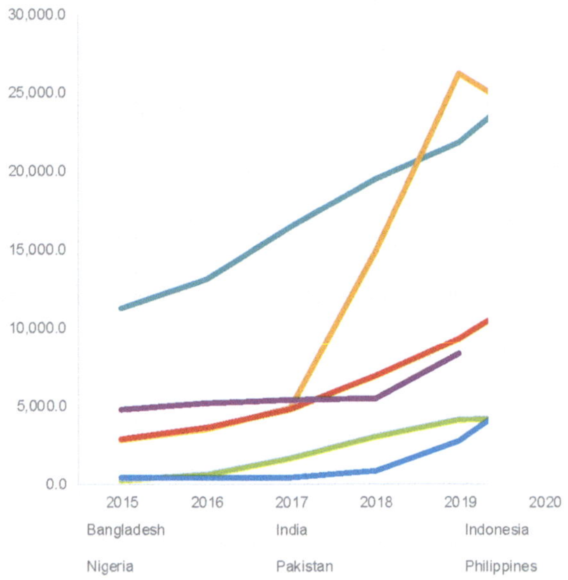

banks (per 1000 adults) compared to the UAE or Saudi Arabia.[6]

(b) MM systems in the MENA region have picked up recently. Figures 21.1, 21.3, and 21.5 all show clear upward trends for the last 3 years. Furthermore, in 2021, MENA experienced the most considerable annual growth in active MM accounts compared to all other regions of the World. New active accounts in MENA increased by 68%, followed by Latin America and the Caribbean with 34% increase (GSMA, 2022). This increase may be related to the recent COVID-19 pandemic and the potential of MM systems to provide resilience to external shocks. Some of the evidence that MM systems have helped societies in MENA tackle the recent challenges that have arisen (and therefore remain resilient in the face of economic uncertainty) is summarized below:

[6]Similarly, for other indicators of the formal banking system, such as "the number of ATMs per 100 K adults."

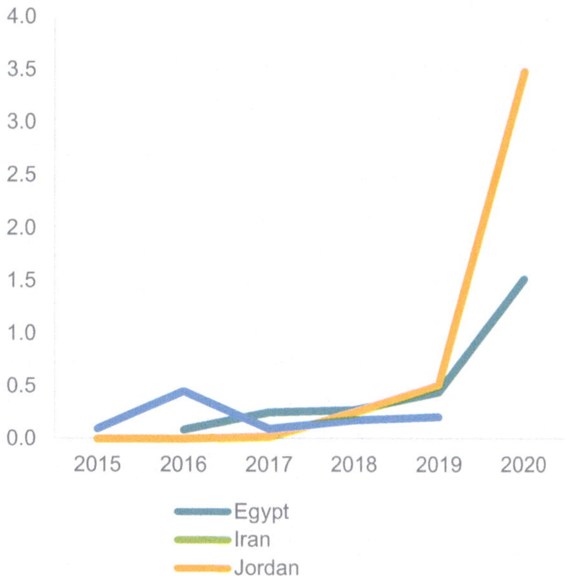

Fig. 21.5 Value of MM transactions (as % of GDP) in selected MENA countries (for 2015–2020)

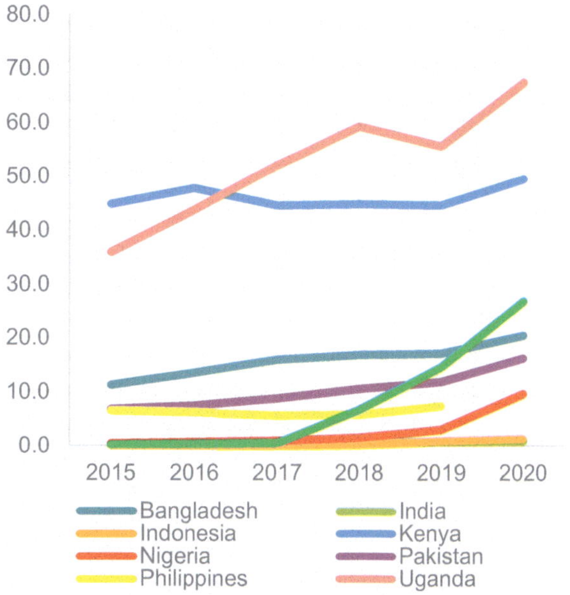

Fig. 21.6 Value of MM transactions (as % of GDP) in selected Rest of the World countries (for 2015–2020)

- Customers and businesses have embraced MM (and other digital payments) as a safer and more accessible transaction option than handling cash (as MM transactions require minimum physical interaction).
- Governments and nongovernmental organizations (NGOs) have utilized MM to securely and quickly distribute humanitarian subsidies, relief funds, and fiscal support to otherwise unbanked population segments. For instance, MM systems were recently used in Jordan to distribute financial assistance to numerous Syrian refugees.[7]
- MM has enabled users (either consumers or firms) to get credit (loans) and access to

[7]Most refugees in Jordan are not allowed to open a bank account; nevertheless, they can get a Jordanian SIM card

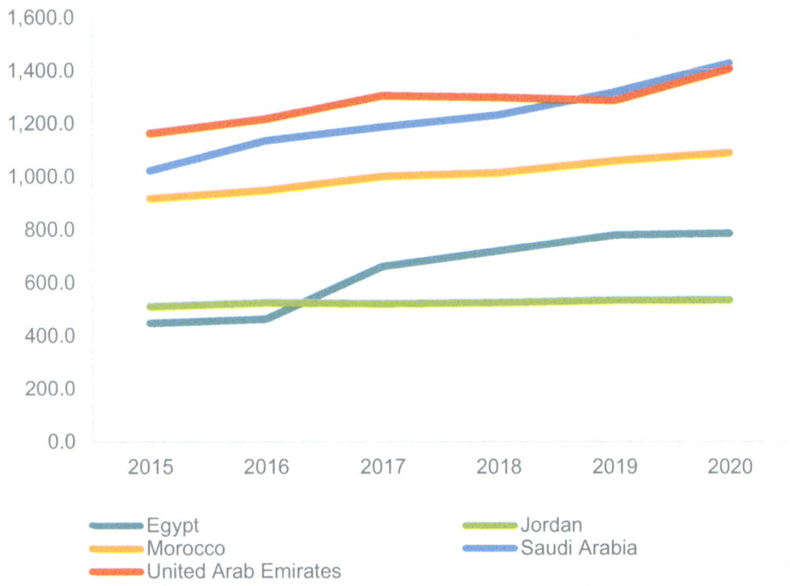

Fig. 21.7 Number of deposit accounts with commercial banks (per 1000 adults) in selected MENA countries (for 2015–2020)

investment and savings products. For instance, in Egypt, there has been a 27% increase in mobile-enabled loans and a 21% increase in mobile-enabled savings compared to before the pandemic (GSMA, 2022).

- Diasporas continued using MM systems for international remittances to relatives who suffered job losses and income cuts during the COVID-19 pandemic. MM systems are much more affordable for remittances than regular bank accounts. For instance, in 2021, the average cost to send $200 internationally through an MM account was 3.90% of the transacted value, almost half of the cost (6.89%) through a regular bank account (GSMA, 2022). In other words, MM systems contribute toward achieving SDG10, which is to "reduce inequality within and among countries."
- Some MENA countries have revised their regulatory regimes and implemented measures to support MM services, which had been pending for a long time. For instance, regulators in Egypt and Jordan have relaxed the Know Your Customer (KYC) requirements for new customer enrolment (to limit contact with agents) and have also allowed for more funds to flow through MM by increasing the transaction limits for MM services per day (GSMA, 2020b). Furthermore, in Jordan, authorities have undertaken a total data cleansing effort (along with other regulatory enhancements) and have eventually paved the way for the opening of approximately 85,000 new mobile wallets (a 20% growth) during the COVID-19 pandemic.[8]

21.3 Micro- and Macroeconomic Determinants of MM Adoption

A review of the relevant literature reveals three main reasons, as summarized below, why MM systems have been increasingly popular in low- and middle-income countries (see also Reppas

(and therefore access to MM services) through their passport or some other residency permit (GSMA, 2020c).

[8]Info is available from: https://www.centerforfinancialinclusion.org/mobile-money-in-times-of-crisis-an-unlikely-hero

and Muschert [2021] for a more thorough discussion):

(a) MM systems represent a digital transformation of exchange relations (in the form of the migration of remittances from cash-based systems to electronic media). MM increases transparency by offering a more secure and convenient alternative to cash, reducing transaction costs of transporting money through intermediaries, facilitating trade and business planning, and offering a safer option for savings.[9]
(b) MM helps include unbanked and financially excluded populations in formal economic relations. In rural areas, where population density falls, the cost of establishing a geographically vast network of bank branches is high. Instead, MNOs may still find it profitable to launch MM platforms by relying on their already available mobile network infrastructure and agents. Furthermore, the typical rural family (in low-income countries) tends to be excluded from conventional banking services because they cannot meet the minimum account balance requirements. Under MM systems, such requirements are typically not in place (instead, mobile network operators typically rely on low commissions for a larger volume of transactions to cover some of their operational costs). Moreover, banks are often reluctant to open accounts and provide other services, such as insurance or loans, to low-income people who may not have any records of their financial transactions. Instead, MNOs can rely on the history of available transactions from mobile usage to build more accessible customer profiles.
(c) MM results in poverty alleviation and provides resilience to financial shocks. MM systems have been viewed as a means of providing self-reliance and security to local communities, particularly as a means of "insulating" local economies either from large exogenous communal shocks (e.g., natural disasters or medical epidemics) or from idiosyncratic financial shocks (see, for instance, Suri et al. [2012]; Suri and Jack [2016]). Intuitively, improved risk-sharing comes from households participating in MM systems with more people to rely upon (whenever a negative shock occurs). In short, MM has the potential to build social relationships of trust, reciprocity, solidarity, mutual aid, and cooperation among communities, resulting in poverty alleviation.

Most of the empirical studies on MM adoption have focused almost exclusively on individual countries in Africa. Instead, an empirical investigation explaining MM's lower levels of market penetration in the MENA region has been largely missing. This chapter fills this gap in the literature by identifying some of the possible micro- and macroeconomic determinants for adopting MM in the MENA region. Therefore, researchers (either from economics, sociology, or some other quantitative social science background) may consider including the following variables when investigating the broader socioeconomic characteristics for the adoption of MM in the MENA region:

- Variables for a nation's formal banking sector (the IMF [2020] suggested that there may be a negative correlation between the adoption of MM and a nation's robust banking system). Examples of such variables include the density of commercial bank branches, the density of ATMs, and domestic credit by banks provided to the private sector.
- Variables for remittances (domestic and international). As mentioned, MM systems have allowed users to spread the risk of adverse financial shocks (such as job losses or income cuts). In 2021, MM-enabled remittances grew globally by 48%. Despite this impressive growth, remittances through MM still represent only about 3% of the total flows to low-income and middle-income countries (GSMA,

[9]For instance, Economides and Jeziorski (2017) find that MM has succeeded in areas with high criminality, that is, where holding cash is risky and mainstream banking services are unavailable.

2022). Nevertheless, this—yet small—percentage indicates MM's considerable potential and advantage over regular bank remittances.

- Examples of variables referring to the institutional economy and governance regime include the GSMA's "Regulatory Index." variables for a country's governance index, that is, the extent to which a country's regulatory regime provides an enabling environment for MM services. For instance, Kenya's M-Pesa MM system became popular once the telecommunications company worked closely with government agencies on legal compliance, product functionality, technical requirements, and consumer protection (Lashitew et al., 2019). Likewise, regulators always need to design interventions (to avoid, for instance, money laundering or protect consumers' rights) in consultation with the MM industry. The sustainability of MM systems in Tanzania, Uganda, Ghana, and Cameroon has recently been threatened by the imposition of government taxes on specific MM transactions (GSMA, 2022).[10] The World Bank's "Governance Indicator"[11] and the Heritage Foundation's "Index of Economic Freedom"[12] comprise six dimensions: authorization, consumer protection, transaction limits, Know Your Customer (KYC), agent networks, and investment.
- Variables for a country's urban population (or rural density). MM systems have been popular among people living in distant/rural areas without access to the formal banking sector. Nguena (2019), for instance, finds that the urban population is the second most significant variable explaining (with a positive coefficient) the development of mobile banking in Africa.
- Variables for infrastructure and technological changes (e.g., quality changes in MNO services). The reasoning for including infrastructure and technological variables is that higher telecommunications infrastructure allows for more mobile phone use.
- Standard socioeconomic variables for education, age, and income (or income inequality). Aker and Mbiti (2010) and Munyegera and Matsumoto (2016) find that more educated and wealthier households are more likely to adopt MM systems. Similarly, Amoah et al. (2020) identify age, income, and education as critical microdeterminants of MM usage in Ghana (i.e., younger, more educated, and technology-savvy cohorts seem to have embraced MM systems). Mouna and Jarboui (2022) also find that poorer and less educated people in the MENA region are disproportionally excluded from digital financial services. Also, knowledge and skill-related barriers (e.g., not knowing how to use a handset, low literacy) prevent both male and female owners from having an account (GSMA, 2022).

21.4 Conclusions

Over the past decade, MM systems have significantly changed the landscape of mobile financial inclusion for millions globally, particularly in low- and middle-income countries. MM systems have been increasingly diverse, meeting the needs of financially excluded and often marginalized populations, and have adapted to various cultural and economic needs. In several countries, MM systems have diversified beyond P2P transfers (e.g., support for bill payments, merchant payments, and international remittances). In short, MM systems advance financial inclusion, as required by SDG 8.10. The global MM market is far from being saturated, and future demand for such systems will likely remain high (GSMA, 2022).

Within the MENA region, there is considerable heterogeneity in the adoption of MM systems, ranging, in some countries, from a handful of live systems operating simultaneously to none in others. Nevertheless, overall, MENA experienced, in 2021, the fastest growth in new active MM accounts globally (about 4%). Furthermore, MENA's well-developed banking system has

[10] https://www.gsma.com/mobilefordevelopment/resources/the-mobile-money-regulatory-index-2/

[11] http://info.worldbank.org/governance/wgi/

[12] https://www.heritage.org/index/

been unable to reach vast segments of the population (particularly the underprivileged ones). In contrast, population growth in the region has been found to increase income inequality (Neaim & Gaysset, 2018). Also pending for years, regulatory policy reforms to support MM services were recently undertaken in Egypt, Angola, and Jordan. At the same time, MNOs in the MENA region have continued investing in commercial 5G services.[13] Overall, the observed trend in MENA toward a more digitized MM ecosystem may have been partly supported by the recent COVID-19 pandemic (as MM transactions require minimum physical interaction) and financial innovation initiatives in parts of the region (such as Morocco, Tunisia, and Algeria). Overall, the main comparative advantage of countries in the Middle East has been the large segments of their populations below the age of 30, who are in their most economically productive phase (Reppas & Temouri, 2023). These technologically competitive age groups are ready to embrace digital transactions as an alternative to traditional banking services and move from a cash economy into a more inclusive digital economy.

Most empirical studies on adopting MM systems have focused almost exclusively on individual countries in Africa. Nevertheless, a comprehensive investigation explaining MM's lower levels of market penetration for the MENA region has been largely missing. This chapter has tried to fill this gap in the research literature by identifying some micro- and macroeconomic determinants for adopting MM in the MENA region. Therefore, it provides some valuable insights to researchers (either from economics, sociology, or other quantitative social science disciplines) wishing to empirically investigate some of the broader socioeconomic characteristics that account for the adoption of MM in the MENA region.

Researchers, nevertheless, should always exercise caution in making inferences from existing studies about the factors that play a role in the adoption and sustained usage of MM systems (Aron, 2017, 2018). This chapter suggests that MM's micro- and macroeconomic drivers must be empirically tested. Similarly, Khan and Blumenstock (2016) conclude that behavioral models for adopting MM do not necessarily apply to all developing countries. In other words, some findings in the literature may emerge from idiosyncratic aspects of the data examined.

[13] For instance, new 5G networks were launched (in 2020) in Oman (Ooredoo), Bahrain (Zain), and Israel (Pelephone). Additionally, Ooredoo reported that its 5G mobile network coverage had expanded to more than 90% of the country's populated areas in Qatar; Zain in Saudi Arabia has increased its 5G footprint to 38 cities; Etisalat in the UAE stated that it anticipates having 5G networks in place throughout the entire nation by the end of 2025 (GSMA, 2020a); and 5G activities should pick up in MENA nations outside of the GCC (where they are currently in a trial phase in Egypt, Iran, and Morocco).

References

Aker, J. C., & Mbiti, I. M. (2010). Mobile phones and economic development in Africa. *Journal of Economic Perspectives, 24*(3), 207–232. Available from: https://doi.org/10.1257/jep.24.3.207

Amoah, A., Korle, K., & Asiama, R. K. (2020). Mobile money as a financial inclusion instrument: What are the determinants? *International Journal of Social Economics, 47*(10), 1283–1299.

Aron, J. (2017). *Leapfrogging: A survey of the nature and economic implications of mobile money*. CSAE Working Paper WPS/2017-02, Center for the Study of African Economies. Available from: https://www.csae.ox.ac.uk/materials/papers/csae-wps-2017-02.pdf

Aron, J. (2018). Mobile money and the economy: A review of the evidence. *The World Bank Research Observer, 33*(2), 135–188. Available from: https://doi.org/10.1093/wbro/lky001

Blumenstock, J. E., Eagle, N., & Fafchamps, M. (2016). Airtime transfers and mobile communications: Evidence in the aftermath of natural disasters. *Journal of Development Economics, 120*, 157–181. Available from: https://doi.org/10.1016/j.jdeveco.2016.01.003

Economides, N., & Jeziorski, P. (2017). Mobile money in Tanzania. *Marketing Science, 36*(6), 815–837. Available from: https://doi.org/10.1287/mksc.2017.1027

GSMA. (2020a). *The mobile economy in the Middle East and North Africa 2020*. Available from: https://www.gsma.com/mobileeconomy/wp-content/uploads/2020/11/GSMA_MobileEconomy2020_MENA.pdf

GSMA. (2020b). *Mobile money recommendations to Central Banks in response to Covid 19*. Available from: https://www.gsma.com/mobilefordevelopment/wp-content/uploads/2020/04/Mobile-money-recommendations-to-central-banks-in-response-to-COVID-19.pdf

GSMA. (2020c). *Humanitarian cash and voucher assistance in Jordan: A gateway to mobile financial services*. Available from: https://www.gsma.com/mobilefordevelopment/wp-content/uploads/2020/01/Jordan_Mobile_Money_CVA_Case_Study_Web_Spreads.pdf

GSMA. (2022). *State of the industry report on mobile money 2022*. Available from: https://www.gsma.com/sotir/

IMF. (2020). *Measuring financial access: 10 Years of the IMF financial access survey*. International Monetary Fund, Departmental Paper No. 20/08. Available from: https://www.imf.org/en/Publications/Departmental-Papers-Policy-Papers/Issues/2020/0/5/12/Measuring-Financial-Access-10-Years-of-the-IMF-Financial-Access-Survey-49298

Lashitew, A. A., van Tulder, R., & Liasse, L. (2019). Mobile phones for financial inclusion: What explains the diffusion of mobile money innovations? *Research Policy, 48*(5), 1201–1215. https://doi.org/10.1016/j.respol.2018.12.010

Mouna, A., & Jarboui, A. (2022). Understanding the link between government cashless policy, digital financial services and socio-demographic characteristics in the MENA countries. *International Journal of Sociology and Social Policy, 42*, 416–433. https://doi.org/10.1108/IJSSP-12-2020-0544

Munyegera, G. K. & Matsumoto, T. (2016). Mobile Money, Remittances, and Household Welfare: Panel Evidence from Rural Uganda. World Development, 79, pp.127-137. Available from: https://doi.org/10.1016/j.worlddev.2015.11.006

Neaim, S., & Gaysset, I. (2018). Financial inclusion and stability in MENA: Evidence from poverty and inequality. *Finance Research Letters, 24*, 230–237. https://doi.org/10.1016/j.frl.2017.09.007

Nguena, C.-L. (2019). On financial innovation in developing countries: The determinants of Mobile banking and financial development in Africa. *Journal of Innovation Economics and Management, 29*, 69–94. https://doi.org/10.3917/jie.029.0069

Reppas, D., & Muschert, G. W. (2021). Mobile money systems as Avant-Garde in the digital transition of financial relations. In S. H. Park, M. A. Gonzalez-Perez, & D. Floriani (Eds.), *The Palgrave handbook of corporate sustainability in the digital era* (pp. 157–168). Palgrave Macmillan. https://doi.org/10.1007/978-3-030-42412-1_8

Reppas, D., & Temouri, Y. (2023). Economic environment in the Middle East: Facts and initiatives. In P. Budhwar, V. Pereira, & K. Mellahi (Eds.), *Doing business in the Middle East: A research-based practitioners' guide*. Routledge.

Suri, T., & Jack, W. (2016). The long-run poverty and gender impacts of mobile money. *Science, 354*(6317), 1288–1292. https://doi.org/10.1126/science.aah5309

Suri, T., Jack, W., & Stoker, T. M. (2012). Documenting the birth of a financial economy. *PNAS, 109*(26), 10257–10262. https://doi.org/10.1073/pnas.1115843109

UN SDG Indicators 8.10 for Measuring Financial Inclusion: An Assessment

22

Mandira Sarma

Abstract

Financial inclusion is an essential pre-condition for sustainable economic development and poverty alleviation. Sub-target 8.10 of the Sustainable Development Goal (SDG) of the United Nations (UN) emphasizes expanding "access to financial services like banking, insurance and financial services for all." To take stock of countries' progress in this sub-target, UN proposed using a set of three indicators: number of commercial bank branches and automated teller machine (ATM) per 100,000 adults (indicators 8.10.1 (a) and (b), respectively) and the proportion of adults with an account at a bank, other financial institution, or with a mobile-money service provider (indicator 8.10.2). This chapter examines these indicators and presents an assessment of whether they sufficiently capture the inclusiveness of a financial system as envisaged under SDG sub-target 8.10. Considering the existing literature on multidimensionality of financial inclusion, it is argued that this limited set of indicators are inadequate and methodologically insufficient. Additional indicators are proposed to augment the set. Further, a multidimensional Index of Financial Inclusion (IFI) is proposed as a measure of financial inclusion.

M. Sarma (✉)
Jawaharlal Nehru University, Delhi, India
e-mail: sarmam@mail.jnu.ac.in

22.1 Introduction

The eighth sustainable development goal (SDG) of the United Nations (UN) aspires for inclusive, sustainable, and sustained economic growth with productive employment and decent work for all. A vital conduit to achieve this goal is an inclusive financial system that allows all economic agents to actively participate in the formal financial system, irrespective of income levels, social status, gender, and other such factors. An inclusive financial system is desirable on many counts. It facilitates efficient intermediation of resources from savers to borrowers for productive investment, thus creating and enlarging livelihood opportunities and capital formation. Providing safe avenues for savings and insurance, an inclusive financial system also helps with financial management, consumption smoothing, and managing unforeseen economic shocks for households and individuals. Thus, financial inclusion is an essential pre-condition for sustainable economic development and poverty alleviation. Because of this, sub-target 8.10 of the SDG emphasizes *expanding "access to financial services like banking, insurance, and financial services for all."* To take stock of countries' progress in this sub-target, the UN proposed using a set of three indicators: the number of commercial bank branches and automated teller machines (ATM) per 100,000 adults (indicators 8.10.1 (a) and (b), respectively) and the proportion of adults with an account at a bank, other financial institution, or a mobile money service provider (indicator 8.10.2).

G. W. Muschert et al. (eds.), *Financial Inclusion*, Sustainable Development Goals Series,
https://doi.org/10.1007/978-3-031-68803-4_22

In this note, we examine whether this limited set of three indicators sufficiently captures the inclusiveness of a financial system as envisaged under SDG sub-target 8.10. In this context, reviewing the existing literature on related issues and given the multidimensionality of financial inclusion, we argue that these indicators are inadequate and methodologically inferior measures of financial inclusion.

We organize this chapter as follows: Sect. 22.2 reviews the literature on measuring financial inclusion in the context of its multidimensionality. Section 22.3 discusses SDG indicators 8.10.1 and 8.10.2 and comments on the adequacy of these indicators with some discussions on them. Section 22.4 suggests an augmented list of indicators and a methodology to construct an index of financial inclusion (IFI). Section 22.5 concludes.

22.2 Multidimensionality and Financial Inclusion Measures

People's exclusion from the formal financial system may occur in various ways, e.g., due to their remoteness from financial institutions, due to unreasonable conditions of financial services, due to the high price of these services, due to targeted marketing of these services, and due to other obstacles such as psychological barriers and financial illiteracy. Thus, the inclusiveness of a financial system should be viewed from several dimensions: accessibility, availability, affordability, and usability, to name a few. The academic community, policymakers, and financial sector practitioners widely acknowledge the multidimensionality of an all-inclusive financial system. Accordingly, financial inclusion is a multidimensional phenomenon. Financial inclusion has been defined as a process that ensures the ease of access, availability, and usage of the formal financial system (Sarma, 2008).

A substantial body of literature has developed on the vital issue of how financial inclusion should be measured. The simplest way has been to use the proportion of the adult population or proportion of households (of an economy) having an account in a bank or other formal financial institution (see, e.g., Honohan, 2008). This measure is similar to using only the SDG indicator 8.10.2 mentioned above. This single-indicator approach is straightforward; however, it incorporates only one dimension of financial inclusion: banking or financial sector penetration or access to the formal financial system or services. If we use only this measure, we will ignore other important aspects of an inclusive financial system, such as availability, quality, affordability, and volume of usage of financial services. As many scholars have documented, merely having a bank account does not necessarily imply that the account is utilized adequately (see, e.g., Kempson et al., 2004). Often, people with bank accounts do not use them enough due to the remoteness of bank branches (Diniz et al., 2012). Even when financial services are available close by, sometimes people feel reluctant to use their bank accounts due to psychological barriers and financial illiteracy. Kempson et al. (2004) defined the notion of "underbanked" or "marginally banked" people as those who do not make adequate use of their bank accounts despite having one. A measure of financial inclusion should consider access and other dimensions like availability and usage. The proportion of "banked" adults measures only one aspect of financial inclusion: access to the financial system (of the adult population). It ignores other important aspects and overlooks the multidimensionality of financial inclusion.

The SDG indicators' set included in 8.10.1 and 8.10.2 falls within the second approach to measuring financial inclusion, the multiple indicators approach. Since a single indicator cannot measure all dimensions of financial inclusion, policymakers use various indicators, each measuring a specific dimension, partially or entirely. The most commonly used indicators are the proportion of adults with a bank or financial account, number of bank accounts (per 1000 adult persons), number of bank branches (per million people), number of automated teller machines (ATM) per 100,000 people, the volume of bank credit disbursed by the financial or banking system, amount of bank deposit mobilized, the volume of financial transactions and remittances, etc. This approach is better than the single-indicator approach because it acknowledges the multidimensionality of financial inclusion and provides measures for different dimensions. However, the major disadvantage of such a multiple-indicator approach is that the multiplicity of indicators makes comparison difficult. For example, an economy may be ahead in one indicator but behind in another indicator vis-à-vis another economy. In such a case, which economy has better financial inclusion? In the next section, we illustrate this problem using the SDG indicators 8.10.1 and 8.10.2.

Thus, despite giving essential information on various dimensions of an inclusive financial system, multiple indicators may fail to provide a clear picture of the overall level of inclusiveness. A more comprehensive picture emerges by aggregating all indicators in a single index, thereby incorporating information on various dimensions of financial inclusion in a single number. This is the third approach to measuring financial inclusion, often called the index-based approach. By aggregating various indicators into a single number, this index-based approach incorporates the multidimensionality of financial inclusion that is absent in the single-indicator approach and, at the same time, mitigates the challenges posed by the multiplicity of indicators in the multiple-indicator approach. Hence, the index approach is superior.

22.3 SDG Indicators 8.10.1 and 8.10.2

The UN SDG indicators website (https://unstats.un.org/sdgs/metadata/) defines the development target 8.10 and associated indicators 8.10.1 and 8.10.2 as follows:

Target 8.10: Strengthen the capacity of domestic financial institutions to encourage and expand access to banking, insurance, and financial services for all.

Indicator 8.10.1 (a): Number of commercial bank branches per 100,000 adults.

Indicator 8.10.1 (b): Number of automated teller machines (ATMs) per 100,000 adults.

Indicator 8.10.2: Proportion of adults (15 years and older) with an account at a bank or other financial institution or with a mobile-money-service provider.

Indicators 8.10.1 (a) and (b) are measures of the physical availability of banking services, and indicator 8.10.2 measures the penetration or accessibility of banking and alternative financial institutions, including mobile money services. The primary data source for indicators 8.10.1 is the Financial Access Survey (FAS), released annually by the International Monetary Fund (IMF). For indicator 8.10.2, data are available from the World Bank's Global Findex Database, which is updated once every three years.

These indicators should be improved, particularly the ones included in 8.10.1, as they provide information on only one type of availability, viz., branch outlets and ATMs of commercial banks. Indicator 8.10.2, a measure of financial sector penetration, includes banking and other kinds of financial institutions and, hence, is a relatively complete measure of financial sector penetration. Even though indicator 8.10.2 is a more comprehensive measure, this limited set of 8.10.1 (a), 8.10.1 (b), and 8.10.2 needs to be revised to measure financial inclusion as envisaged under SDG sub-target 8.10. The inadequacy is at various levels, and we discuss each count.

The SGD Indicators 8.10.1 (a) and (b) include only commercial banks: The availability indicators defined under 8.10.1 (a) and (b) include only commercial bank branches and ATMs; thus, they measure the availability of physical outlets of only one type of financial service provider and are limited in nature. The availability of physical outlets of alternative financial institutions, such as cooperative banks, microfinance institutions, postal banks, and mobile money service outlets, should be taken into account. It is well known that these alternative institutions comprise a substantial part of the formal financial systems in many countries, particularly in developing and low-income economies. For example, as per the FAS data of the IMF, in the years 2014 and 2017, the numbers of registered mobile money agent outlets in Bangladesh per 100,000 adults were 501 and 687, respectively, which far exceeded the number of commercial bank branches at about 9 branches per 100,000 adults in both years. Similarly, in Cambodia, approximately 8 commercial bank branches were available for 100,000 adults in 2017, while approximately 12 branches of microfinance institutions (MFI) were available per 100,000 adults in the same year. In Cameron, only 2 branches of commercial banks and more than 8 branches of MFI branches were available per 100,000 adults in 2017. Thus, it is clear that large segments of the population have access to alternative formal financial service providers in many countries, particularly low-income and developing economies. In other words, alternate financial institutions supplement the mainstream banking segments in a significant way in some countries. We do not include a measure of the outlets of these alternative institutions in our measure of availability. In that case, a full measure of this dimension cannot be achieved. Thus, either more indicators must be added to the list of indicators 8.10.1 (a) and (b), or the indicators themselves should be modified to take other institutions into account.

Internet banking is an essential aspect of financial inclusion. Financial service provisioning worldwide has moved considerably toward digital forms with the widespread usage of the Internet and other digital platforms. There is no doubt that these technological innovations have, to a large extent, reduced the importance of physical bank branches. In many countries, bank branches are being closed down as they become redundant due to widely available, less expensive digital and electronic platforms, particularly in high-income economies. Thus, in this situation, the importance of physical branches is significantly reduced. However, this reduction in bank branches is not necessarily an indication of a decline in the financial inclusiveness of the system. For example, if commercial banks provide Internet banking facilities, customers need not visit a bank branch or ATM and can do most of their online financial transactions. In other words, when the closure of physical branches is supplemented with the electronic provision of financial services, there is little impact on financial inclusion on this count.

In the current set of indicators, no measure of electronic or Internet-based availability of financial services appears.

The usage dimension is missing from the SDG indicator set. A critical dimension missing from the set of SDG indicators is the use of the financial system. It has been argued that an inclusive financial system has many account holders and is widely available and adequately utilized. Simply having an account in a bank does not imply that accounts are being used. For example, in an extensive survey of low- and moderate-income (LMI) households in three major cities in the United States, Seidman et al. (2005) found that nearly two-thirds of the population with a bank account used informal non-bank services despite having access to their branches. Similarly, according to multiple sources, many bank accounts in India lie dormant.[1] Only if accounts are used adequately for saving, borrowing, payment, remittances, and other transactions is the potential of the financial system fully realized, and the benefits of the system accrue to the user. Such a usage dimension can be measured in terms of outstanding deposits, the volume of general credit and SME loans, the volume of remittances, internet-based transactions, mobile money transactions, and so on. The proposed set of SDG indicators does not include any indicators to measure usage dimensions. Both IMF-FAS and WB-Findex have many indicators to indicate the number and volume of usage of bank accounts and

[1] See, e.g., report in Business Standard, March 22, 2018, at https://www.business-standard.com/article/news-ians/20-jan-dhan-accounts-lying-dormant-minister-118032200961_1.html, last accessed in August 2021.

Table 22.1 Illustrative table with SDG indicators 8.10.1 and 8.10.2 for a select set of countries, 2017

	Indicator 8.10.1 (a)	Indicator 8.10.1 (b)	Indicator 8.10.2
Country name↓	No. of commercial bank branches per 100,000 adults	No. of ATMs per 100,000 adults	% of adults with an account
Australia	29.6	160.3	99.5
Austria	12.1	165.0	98.2
Bangladesh	8.8	8.4	50.0
Botswana	8.4	35.4	51.0
Cambodia	7.5	16.7	21.7
Cameron	1.9	4.4	34.6
Chile	14.8	51.6	74.3
China	8.8	84.4	80.2
Czech Republic	21.3	55.6	81.0
Estonia	10.1	67.8	98.0
India	14.5	22.0	79.9
Jordan	14.4	27.5	42.5
Malaysia	10.2	50.9	85.3
Mauritius	18.6	43.5	89.8
Nepal	12.1	11.0	45.4
Pakistan	10.2	10.0	21.3
Paraguay	11.7	27.0	48.6
Peru	7.6	107.5	42.6
The Philippines	9.0	28.1	34.5
Sweden	16.2	32.0	99.7
Switzerland	40.8	97.9	98.4
Thailand	11.9	117.8	81.6

Source: FAS IMF database, WB-Global Findex database, both accessed in August 2021

other financial services, some of which should be included in the SDG indicators.

Insurance indicators are missing from the SDG set: Expansion of access to insurance services for all has been a stated objective of SDG Target 8.10. However, the proposed set of indicators does not have any for measuring insurance penetration or insurance service availability. Thus, some measures for availability, penetration, and usage of insurance services should also be included in the SDG 8.10 set of indicators.

Lack of comparability: As we discussed in the last section, one major limitation of the multiple indicator approach, like the three SDG 8.10 indicators, is that it can make it hard to make sense of how different countries are doing overall regarding financial inclusion. An illustrative example is deduced from the data on the three SDG 8.10 indicators presented in Table 22.1 for a set of 24 countries for 2017, the latest year for which all three indicators are available at the time of writing. By themselves, the individual indicators convey precise achievement measures in specific dimensions. However, if we want to compare overall achievement by combining all dimensions, we may not get a clear picture. For example, consider Australia and Austria in Table 22.1.

Regarding indicator 8.10.1 (a), i.e., availability of bank branches, Australia is ahead of Austria, while in terms of 8.10.1 (b), indicating the availability of ATMs, Austria is ahead of Australia. In terms of financial penetration, both countries are at similar levels. A similar comparison of these indicators between Austria and Sweden reveals that despite the fact that both countries have a similar level of financial penetration (98 and 99.7%, respectively), Sweden outperforms Austria in terms of branch availability. In contrast, Austria ranks way ahead of Sweden in terms of ATM availability.

In taking stock of overall progress in policies and schemes for promoting financial inclusion across countries and over time, the multiple indicator approach is particularly challenging as it makes comparison difficult. Such difficulty for comparison is more challenging for developing and low-income economies. Many developing and low-income countries are home to a large number of financially excluded people, and financial inclusion is a policy priority in these countries. Using the examples of financial inclusion achievements from Table 22.1, e.g., Bangladesh, Pakistan, and Nepal, we can see that Nepal has better availability of bank branches and ATMs. In contrast, Bangladesh has the least availability of these physical outlets among the three countries, with Pakistan in the middle. However, regarding the third SDG indicator, viz., the proportion of adults having an account, Bangladesh ranks first, Pakistan ranks third, and Nepal remains in between. A similar difficulty arises when we compare Thailand, India, Malaysia, and Chile (Table 22.1). Thus, the use of multiple indicators, though more informative, may make it difficult to compare the overall level of financial inclusion across countries and over time. This is a drawback of the multiple-indicator approach to measuring financial inclusion. As a remedy, the index-based approach appears in the literature. Individual indicators are aggregated using an appropriate method to create an index that captures information on all dimensions in a single number.

22.4 Recommendations

As discussed above, the current SDG indicators for measuring financial inclusion are restrictive due to their narrow focus and due to the need for an index-based approach. This limited set has no indicators to measure insurance coverage or the usage dimension of financial inclusion. For the availability dimension, outlets of financial institutions other than commercial bank branches and ATMs are excluded. Thus, a more extensive set of indicators is necessary. After having a more exhaustive set of indicators, an appropriate methodology should be used to construct an index of financial inclusion so that the ultimate measure incorporates information on countries' achievements in various dimensions in a single number in a meaningful manner. Here, we discuss these two points.

Additional Indicators A list of additional indicators that can be added to augment the SDG 8.10 indicators is provided in Table 22.2. This list is, however, only suggestive and not exhaustive. Out of the three SDG 8.10 indicators, only indicator 8.10.2 is complete, as it covers banks and other forms of financial institutions in measuring financial penetration or accessibility. The other two indicators, 8.10.1 (a) and 8.10.1 (b), are partial measures of the availability dimension of the financial system as they include only commercial bank branches and ATMs. To complete the availability dimension in measuring financial inclusion, it is essential to include alternative types of financial institutions, such as microfinance institutions, mobile money service providers, insurance corporations, cooperative banks, and other deposit-taking institutions. To augment the set of indicators for the availability dimension, we suggest five more indicators. These include the physical outlets of cooperative credit unions, microfinance institutions, mobile money service providers, other deposit-taking institutions (like postal banks), and insurance service providers. It is important to note that these alternative financial institutions have varying degrees of importance in different countries. For example, microfinance institutions or mobile money services may not exist in Sweden or Austria. However, such services may form a substantial part of Bangladesh's and Cameroon's financial systems. Therefore, the overall measure of availability should include information on all alternative forms of financial institutions that are available and relevant to a country.

The SDG indicators should also include indicators to measure the usage dimension because an inclusive financial system should be one that is adequately utilized. At the moment, there is no indicator for the usage dimension. In Table 22.2,

Table 22.2 Suggestive list of additional indicators to augment the SDG 8.10 indicators

Indicator	Description	Dimension measured
8.10.2	% of adults with an account at a bank, other financial institution, or mobile money service provider	Financial sector penetration
8.10.1 (a)	No. of commercial bank branches per 100,000 adults	Availability of financial services
8.10.1 (b)	No. of commercial ATMs per 100,000 adults	
Additional indicators to augment Availability Dimension (suggestive list)	1. Number of credit union and credit cooperative branches per 100,000 adults	Availability of financial services
	2. Number of registered mobile money agent outlets per 100,000 adults	
	3. Number of insurance corporations per 100,000 adults	
	4. Number of all microfinance institution branches per 100,000 adults	
	5. Number of other deposit-taking branches per 100,000 adults	
Additional indicators to measure the Usage Dimension (suggestive list)	6. Outstanding loans from commercial banks (% of GDP)	Usage of financial services
	7. Outstanding deposits with commercial banks (% of GDP)	
	8. Outstanding loans from credit unions and credit cooperatives (% of GDP)	
	9. Outstanding deposits with credit unions and credit cooperatives (% of GDP)	
	10. Outstanding loans from all microfinance institutions (% of GDP)	
	11. Value of mobile and internet banking transactions (during the reference year) (% of GDP)	
	12. Outstanding Deposits, Insurance corporations (% of GDP)	

we list seven indicators to measure the usage dimension. These include the amount of deposits mobilized through financial institutions, the number of mobile money transactions, the amount of credit disbursed through different institutions, and so on. All of these indicators are shown as a percentage of an economy's GDP so that they can be compared.

In Table 22.3, we present data on the augmented list of indicators for a select set of countries for illustration. As we can see, data on various indicators for the augmented SDG 8.10 indicator list are available from the Financial Access Survey (FAS) database of the IMF, the same database suggested for indicators 8.10.1 (a) and (b). The availability of relevant data for many countries makes it pertinent to utilize such data to construct a complete set of indicators of financial inclusion, as suggested here.

A Comprehensive Measure Based on Multiple Indicators

Multiple indicators are necessary to incorporate various aspects of the inclusiveness of a financial

Table 22.3 Augmented list of indicators for select countries, 2017

Indicators		Australia	Austria	Bangladesh	Botswana	Cambodia	Cameroon	Chile	China	Czech Republic	Estonia	India	Jordan
Indicator 8.10.2	% of adults with an account	99.5	98.2	50.0	51.0	21.7	34.6	74.3	80.2	81.0	98.0	79.9	42.5
Indicator 8.10.1 (a)	No. of commercial bank branches per 100,000 adults	29.6	12.1	8.8	8.4	7.5	1.9	14.8	8.8	21.3	10.1	14.5	14.4
Indicator 8.10.1 (b)	No. of ATMs per 100,000 adults	160.3	165.0	8.4	35.4	16.7	4.4	51.6	84.4	55.6	67.8	22.0	27.5
Additional indicators to augment Availability Dimension	Credit union/credit cooperative branches	2.91	29.70	174.88	4.64		7.41	1.35	2.19	0.23		22.79	
	Registered mobile money agent outlets			686.84	93.45	309.62	588.28						8.19
	Insurance corporations	0.67	0.49	0.08	1.66	0.11	0.19	0.46	0.02	0.57	1.72	0.01	0.38
	Microfinance institution branches			16.93		12.19	8.15					1.02	
	Other deposit takers' branches		14.53	11.45	9.61		0.08	0.09	8.33		2.17		
Additional indicators to measure Usage Dimension (outstanding value as % of GDP)	Loans from commercial Banks	137.44	24.65	42.69	28.89	72.14	15.94	90.08	100.80	50.14	63.27	46.33	83.75
	Deposits with commercial Bank	79.78	27.86	50.22	25.77	77.56	19.78	59.98	149.97	66.97	54.38	62.79	112.92
	Loans from credit unions/credit cooperatives	1.94	31.55	0.26	0.25		0.91	1.08	4.24	0.17		4.95	
	Deposits credit unions/ credit cooperatives	2.25	34.47	0.62	0.24		1.24	0.47	6.77	0.22		5.94	
	Loans from microfinance institutions			3.24		19.19	1.37						
	Mobile/internet banking transactions		656.68	1.87	3.07	90.52	2.40	626.99		799.23	539.35	7.67	27.11
	Deposits, insurance	20.32		1.45		14.08	1.42	2.10	1.46	5.74	18.72	16.80	
	Deposits, other deposit takers		32.57	4.90	2.09		0.07	7.43	44.24	0.03	0.35	0.14	
	Loans, other deposit takers				2.66		0.03		41.01	0.00	0.36	1.43	

Source: FAS (IMF), Global Findex (WB), accessed in August 2021

system. As discussed in the previous section, the challenges of multiplicity can be mitigated by constructing an index incorporating all indicators into a single number using a robust methodology. Such a methodology should generate a meaningful index, amenable for comparison across economies and over time. Several attempts have been made to devise multidimensional indices for financial inclusion (Amidzic et al., 2014; Cámara & Tuesta, 2014; Sarma, 2008, 2012, 2015; Sarma & Pais, 2011; Zulaica Pineyro, 2013). This section illustrates the index of financial inclusion, proposed and developed by Sarma (2015, 2016). In Sarma (2016), various methodologies are reviewed against the backdrop of conceptual considerations that should be accounted for while developing an index of financial inclusion.

Index of Financial Inclusion (IFI)

The IFI is a multidimensional measure of financial inclusion that incorporates information on various dimensions of an inclusive financial system in a single number in the interval [0,1]. The IFI is a unit-free and monotonic measure. The bounds of IFI are well defined, where 0 indicates an entirely exclusionary system and 1 indicates a fully inclusive financial system. The monotonicity of IFI implies that the methodology generates IFI values such that higher achievement in any given dimension, ceteris paribus, gives rise to a higher value of IFI. In addition, the IFI methodology is stable, which makes it amenable for comparison over time and across economies. Further, it does not suffer from the "perfect substitutability" that some other indexes display (Sarma, 2016).

In Table 22.4, the IFI values are computed for a select set of countries as an illustration. The methodology and the mathematical steps for IFI computation are elaborated in Sarma (2015, 2016), and here we do not delve into the technical details for brevity.

In Table 22.4, columns 2, 3, and 4 show how well each country (shown in Column 1) fared in the three important areas of financial inclusion we consider: financial penetration (or accessibility), availability of financial services, and use of the financial system. These achievements are normalized to lie within [0,1] by comparing them to some benchmarks. The normalized achievement in a particular dimension is called the dimension index. In columns 5, 6, and 7, we present the dimension index for each dimension considered in this example. The dimension indexes are measures of achievement in specific dimensions vis-à-vis a particular benchmark that measures the desirable (maximum or optimal) outcome in a particular dimension. For instance, the proportion of adults who have an account with a formal financial service provider is a proxy for the accessibility (penetration) dimension in this context. The benchmark here can be fixed at 100, indicating that all adults should have an account. For the other two dimensions of financial inclusion considered here, viz., availability and usage, it is, however, not easy to find benchmarks as there is no straightforward way of deciding the maximum or optimal value of how many outlets per 100,000 adults and how much volume of credit+deposit+transaction to GDP are most desirable from the point of view of inclusive finance. Sarma (2016) suggested that one can fix the benchmark for these dimensions at the 90th percentile on the statistical distribution of values for all countries. Accordingly, in this illustrative example, a value of 150 has been used for availability, and a value of 400 is used for the usage dimension. If the achievement of a country is more than the benchmark(s), then the achievement is set equal to the benchmark.

After deciding on the benchmarks, the following formula calculates the dimension index for a particular dimension:

$$D_i = \frac{A_i}{M}, \text{if } A_i < M$$

$$D_i = 1, \text{if } A_i \geq M$$

where,

$$D_i = \text{dimension index for } i^{\text{th}} \text{ dimension}$$

$$A_i = \text{actual value of the } i^{\text{th}} \text{ dimension}$$

$$M = \text{benchmark for } i^{\text{th}} \text{ dimension}$$

Table 22.4 Computing index of financial inclusion from indicators (illustrative example), data pertain to 2017

Country	Financial penetration (% of adults with an account)	Availability of financial services (no. of physical outlets per 100,000 adults)	Usage of financial services (value of credit, deposit, and transaction as % of GDP)	Dimension index for financial penetration	Dimension index for availability	Dimension index for usage	Index of financial inclusion (IFI)
Australia	99.52	193.54	241.73	1.00	1.00	0.60	0.88
Austria	98.16	221.81	807.78	0.98	1.00	1.00	0.99
Bangladesh	50.05	907.35	105.26	0.50	1.00	0.26	0.68
Botswana	51.03	153.15	62.97	0.51	1.00	0.16	0.66
Cambodia	21.67	346.15	273.49	0.22	1.00	0.68	0.71
Cameroon	34.59	610.38	43.16	0.35	1.00	0.11	0.62
Chile	74.35	68.30	788.14	0.74	0.46	1.00	0.78
China	80.23	103.78	348.48	0.80	0.69	0.87	0.83
Czech Republic	80.99	77.73	922.50	0.81	0.52	1.00	0.81
Estonia	97.99	81.73	676.42	0.98	0.54	1.00	0.86
India	79.88	60.32	146.04	0.80	0.40	0.37	0.63
Jordan	42.49	50.49	223.78	0.42	0.34	0.56	0.56

Source: Author's computation based on data from Table 22.3 and methodology from Sarma (2015, 2016)

A country's dimension indexes are considered a point in the Cartesian plane, known as the achievement point of the country. For example, in Table 22.4, the achievement point of Australia is (1, 1, 0.6) and that of Jordan is (0.42, 0.34, 0.56). In this representation, the first coordinate is for financial penetration, the second is for availability, and the third is for the usage dimension. In the three-dimensional Cartesian plane, the point (0,0,0) and the point (1,1,1) are unique reference points because (0,0,0) indicates the worst performance, whereas (1,1,1) indicates the ideal performance in all three dimensions. The average distance of a country's achievement point from the worst point (0,0,0) and its proximity (inverse distance) to the ideal point (1,1,1) serve as indicators of the country's overall success in promoting financial inclusion. A simple Euclidean distance formula is used to compute these distances. A simple average of the achievement point's distance from the worst point and the inverse distance from the ideal point gives us the country's financial inclusion index (IFI). The last column of Table 22.4 presents the IFI values for a select set of countries as an illustration; here, a high value of IFI indicates high financial inclusion, and a low value indicates low inclusion.

22.5 Conclusion

The set of SDG indicators proposed for measuring the sub-target 8.10 involving financial inclusion is limited on many counts. In this note, we discuss the restrictive nature of these indicators and suggest an augmented set of indicators to measure various dimensions of an inclusive financial system fully. We also suggest additional indicators to deal with insurance availability and usage. As pointed out in the literature, an index-based approach to measuring financial inclusion is methodologically superior to having multiple indicators. Here, we suggest the index of financial inclusion (IFI) approach of Sarma (2015, 2016) as a method for index construction.

References

Amidzic, G., Massara, A., & Mialou, A. (2014). *Assessing countries' financial inclusion standing—A new composite index*. IMF Working Paper WP/14/36.

Cámara, N., & Tuesta, D. (2014). *Measuring financial inclusion: A multidimensional index*. BBVA Working Paper, N° 14/26.

Diniz, E., Birochi, R., & Pozzebon, M. (2012). Triggers and barriers to financial inclusion: The use of ICT-based branchless banking in an Amazon county. *Electronic Commerce Research and Applications, 11*(5), 484–494.

Honohan, P. (2008). Cross-Country Variation in Household Access to Financial Services. *Journal of Banking and Finance, 32*, 2493–2500.

Kempson, E., Atkinson, A., & Pilley, O. (2004). *Policy level response to financial exclusion in developed economies: lessons for developing countries*. Report of Personal Finance Research Centre, University of Bristol.

Sarma, M. (2008). *Index of financial inclusion*. ICRIER Working Paper 215.

Sarma, M. (2012). *Index of financial inclusion–A measure of financial sector inclusiveness, Working Paper No. 07/2012*. Berlin Working Papers on Money, Finance, Trade and Development.

Sarma, M. (2015). Measuring financial inclusion. *Economics Bulletin, 35*(1), 604–611.

Sarma, M. (2016). Measuring financial inclusion using multidimensional data. *World Economics, 17*(1), 15–40.

Sarma, M., & Pais, J. (2011). Financial inclusion and development. *Journal of International Development, 23*, 613–628.

Seidman, E., Hababou, M., & Kramer, J. (2005). *Getting to know underbanked consumers: A financial services analysis*. Report of the Center for Financial Services Innovation.

Zulaica Pineyro, C. M. (2013). Financial inclusion index: Proposal for a multidimensional measure for Mexico. *Revista Mexicana de Economıa y Finanzas, 8*(2), 157–180.

GPSR Compliance

The European Union's (EU) General Product Safety Regulation (GPSR) is a set of rules that requires consumer products to be safe and our obligations to ensure this.

If you have any concerns about our products, you can contact us on ProductSafety@springernature.com

In case Publisher is established outside the EU, the EU authorized representative is:

Springer Nature Customer Service Center GmbH
Europaplatz 3
69115 Heidelberg, Germany

The manufacturer's authorised representative in the EU is Springer Nature Customer Service Centre GmbH, Europaplatz 3, 69115 Heidelberg, Germany. If you have any concerns regarding our products, please contact ProductSafety@springernature.com

Printed and bound by CPI Group (UK) Ltd, Croydon, CR0 4YY
30/09/2026
02225598-0019